The regulation of the adipose tissue mass

The regulation of
the adipose tissue mass

Proceedings of the IV International Meeting of
Endocrinology, Marseilles, July 10-12, 1973

Editors

J. VAGUE and J. BOYER

Clinique Endocrinologique,
Hôpital de la Timone,
Marseilles, France

Co-editor
G. M. ADDISON
Department of Medical Biochemistry,
University Hospital of Wales,
Cardiff, U.K.

 1974

Excerpta Medica Amsterdam
American Elsevier Publishing Co., Inc. New York

International Congress Series No. 315

ISBN Excerpta Medica 90 219 0225 7
ISBN American Elsevier 0 444 15054 4
Library of Congress Catalog Card Number 74-77376

Publisher
Excerpta Medica
335 Jan van Galenstraat
Amsterdam
P.O. Box 1126

Sole Distributor for the U.S.A. and Canada
American Elsevier Publishing Company, Inc.
52 Vanderbilt Avenue
New York, N.Y. 10017

Set by Blankenstein bv, Amsterdam
Printed in The Netherlands by Drukkerij Bakker, Zaandijk

Organizing Committee

President	J. Vague
Secretary General	J. Boyer
Treasurer	J.-L. Codaccioni
Executive Committee	P. Bernard, P. Jaquet, J. Jubelin, A. Mattei, G. Miller, J. Nicolino, P. Rubin, R. Simonin, M. Teitelbaum, P. Vague

Contents

ADIPOSE TISSUE CELLULARITY AND METABOLISM

ADIPOSE MASS AND OBESITY

ADIPOSE MASS AND BODY COMPOSITION

ADIPOSE MASS AND BLOOD LIPIDS

Contributors

(First authors and participants in discussions)

G. Ailhaud
Centre de Biochimie,
Faculté des Sciences,
University of Nice,
Parc Valrose, 0634 Nice, France

A. Angel
Department of Medicine,
Room No. 7368,
Medical Sciences Building,
University of Toronto, Toronto, Canada

M. Apfelbaum
Unité de Recherches Diététiques,
Hôpital Bichat, 170 Boulevard Ney,
75877 Paris, France

M. Berger
Joslin Research Laboratory,
Harvard Medical School,
170 Pilgrim Road,
Boston, Mass. 02166, U.S.A.

R.S. Bernstein
St. Luke's Hospital Center,
Amsterdam Avenue at 114th Street,
New York, N.Y. 10025, U.S.A.

P. Björntorp
First Medical Service,
Sahlgren's Hospital,
S-413 45 Gothenburg, Sweden

G.A. Bray
Harbor General Hospital,
Carson Street, Torrance, Calif., U.S.A.

H. Breidahl
Alfred Hospital, 421 St. Kilda Road,
Melbourne, Victoria 3004, Australia

F. Ceresa
Istituto di Patologia Speciale Medica,
University of Turin,
Via Legano 40, Turin, Italy

C. Chlouverakis
School of Medicine,
State University of New York at Buffalo,
E.J. Meyer Memorial Hospital,
462 Grider Street,
Buffalo, N.Y., 14215, U.S.A.

G. Copinschi
Laboratoire de Médecine Expérimentale,
Université Libre,
4, Rue Héger-Bordet,
Brussels, Belgium

E. Couturier
Laboratoire de Médecine Expérimentale,
Université Libre,
115 Boulevard de Waterloo,
Brussels, Belgium

G. Debry
Department of Nutrition and Metabolic
Diseases,
University of Nancy, 40 Rue Lionnois,
54000 Nancy, France

P. de Gasquet
Laboratoire de Nutrition Humaine,
Hôpital Bichat, 170 Boulevard Ney,
75018 Paris, France

J.L. de Gennes
Clinique Endocrinologique,
Centre Hospitalier Universitaire
Pitié Salpêtrière,
83 Boulevard de l'Hôpital,
Paris, France

P.A. Delwaide
Chimie Médicale,
University of Liège,
1 Rue des Bonnes Villes,
B-4000 Liège, Belgium

M. DiGirolamo
  Department of Medicine,
  Woodruff Medical Center,
  Emory University,
  Thomas K. Glenn Memorial Building,
  69 Butler Street, S.E.,
  Atlanta, Ga. 30303, U.S.A.

J. Durnin
  Institute of Physiology,
  University of Glasgow,
  Glasgow G12 8QQ, U.K.

G. Enzi
  Clinica Medica Generale,
  University of Padua,
  Via Giustiniani 1, Padua, Italy

E. Eschwege
  Unité de Recherches Statistiques, INSERM,
  16 bis Avenue Paul Vaillant Couturier,
  Villejuif, France

M. Faggiano
  Division of Endocrinology,
  I Faculty of Medicine,
  University of Naples, Naples, Italy

P.P. Foa
  Sinai Hospital,
  Detroit, Mich. 48235, U.S.A.

M. Fredman
  Department of Anatomy,
  University of Cape Town
    Medical School, Observatory,
  Cape, South Africa

D. Galton
  Department of Medicine,
  St. Bartholomew's Hospital,
  West Smithfield,
  London EC1A 7BE, U.K.

G. Giordano
  Cattedra di Endocrinologia,
  Istituto Scientifico di Medicina Interna,
  University of Genoa, Genoa, Italy

H. Giudicelli
  Laboratoire Clinique Endocrinologique,
  Hôpital de la Conception,
  144 Rue Saint Pierre,
  Marseilles, France

R.B. Goldrick
  Department of Chemical Pathology,
  St. Bartholomew's Hospital,
  West Smithfield,
  London EC1A 7BE, U.K.

F.A. Gries
  Diabetes Research Institute,
  Auf dem Hennekamp 65, 4 Düsseldorf,
  Federal Republic of Germany

M. Gurr
  Unilever Research Laboratory,
  Sharnbrook, Bedford MK 44 KQ, U.K.

B. Guy-Grand
  Clinique Médicale de l'Hôtel Dieu,
  1 Impasse du Parvis Notre Dame,
  Paris, France

M. Harter
  Département Endocrinologie Nutrition
    Métabolique,
  Centre Hospitalier de Recherche,
  Nice, France

J.O. Holloszy
  Department of Preventive Medicine
    and Public Health,
  Washington University School of
    Medicine,
  4566, Scott Avenue,
  St. Louis, Mo. 63110, U.S.A.

B. Jeanrenaud
  Laboratoires de Recherches Médicales,
  University of Geneva,
  Avenue de la Roseraie 64,
  1211 Geneva 4, Switzerland

R.H. Johnson
  Institute of Neurological Sciences,
  Southern General Hospital,
  1345 Govan Road,
  Glasgow G-51 4TF, U.K.

R.K. Kalkhoff
  Clinical Research Center, Milwaukee
    County General Hospital,
  8700 W. Wisconsin Avenue,
  Milwaukee, Wis. 53226, U.S.A.

L. Kazdová
  Metabolism and Nutrition Research Centre,
  Institute of Clinical and Experimental
    Medicine,
  Budejovicka 800,
  Prague, Czechoslovakia

H.M.J. Krans
Department of Endocrinology
and Metabolic Diseases,
University Hospital,
Rijnsburgerweg 10, Leiden,
The Netherlands

P. Laudat
Hôpital Cochin,
27 Rue du Faubourg Saint-Jacques,
75674, Paris, France

P. Lefebvre
Institut de Médecine,
University of Liège,
Hôpital de Bavière,
B4000 Liège, Belgium

F. Legros
Laboratoire de Physiopathologie,
Faculté de Medecine,
Université Libre,
115 Boulevard de Waterloo,
Brussels, Belgium.

D. Lemonnier
Laboratoire de Nutrition Humaine,
Institut Scientifique et Technique
de l'Alimentation,
292 Rue Saint-Martin, Paris, France

J.E. Lindsay Carter
Department of Physical Education,
San Diego State University,
San Diego, Calif. 92115, U.S.A.

J.N. Livingston
Taylor Building, The Johns Hopkins
Medical School,
720 Rutland Avenue,
Baltimore, Md. 21205, U.S.A.

E. Martino
Centro per la Prevenzione del Gozzo,
Istituto di Medicina del Lavoro,
University of Pisa,
Via Bonanno 48, Pisa, Italy

S. Matsuki
Department of Internal Medicine,
Keio University School of Medicine,
35 Shinanomachi, Shinjuku-ku,
Tokyo, Japan

K.I. McLeod
Department of Clinical Science,
John Curtin School of Medical Research,
Australian National University,
P.O. Box 4, Canberra, Australia

J.H. Nielsen
Research Laboratory,
Steno Memorial Hospital,
Gentofte, Denmark

Z.L. Ostrowski
ENSP,
3 Square du Rhône,
Paris 17, France

I. Perlstein
3333 Bardstown Road,
Louisville, Ky., U.S.A.

W.J. Poznanski
Department of Metabolism,
Ottawa Civic Hospital,
Carling Avenue,
Ottawa, Canada

L. Rakow
II Department of Pathology,
Centre for Biology and
Theoretical Medicine,
Oberer Eselsberg, D79 Ulm,
Federal Republic of Germany

J. Raulin
Hôpital de Bicêtre,
Le Kremlin,
94270 Bicêtre, France

D.A.K. Roncari
Medical Sciences Building,
University of Toronto,
Toronto, Canada

S. Rössner
Department of Internal Medicine,
Karolinska Sjukhuset,
Stockholm 60, Sweden

D. Samuel
Serrier Laboratories Ltd.,
Pinner Road, North Harrow,
Middlesex, U.K.

R. *Schemmel*
    Department of Food Science
        and Human Nutrition,
    Michigan State University,
    East Lansing, Mich. 48823, U.S.A.

P. *Schwandt*
    First University Medical Clinic,
    Ziemssenstrasse 1, 8 Munich 15,
    Federal Republic of Germany

B. *Shapiro*
    Department of Biochemistry,
    Hebrew University – Hadassah
        Medical School,
    Jerusalem, Israel

D. *Steinberg*
    Division of Metabolic Disease,
    Department of Medicine,
    School of Medicine,
    Basic Science Building,
    University of California, San Diego,
    La Jolla, Calif. 92037, U.S.A.

M.P. *Stern*
    Stanford University Medical Center,
    Stanford, Calif. 94305, U.S.A.

T. *Tomita*
    Shizuoka College of Pharmacy,
    160 Oshika, Shizuoka-shi, Japan

C. *Tordet-Caridroit*
    Centre des Recherches Biologiques
        Néonatales,
    Hôpital Port Royal,
    123 Boulevard de Port-Royal,
    75674 Paris, France

J. *Vague*
    Endocrinology Clinic,
    Hôpital de la Timone,
    Chemin de l'Armée d'Afrique,
    13005 Marseilles, France

P. *Vague*
    Endocrinology Clinic,
    Hôpital de la Timone,
    Chemin de l'Armée d'Afrique,
    13005 Marseilles, France

A. *Vezinhet*
    Station de Physiologie Animale,
    École Nationale Supérieure
        Agronomique,
    Place Viala, Montpellier, France

G. *Walldius*
    King Gustaf V Research Institute,
    Karolinska Sjukhuset,
    Stockholm 60, Sweden

K.M. *West*
    University of Oklahoma
        College of Medicine,
    800 N. E., 13th Street, Oklahoma City,
    Okla. 73190, U.S.A.

J.H. *Willmore*
    Human Performance Laboratory,
    Department of Physical Education,
    University of California at Davis,
    Davis, Calif. 95616, U.S.A.

J. *Winand*
    Laboratory of Biochemistry and
        Nutrition,
    Medical School, Université Libre,
    115 Boulevard de Waterloo,
    Brussels, Belgium

T.T. *Yen*
    Eli Lilly and Co.,
    Indianapolis, Ind., U.S.A.

N. *Zaragoza-Hermans*
    Département de Biochimie Clinique,
    Service de Médecine,
    Hôpital Cantonal Universitaire,
    1101 Lausanne, Switzerland

Introductory remarks

J. Vague

Endocrinology Clinic, Hôpital de la Timone, Marseilles, France

These Proceedings of the Fourth International Meeting of Endocrinology of Marseilles are a continuation of *Physiopathology of Adipose Tissue* (J. Vague (Ed.), Excerpta Medica, 1969), which contained the Proceedings of our Third Meeting held in 1968.

The novelty of the present volume lies perhaps in the different points of view from which the wide problem of the adipose mass is considered. These include the notion of a determined weight, the mechanism of its regulation and the means of modifying it as physiologically as possible when it is maladapted to environment and activity.

We know today that mass and energy are mutually convertible. Both are measurable from the electron to the giant star. Perpetual movement is the law of matter, even when it looks inert. This movement does not preclude the relative stability of minerals and it keeps pace with the renewal of living molecules, the rapidity of which is a function of metabolic activity. So life interposes at first as a transitory brake, afterwards as an accelerator in the degradation of energy which goes from sunlight to inutile heat and to that constant increase of entropy of the universe which seems to be its inescapable law.

Meanwhile permanent energy transfers result in the formation of masses of living matter, the importance of which is linked to autoregulated chemical reactions.

Of course mass as measured with very high accuracy by physicists is not necessary for our purpose. As biologists we can consider the notion of adipose mass in three ways:

1. The easily measurable weight of triglyceride, complex lipids not being a part of reserve substances.
2. The number of molecules of triglyceride per unit weight of adipose tissue and consequently the total depot fat. This notion, which has some interest owing to the biological consequences of the fatty acid chain length, loses this interest on energetic grounds, triglycerides being insoluble and lipases being constrained to act in heterogeneous systems.
3. The external surface of the triglyceride droplet in the adipocyte, a factor leading to the different progression of surface and volume and the relationship between the surface of adipocytes and total adipose mass.

Two aspects of the adipose mass will be met constantly, the weight of this mass and its surface, along with the molecules which determine its value and activity.

Adipose mass, which for a long time was considered to be the passive result of food intake and energy expenditure, looks today much more complex and interesting. The metabolism of the nervous system is critical and in normal conditions relies almost entirely on the oxidation of glucose. Increase in brain size being a corollary of zoological progress, man, with the largest brain, has the ability to ensure an adequate glucose supply, i.e. 100 g daily, which is a quarter of the basal metabolic requirement for only one fiftieth of the body weight. Gluconeogenesis from protein and fatty acid utilization by muscle together with inhibition of glycolysis by these fatty acids produce and at the same time spare the glucose needed by the brain.

In contrast to plants and various micro-organisms, animal tissues lack isocitrate and malate synthetase and consequently are unable to synthesize carbohydrates from the carbon atoms of acetyl-CoA produced by fatty acid oxidation. So the sequence glucose to fatty acid is irreversible in animals, which must exclusively use proteins for gluconeogenesis.

Man, in whom the brain is an important organ, will lose most of his protein stores during fasting even with partial utilization of ketone bodies by the nervous system. The sparing of these stores during exercise and fasting is provided by the adipose mass which, useless for gluconeogenesis, at least provides the musculature with fatty acids.

Adipose tissue triglycerides come from lipids and carbohydrates. The proportion of the latter allotted respectively to muscle glycogen synthesis and to adipocyte lipid synthesis determines the conditions under which te organism faces fasting and exercise. The need for gluconeogenesis from protein and adipose mass utilization is as much delayed as there is available glycogen mass. Generally speaking the adipose mass is inversely proportional to the intensity of normal strain and directly related to the foreseen energy needs.

Fed by maternal blood the human foetus does not need any energy reserves. Adipose mass appears at the end of the fifth month but develops only during the ninth month, reaching 17% of the body weight at birth, which allows the best economy of glucose for the brain when glycogenolysis is still immature. Later on adipose mass increases in the unexercised baby, after which it is decreased in males by the action of testosterone, aggressiveness and higher muscular activity. Adipose mass remains elevated in the female, who is responsible for pregnancy. Sexual attractiveness and the aesthetic considerations which follow it are to a large extent the result of these differences.

The adipose mass increases in women (as in female animals) during the first part of pregnancy, before the foetus, by rapid growth during the second part of pregnancy, has become a greedy consumer of the accumulated energy. This adipose mass growth before utilization is as rational as the filling of a tank by fuel and is observed in migratory birds before flight, in hibernating mammals before sleep and in aquatic mammals before contact with cold water. Be it directed or stochastic, the mechanism of such a 'preprogrammed' adaptative phenomenon is not today within the field of science and should not hold our attention, but its reality is evident. It is the cause of the remarkable stability of adipose mass when internal and external conditions of life are not modified. The amount of the adipose mass is determined by a neuro-hormono-enzymatic system which acts as a function of circumstances and energy needs. The level of this regulation induces the need for a determined and apparently necessary adipose mass, rapid modification of which causes suffering and stress.

Food intake and spontaneous muscular activity, until recently considered to be the main determinants of the adipose mass, now appear to be the consequences of the need for this mass, which regulates in turn hunger and activity. Obese and more especially people increasing their weight are hungry and unconsciously reduce their activity until they have accumulated the adipose mass they need in the conditions in which they are living. Anxious and/or aggressive hunger and difficulty of strain as long as this need is not satisfied, or, alternatively, premature repletion and a need to move are phenomena in which we see the competition for energy uptake between adipose mass, muscle mass and brain.

Besides the necessary adipose mass under particular conditions (which must be preserved as long as these conditions are not modified), we have to consider the pathogenic adipose mass, in other words the threshold above which adipose mass is dangerous and must be decreased. Fat excess leads to two types of complications: mechanical (in direct relation to overload) and metabolic (diabetes, hyperlipoproteinaemia, particularly of the pre-β carbohydrate-induced type, hyperuricaemia, hypertension, atherosclerosis). The latter are more complex and not in direct relation to the value of the total adipose mass but to the predominance of fat in the upper part of the body, the increased volume of these adipocytes and the relatively reduced number of lower body adipocytes, giving rise in both sexes to android obesity, and the degree of genetic predisposition to one or usually more of these metabolic disturbances.

When the threshold above which adipose mass becomes pathogenic is higher than the necessary adipose mass, treatment and prophylaxis are easy and efficient. On the other hand, treatment and prophylaxis of metabolic disturbances induced by obesity may be at the price of depression and reduction of functional capacity. It is not the excess of adipose mass itself we must treat (that would be at the best a short-lived and unprofitable success) but the need for this excess, by an improvement of the physiological conditions of life, which is a prerequisite for reeducation regarding food intake and muscular activity.

Lipid biosynthesis

Enzymatic aspects of neutral lipid biosynthesis

G. Ailhaud, D. Samuel and R. Négrel

Centre de Biochimie, Faculté des Sciences, Université de Nice, Nice, France

The role of neutral lipids as an energy source in eukaryotic cells is well established, particularly in mammals, e.g. heart tissue (Masoro, 1968; Crass, 1972). On the other hand, the structural role of the complex glycerolipids in many biological activities of different membranes has recently been well substantiated (Steim et al., 1969; Overath et al., 1970; Raison and Lyons, 1971; Mavis and Vagelos, 1972; Mavis et al., 1972).

Neutral lipids are almost absent from prokaryotic cells. However, lipids can be used and degraded by β-oxydation when present in the form of free fatty acids as the unique carbon source in the culture medium of *Escherichia coli* or *Pseudomonas putida* (Overath et al., 1967; Samuel and Ailhaud, 1969 and unpublished data).

Thus while the mechanisms and the regulation of neutral lipid biosynthesis primarily occur in mammals, nevertheless bacteria will offer some interesting features in the study of fatty acid activation.

We should like to review the current understanding of neutral lipid biosynthesis from an enzymological point of view.

FATTY ACID TRANSPORT AND ACTIVATION

The transport of fatty acids from exogenous sources into the cell represents the primary important event in enterocytes and adipocytes. The data regarding the mode of transport are still inconclusive. While passive diffusion through the plasma membrane seems likely (Spector, 1968; Mishkin et al., 1972; Lee et al., 1971), other authors have shown the influence of blood supply on lipid uptake from micellar solutions by the rat small intestine, which suggests an active process for oleic acid and monoolein uptake (Sylvèn, 1970). A facilitated diffusion has also been implied (Lyon, 1968). In hepatocytes it has been proposed that the fatty acid uptake is related to the deacylation-reacylation cycle of phospholipids in plasma membrane (Wright and Green, 1971).

In bacteria the isolation of *E. coli* mutants defective in acyl-CoA synthetase (acid:CoA ligase (AMP)) has shown that fatty acid uptake is strictly dependent on the presence of this enzyme (Klein et al., 1971; Samuel et al., 1972). Table 1 indicates that acyl-CoA synthetase is loosely bound to the periplasmic membrane of *E. coli*; its location is thus compatible with a role in an active transport of fatty acids across the bacterial membrane. Acyl-CoA synthetase has been purified from *E. coli* cells grown on fatty acids and derepressed for the enzymes of β-oxidation. The purification scheme is given in Table 2. The enzyme is an acid:CoA ligase (AMP), active on C_4-C_{18} fatty acids. The ratio of hexanoate activity versus palmitate activity remains constant throughout the purification procedure.

Further purification of the enzyme (mol. wt. 120,000) was obtained on a hydroxyapatite column as shown in Figure 1. The separation of two activities was observed. Peak A contained fractions active on C_6 and C_{12} fatty acids. Peaks B and C were active on palmitate. An overall purification of 770-fold was achieved in both cases. Examination of peaks A and C by poly-acrylamide disc-gel electrophoresis indicated a degree of purity of approximately 50%. Fast inactivation of separated A and C activities was observed. The results of thermal inactivation and kinetic experiments also support the idea that two enzymes exist for the activation of short

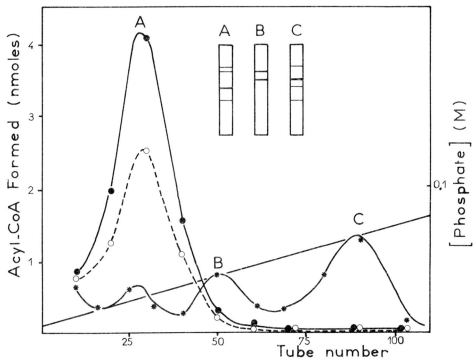

Fig. 1   Chromatography of fraction V (see Table 2) on hydroxyapatite. The hydroxyapatite was prepared by suspending dry hydroxyapatite in 0.5 M potassium phosphate buffer pH 7.3 and washing it with 0.005 M of the same buffer containing 0.01 M 2-mercaptoethanol. The enzyme (135 mU, 1.4 mg of protein) was dialyzed against the same buffer and put on the column (1×6 cm). Elution was performed with 150 ml of a linear gradient from 0.005 to 0.08 M of potassium phosphate buffer pH 7.3 containing 0.010 M of 2-mercaptoethanol (1.4 ml/tube). The substrates used for assays were: ●, [1-¹⁴C]hexanoate; ○, [1-¹⁴C] dodecanoate; *, [1-¹⁴C]-hexadecanoate. 0.020 ml aliquots were used to assay the activity. Disc-gel electrophoresis patterns are given in the upper part of the figure. (From *Europ. J. Biochem.*, 1970, *12*, 576; courtesy of the editors.)

TABLE 1   Subcellular localization of acyl-CoA synthetase

Fraction	Total protein (mg)	Total activity (mU)	Specific activity (mU/mg)	Proportion of total (%)
Lysed protoplasts	55	24	0.44	100
Membranes	20	3.4	0.17	14
Supernatant	39	8.7	0.22	36
Membrane washes	2.8	15.5	5.6	65

Protoplasts were obtained in the following manner: to cells, grown in 300 ml medium containing oleate, were added 900 ml of a solution containing 20% sucrose, 0.4% MgSO₄, 17.5% Difco Penassay broth and 1.3 × 10⁶ penicillin units. After 3 hours without shaking the protoplasts were centrifuged and washed with a solution devoid of penicillin but supplemented with 1% serum albumin. Lysis of protoplasts was in 10 mM 2-mercaptoethanol at 0° C. 5 minutes later potassium phosphate buffer pH 7.3 up to 10 mM and MgSO₄ up to 5 mM were added. Whole cells were eliminated by centrifugation at 600×g for 10 minutes. Lysed protoplasts, after treatment with DNase for 30 minutes at 10° C were separated, by centrifugation, into membrane and supernatant fractions. Membranes were washed with twice 5 ml of 0.02 M potassium phosphate buffer pH 7.3 containing 10 mM 2-mercaptoethanol and are referred to as 'membrane washes'. (From *Europ. J. Biochem.*, 1970, *12*, 576; courtesy of the editors.)

TABLE 2 Purification of acyl-CoA synthetase

Fraction	Step	Total protein (mg)	Total activity (mU)	Specific activity (mU/mg)	Yield (%)	Purification (-fold)	Hexanoate activity / Hexadecanoate activity
I	Streptomycin sulfate supernatant	4000	5150	1.3			3.7
II	0—0.4 satd. ammonium sulfate fraction	1000	1530	1.5	30		—
III	First DEAE-cellulose column	112	1470	12.8	28.6	9.8	4.3
IV	Second DEAE-cellulose column	7.9	860	110	16.7	84.5	—
V	Sephadex G-200 column	4	425	106	8.2	81.5	3.7

The conditions of assay were as follows: 100 mM potassium phosphate pH 7.6, 8 mM $MgCl_2$, 10 mM ATP, 0.5 mM CoA and 0.7 mM [1-^{14}C]hexadecanoic acid (specific activity ranging from 3000 to 7000 d.p.m./nmole); 10 minutes incubation at 30° C, final volume 0.1 ml. Control experiments lacking CoA were systematically included and their values subtracted from those obtained with complete medium. CoA derivatives were extracted and counted as previously described (Samuel and Ailhaud, 1969). (From *Europ. J. Biochem.*, 1970, *12*, 576; courtesy of the editors.)

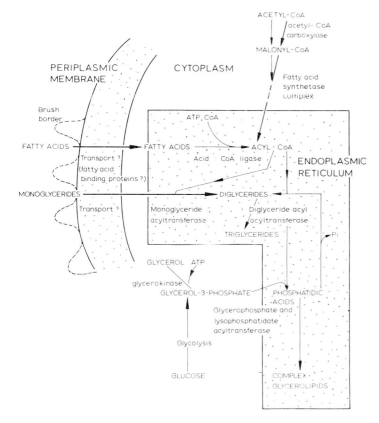

Fig. 2. General scheme of neutral lipid biosynthesis.

and medium chain fatty acids on one hand and long chain fatty acids on the other. These enzymes are present as a bi-enzymatic complex whose dissociation leads to inactivation (Samuel et al., 1970). In eukaryotic cells fatty acid activation is primarily located in the endoplasmic reticulum (see below and Fig. 2). The mode of fatty acid transport from the periplasmic membrane to the endoplasmic reticulum is still not clear (see above). Two fatty acid-binding proteins have been recently described (Ockner et al., 1972; Ritter and Dempsey, 1973). Both are of cytoplasmic origin, although an association with the cytoplasmic membrane cannot be excluded. The first one (mol. wt. 12,000) has been found in the cytosol of intestinal mucosa, liver, myocardium, kidney and adipose tissue of the rat, while the second one (mol. wt. 150,000) was purified to homogeneity from rat liver cytoplasm. Their role in fatty acid transport remains speculative.

TABLE 3   Comparative properties of *E. coli* and rat liver acid:CoA ligase (AMP)

	Molecular weight	Purification (-fold)	K_m (μM)			V_{max} (nmoles/min/mg)		
			ATP	CoA	Palmitate	Octanoate	Palmitate	Oleate
E. coli (Samuel et al., 1970)	120,000	81	3000	90	20	150	106	40
Rat liver (Bar-Tana et al., 1971)	168,000	14	4650	50	42	6	300	94

Acyl-CoA synthetase has successfully been solubilized and partially purified from rat liver microsomal fraction (Bar-Tana et al., 1971). Table 3 shows the main kinetic properties of this enzyme, which are similar to those of the enzyme purified from *E. coli*. However, the specificity of the rat liver enzyme toward short chain fatty acids is very low as compared to that of the *E. coli* enzyme.

ENZYMES OF GLYCEROL-3-PHOSPHATE ESTERIFICATION

Phosphatidic acid biosynthesis is a key step in the formation of neutral lipids starting from glycerol-3-phosphate. Two enzymes are involved in this process. In prokaryotic cells direct proof was obtained by the isolation of thermosensitive mutants of *E. coli* defective either in acyl-ACP(CoA):glycerol-3-phosphate acyltransferase or in acyl-ACP(CoA):lysophosphatidate acyltransferase (Cronan et al., 1970; Hechemy and Goldfine, 1971). In eukaryotic cells, solubilization and partial purification of the two separate activities have been recently achieved (Yamashita and Numa, 1972; Yamashita et al., 1972). For all cells the major pathway of phosphatidic acid synthesis involves the following sequence:
sn-glycerol-3-phosphate+saturated acyl-CoA → l-acylglycerol-3-phosphate+CoA
l-acylglycerol-3-phosphate+unsaturated acyl-CoA → phosphatidic acid + CoA.

In prokaryotic cells acyl-CoA can be replaced by acyl-ACP (Goldfine et al., 1967; Van den Bosch and Vagelos, 1970). Interestingly enough, the asymmetric distribution of saturated and unsaturated fatty acids is reversed in *Clostridium butyricum* phospholipids. Using a mixture of chemically synthesized oleyl- and palmityl-ACP, the phosphatidic acids formed by a membrane fraction of these cells mimic the distribution observed in vivo (Goldfine and Ailhaud, 1971). Thus the asymmetric distribution apparently arises at the level of phosphatidic acid synthesis. It must be pointed out that the preferential esterification of 1-acylglycerol-3-phosphate by unsaturated fatty acids is observed with mammalian enzymes providing that the lysophosphatidate is present in vitro at low concentrations. This situation is likely to be true in vivo, where this intermediate does not accumulate during phosphatidic acid formation (Okuyama and Lands, 1972).

ENZYMES OF MONOGLYCERIDE ESTERIFICATION

The esterification of monoglycerides is quantitatively important in mammals. The hydrolysis of triglycerides either in the intestinal lumen by pancreatic lipase or in the blood by lipoprotein lipase mainly leads to formation of 2-monoglycerides, owing to the positional specificity of both enzymes (Entressangles et al., 1966; Morley and Kuksis, 1972). Different authors (Ailhaud et al., 1964; Johnston and Rao, 1965) have shown that the main pathway of monoglyceride esterification in the enterocytes of different species follows the sequence shown in Figure 3.

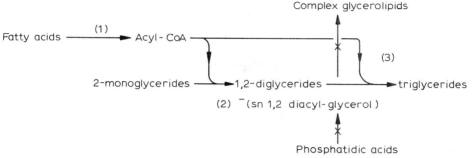

Fig. 3. The monoglyceride pathway in intestinal mucosa.

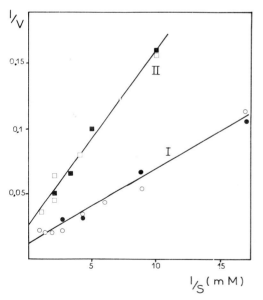

Fig. 4. Existence of two separate enzymes for the esterification of monoglycerides to triglycerides. The Lineweaver-Burk plots correspond to experiments with (●–● ■–■) or without competitor (○–○ □–□). All assays were performed at 37°C and pH 6.0. The final volume (1 ml) contains 0.22 mM palmityl-CoA, 12 mg of serum-albumin, 10% acetone and 0.35 mg of proteins of a microsomal fraction of rat enterocytes. After 5 minutes (zero-order kinetics in this assay), the reaction is stopped by adding ethyl dioxide and the different glycerides separated by thin-layer chromatography as previously described (Ailhaud et al., 1964).

Curve I: substrate 1-monoolein; competitor 1,2-diolein (0.75 mM). The reaction rate (V) is given by the amount of 1,3-diglyceride formed in 5 minutes.

Curve II: substrate 1,2-diolein; competitor 1-monoolein (0.5 mM). The reaction rate (V) is given by the amount of triglyceride formed in 5 minutes. (From G. Ailhaud et al., 1964, *Biochim. biophys. Acta (Amst.)*: courtesy of the editors.)

A 70 to 100-fold purification of a multi-enzyme complex (triglyceride synthetase complex) from a microsomal fraction of hamster intestinal mucosa has been achieved (Rao and Johnston, 1966). It catalyses reactions 1, 2 and 3 shown in Figure 3. No physical separation of the monoglyceride:acyl-CoA acyltransferase from the diglyceride:acyl-CoA acyltransferase was obtained throughout the purification procedure. However, indirect evidence supports the idea of two separate enzymes, as shown in Figure 4 (using a microsomal fraction of rat enterocytes).

No inhibition whatsoever of 1,3-diglyceride synthesis from 1-monoolein and palmityl-CoA as substrates was observed by adding 1,2-diolein as competitor. Similar results were found by adding 1-monoolein as competitor when 1,2-diolein and palmityl-CoA were the substrates of the diglyceride:acyl-CoA acyltransferase activity. Other competition experiments performed with 1- and 2-monoglycerides have shown that a single monoglyceride acyltransferase activity is specific for both substrates (Ailhaud et al., 1964).

LOCALIZATION OF THE TRIGLYCERIDE SYNTHETASE COMPLEX IN SUBCELLULAR FRACTIONS OF ENTEROCYTES

In order to understand better the relationships between fatty acid transport, fatty acid activation and monoglyceride esterification, a careful reexamination concerning the enzymes involved was undertaken, using a technique recently described (Louvard et al., 1973) for the purification of closed vesicles from the brush border of the jejunum pig enterocytes.

As shown in Table 4, the bulk of the glyceride transacylases sediment with the endoplasmic reticulum, where the highest specific activities are found. Figures 5 and 6 give additional support in that, whilst the monoglyceride and diglyceride acyltransferase activities coincide with

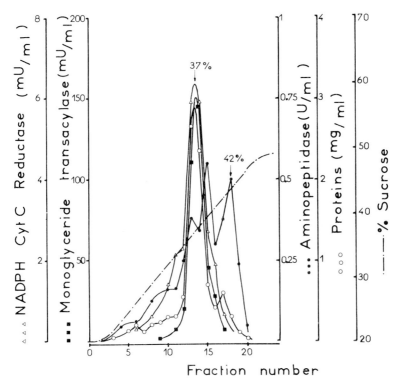

Fig. 5.   Sucrose gradient centrifugation of the endoplasmic reticulum fraction. This fraction was obtained according to Louvard et al. (1973). 12 mg of proteins of fraction I were layered on 32 ml of a 20 to 50% (w/w) sucrose gradient. After 18 hours at 22,000 r.p.m. in a swinging bucket rotor, 1.2 ml fractions were collected and assayed for proteins and for the different enzymatic activities as described in Table 4.

TABLE 4 Subcellular localization of the triglyceride synthetase in the epithelial cells of the pig jejunum

	Aminopeptidase			NADPH-cytochrome c reductase			Acyl-CoA synthetase			Glyceride transacylase		
	Duodenum	Jejunum	Ileum	Duodenum	Jejunum	Ileum	Duodenum	Jejunum	Ileum	Duodenum	Jejunum	Ileum
Homogenate	100 (100)	100 (150)	100 (106)	100 (16)	100 (13.5)	100 (6.1)	100 (4.5)	100 (3)	not determined	100 (14)	100 (17)	100 (11.4)
Endoplasmic reticulum	10 (30)	15 (60)	15 (30)	75 (23)	80 (18.3)	83 (9.5)	81 (4.2)	85 (3.5)	not determined	68 (40)	86 (27)	87 (25)
Brush border	10 (1000)	25 (2000)	18 (1300)	1 (10)	0.1 (0.75)	0.55 (2.2)	no detectable activity	no detectable activity	not determined	2 (12)	1 (8.5)	1.5 (11.3)

Specific activities in parentheses are expressed in nmoles/min/mg protein.
The conditions of assay for aminopeptidase and NADPH-cytochrome c reductase were those used by Louvard et al. (1973). The acyl-CoA synthetase was assayed according to Ailhaud et al. (1964) and the monoglyceride and diglyceride transacylases according to Rodgers (1969).

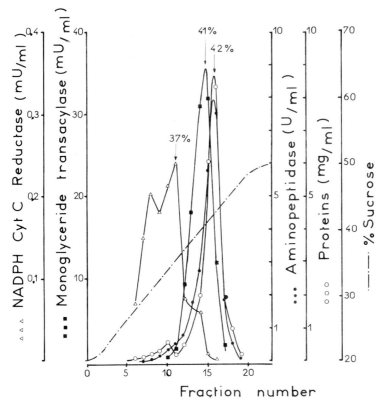

Fig. 6.   Sucrose gradient centrifugation of the brush border fraction. This fraction was obtained according to Louvard et al. (1973). 30 mg of proteins of the brush border fraction were layered and centrifuged under the same conditions as in Figure 5.

NADPH-cytochrome *c* reductase as marker for endoplasmic reticulum, the activities present in the brush border (which represent around 1% of the total activity found in homogenates) follow neither aminopeptidase and alkaline phosphatase nor NADPH-cytochrome *c* reductase. A small contamination of the purified closed vesicles by lateral membranes of the enterocytes and/or periplasmic membranes of the crypt cells could account for this result. Thus the triglyceride synthetase complex seems to be located in the endoplasmic reticulum and absent from the brush border. Electron microscopic studies on the fine structural localization of the glyceride acyltransferase of rat duodenum and jejunum agree well with our results (Higgins and Barnett, 1971).

REGULATORY ASPECTS OF NEUTRAL LIPID BIOSYNTHESIS

Until now no effectors have been found in vitro which will modulate the catalytic activity of acyl-CoA synthetase on the one hand, and monoglyceride and diglyceride acyltransferase on the other hand. However, a modulation of fatty acid biosynthesis, that is before the esterification process, has been postulated both in vitro and in vivo (Bortz and Lynen, 1963). The most recent studies both on prokaryotic cells and on cultured eukaryotic cells, e.g. fibroblasts and hepatocytes, have shown that, in the presence of an exogenous supply of fatty acids, the cell will adjust by lowering the rate of endogenous fatty acid synthesis from acetate. The results shown in Table 5 indicate that, in *E. coli* wild type, the total acetate incorporation is substantially decreased by adding 0.175 to 1.92 mM oleate in the culture medium. Comparatively the rate of unsaturated fatty acid biosynthesis is more reduced than the rate observed for saturated fatty acids (last column of Table 5). A regulation of the rate of fatty acid synthesis is also observed on a short term basis (within 10 minutes) by adding 0.12 mM of a mixture of fatty acids to fibroblasts cultured in a lipid-depleted medium (Jacobs et al., 1973). The exact nature of the inhibition, whether free fatty acid, acyl-CoA or some other intermediate is at present unknown.

The addition of free fatty acids to the medium not only inhibits fatty acid biosynthesis on a short term basis but will also decrease on a long term basis (one day or more) the acetyl-CoA carboxylase content of the fibroblasts as measured by precipitation of the enzyme with specific antibodies. Similar and dramatic decreases of the acetyl-CoA carboxylase levels brought about by nutritional factors (from a fat-free diet to a high-fat diet) have also been demonstrated (Majerus and Kilburn, 1969). In that case the compound(s) which induces the acetyl-CoA carboxylase after fat-free feeding is also unknown.

TABLE 5

Experiment	Culture medium	[1-^{14}C] Acetate incorporation into total lipids (d.p.m.) (10^6)	Ratio unsaturated fatty acids: saturated fatty acids
I	Glycerol	6.2	2.2
	Glycerol + oleate 0.175 mM	3.7	1.24
	Glycerol + oleate 0.525 mM	2.1	0.56
	Glycerol + oleate 1.92 mM	3.6	0.32
II	Glycerol	6.7	1.5
	Glycerol + oleate 0.525 mM	5.7	0.74
III	Glycerol	7.85	0.96
	Glycerol + elaidate 0.525 mM	4.6	0.54
IV	Glycerol	10	1.0
	Glycerol + pentyl-4-undecanoate 0.525 mM	14.5	0.95
	Glycerol + elaidate 0.525 mM	8	0.75

E. coli ML-308 (optical density at 410 nm : 0.25 to 0.3) was grown for one generation (75 minutes) at 37° C with glycerol (2 g/l). The culture was then divided into two 50 ml aliquots. To one aliquot was added the fatty acid. After 75 minutes 0.5 mCi of [1-^{14}C] acetate was added to both cultures. After an additional hour at 37° C with shaking, the cells were centrifuged and carefully washed. Total free fatty acids were obtained after saponification of the cells. Unsaturated and saturated fatty acids were separated by thin-layer chromatography. (From J. Estroumza and G. Ailhaud, 1971, *Biochimie*; courtesy of the editors.)

It has been recently proposed that stearate (or stearyl-CoA) could be the potential negative regulator since, as observed with chick embryos before and after hatching, it is the only metabolite which increases when acetyl-CoA carboxylase and fatty acid synthetase decrease, and vice versa. None of the other metabolites were correlated with enzyme concentration before and after hatching. Furthermore, glycerol-3-phosphate changes in response to stimulation of glucose metabolism could indirectly stimulate fatty acid synthesis by acting as a trap for stearyl-CoA (Goodridge, 1973).

No common regulatory mechanisms of complex glycerolipids and neutral lipids, as shown in Figure 3, can apparently be envisaged to occur in intestinal mucosa. (sn) 1,2-diacyl-glycerol arising from 2-monoglyceride acylation will form a separate pool from the (sn) 1,2-diacyl-glycerol arising from (sn)-glycerol-3-phosphate esterification and dephosphorylation (Johnston et al., 1970). Electron microscopic studies support this idea since, while the glycerophosphate pathway is mainly localized to the rough endoplasmic reticulum, the monoglyceride pathway is found to be located mainly on the inner face of the smooth endoplasmic reticulum (Higgins and Barnett, 1971).

From the above comments, it is clear that the refined regulatory mechanisms of enzyme activities involved in neutral lipid biosynthesis are still largely unknown. It is hoped that the standardization of the procedures for membranous enzyme and receptor solubilization will in future allow a better understanding of these problems, with the use of purified enzymes.

REFERENCES

Ailhaud, G., Samuel, D., Lazdunski, M. and Desnuelle, P. (1964): Quelques observations sur le mode d'action de la monoglycéride transacylase et la diglycéride transacylase de la muqueuse intestinale. Biochim. biophys. Acta (Amst.), 84, 643.

Bar-Tana, J., Rose, G. and Shapiro, B. (1971): The purification and properties of microsomal palmitoyl-CoA synthetase. Biochem. J., 122, 353.

Bortz, W. M. and Lynen, F. (1963): Elevation of long-chain acyl-CoA derivatives in livers of fasted rats. Biochem. Z., 337, 505.

Crass, M. F. (1972): Exogenous substrate effects on endogenous lipid metabolism in the working rat heart. Biochim. biophys. Acta (Amst.), 280, 71.

Cronan Jr, E. J., Ray, T. K. and Vagelos, P. R. (1970): Selection and characterization of an E. coli mutant defective in membrane lipid biosynthesis. Proc. nat. Acad. Sci. (Wash.), 65, 737.

Entressangles, B., Sari, H. and Desnuelle, P. (1966): On the positional specificity of pancreatic lipase. Biochim. biophys. Acta (Amst.), 125, 597.

Estroumza, J. and Ailhaud, G. (1971): Effet des acides gras exogènes sur la vitesse de biosynthèse des acides gras chez E. coli. Biochimie, 53, 837.

Goldfine, H. and Ailhaud, G. (1971): Fatty acyl-acyl carrier protein and fatty acyl-CoA as acyl donors in the biosynthesis of phosphatidic acid in Clostridium butyricum. Biochem. biophys. Res. Commun., 45, 1127.

Goldfine, H., Ailhaud, G. and Vagelos, P. R. (1967): Involvement of acyl carrier protein in acylation of glycerol-3-phosphate in Clostridium butyricum. J. biol. Chem., 242, 4466.

Goodridge, A. G. (1973): Regulation of fatty acid synthesis in the liver of prenatal and early postnatal chicks. J. biol. Chem., 248, 1939.

Hechemy, K. and Goldfine, H. (1971): Isolation and characterization of a temperature-sensitive mutant with a lesion in the acylation of lysophosphatidic acid. Biochem. biophys. Res. Commun., 42, 245.

Higgins, J. A. and Barnett, R. J. (1971): Fine structural localization of acyltransferase. J. Cell Biol., 50, 102.

Jacobs, R. A., Sly, W. S. and Majerus, P. W. (1973): The regulation of fatty acid biosynthesis in human skin fibroblasts. J. biol. Chem., 248, 1268.

Johnston, J. M. and Rao, G. A. (1965): Triglyceride biosynthesis in the intestinal mucosa. Biochim. biophys. Acta (Amst.), 106, 1.

Johnston, J. M., Paltauf, F., Schiller, C. M. and Schultz, L. D. (1970): The utilization of the α-glycero-phosphate and monoglyceride pathways for phosphatidylcholine biosynthesis in the intestine. Biochim. biophys. Acta (Amst.), 218, 124.

Klein, K., Steinberg, R., Fiethen, B. and Overath, P. (1971): Fatty acid degradation in Escherichia coli. An inducible system for the uptake of fatty acids and further characterization of old mutants. Europ. J. Biochem., 19, 442.

Lee, K. Y., Simmonds, W. J. and Hoffman, N. E. (1971): The effects of partition of fatty acid between oil and micelles on its uptake by everted intestinal sacs. Biochim. biophys. Acta (Amst.), 249, 548.

Louvard, D., Maroux, S., Baratti, J., Desnuelle, P. and Mutaftschiev, S. (1973): On the preparation and some properties of closed membrane vesicles from hog duodenal and jejunal brush border. Biochim. biophys. Acta (Amst.), 291, 747.

Lyon, I. (1968): Studies on transmutal potentials in vitro in relation to intestinal absorption. Biochim. biophys. Acta (Amst.), 163, 75.

Majerus, P. W. and Kilburn, E. (1969): Acetyl-CoA carboxylase. The roles of synthesis and degradation in regulation of enzyme levels in rat liver. J. biol. Chem., 244, 6254.

Masoro, E. J. (1968): In: Physiological Chemistry of Lipids in Mammals, p. 222. B. Saunders, Philadelphia.

Mavis, R. D. and Vagelos, P. R. (1972): The effect of phospholipid fatty acid composition on membranous enzymes in Escherichia coli. J. biol. Chem., 247, 652.

Mavis, R. D., Bell, R. M. and Vagelos, P. R. (1972): Effect of phospholipase C hydrolysis of membrane phospholipids on membranous enzymes. J. biol. Chem., 247, 2835.

Mishkin, S., Yalovsky, M. and Kessler, J. I. (1972): Stages of uptake and incorporation of micellar palmitic acid by hamster proximal intestinal mucosa. J. Lipid Res., 13, 155.

Morley, N. and Kuksis, A. (1972): Positional specificity of lipoprotein lipase. J. biol. Chem., 247, 6389.

Ockner, R. K., Manning, J. A. , Poppenhausen, R. B. and Ho, W. K. L. (1972): A binding protein for fatty acid in cytosol of intestinal mucosa, liver, myocardium and other tissues. Science, 177, 56.

Okuyama, H. and Lands, W. E. M. (1972): Variable selectivities of acyl-CoA:monoacylglycerophosphate acyltransferases in rat liver. J. biol. Chem., 247, 1414.

Overath, P., Raufuss, E. M., Stoffel, W. and Ecker, W. (1967): The induction of the enzymes of fatty acid degradation in *Escherichia coli*. Biochem. biophys. Res. Commun., 29, 28.

Overath, P., Schairer, H. U. and Stoffel, W. (1970): Correlation of in vivo and in vitro phase transitions of membrane lipids in *Escherichia coli*. Proc. nat. Acad. Sci. (Wash.), 67, 606.

Raison, J. K. and Lyons, J. M. (1971): Hibernation: Alteration of mitochondrial membranes as a requisite for metabolism at low temperature. Proc. nat. Acad. Sci. (Wash.), 68, 2092.

Rao, G. A. and Johnston, J. M. (1966): Purification and properties of triglyceride synthetase from the intestinal mucosa. Biochim. biophys. Acta (Amst.), 125, 465.

Ritter, M. C. and Dempsey, M. E. (1973): Squalene and sterol carrier protein: structural properties, lipid-binding and function in cholesterol biosynthesis. Proc. nat. Acad. Sci. (Wash.), 70, 265.

Rodgers Jr, J. B. (1969): Assay of acyl-CoA monoglyceride acyltransferase from rat small intestine using continuous recording spectrophotometry. J. Lipid Res., 10, 427.

Samuel, D. and Ailhaud, G. (1969): Comparative aspects of fatty acid activation in *Escherichia coli* and *Clostridium butyricum*. F.E.B.S. Letters, 2, 213.

Samuel, D., Estroumza, J. and Ailhaud, G. (1970): Partial purification and properties of acyl-CoA synthetase of *Escherichia coli*. Europ. J. Biochem., 12, 576.

Samuel, D., Paris, S. and Ailhaud, G. (1972): Isolation and characterization of mutants of *Escherichia coli* with cellular division selectively affected by growth on fatty acids. J. Bact., 112, 480.

Spector, A. A. (1968): The transport and utilization of free fatty acid. Ann. N.Y. Acad. Sci., 149, 768.

Steim, J. M., Tourtelotte, M. E., Reinert, J. C., McElnakey, R. N. and Rader, R. L. (19): Calorimetric evidence for the liquid-crystalline state of lipids in a biomembrane. Proc. nat. Acad. Sci. (Wash.), 63, 104.

Sylvèn, C. (1970): Influence of blood supply on lipid uptake from micellar solutions by the rat small intestine. Biochim. biophys. Acta (Amst.), 203, 365.

Van den Bosch, H. and Vagelos, P. R. (1970): Fatty acyl-CoA and fatty acyl-ACP as acyl donors in the synthesis of lysophosphatidic acids and phosphatidic acids in *Escherichia coli*. Biochim. biophys. Acta (Amst.), 218, 233.

Wright, J. D. and Green, C. (1971): The role of the plasma membrane in fatty acid uptake by rat liver parenchymal cells. Biochem. J., 123, 837.

Yamashita, S. and Numa, S. (1972): Partial purification and properties of glycerophosphate acyltransferase from rat liver. Europ. J. Biochem., 31, 565.

Yamashita, S., Hosaka, K. and Numa, S. (1972): Resolution and reconstitution of the phosphatidate-synthesizing system of rat liver microsomes. Proc. nat. Acad. Sci. (Wash.), 69, 3490.

The effects of protein deficiency in utero on the morphology and metabolism of adipose tissue and liver in the rat

K. I. McLeod, R. B. Goldrick, P. J. Nestel and **H. M. Whyte**

Department of Clinical Science, John Curtin School of Medical Research, Australian National University, Canberra, Australia

Female rats maintained on a protein-deficient diet (6% protein by weight) for two weeks previous to and during gestation give birth to small litters of stunted offspring (test progeny) that exhibit deficits in body and organ weights which persist into maturity despite adequate postnatal feeding. Carcass composition expressed in terms of % by weight of protein, fat, ash and water is, however, normal (McLeod et al., 1972a).

The stunting of growth is associated with a normal food intake per unit body weight but an excessive urinary nitrogen loss as urea, creatinine and α amino acid (McLeod et al., 1972 b). Amino acid incorporation into liver and muscle protein is also significantly reduced.

In most organs, including the liver, the weight deficit is associated with a reduction in cell numbers, cell size being unaffected. However, in epididymal adipose tissue both the number and size of cells are significantly reduced. As adipocyte size has been related to many facets of adipose tissue metabolism (Girolamo and Mendlinger 1969; Goldrick and McLoughlin, 1970; Vintent and Gliemann, 1970; Zinder and Shapiro, 1971) and as the test progeny had already been shown to exhibit deranged protein metabolism, we examined the possibility that the presence of significantly smaller adipocytes in test progeny might lead to deficient adipose tissue lipogenesis. To accentuate any deficiency that may exist, some of the test and control progeny were given a 20% fructose supplement in their drinking water for two weeks before the experiments. Fructose is a potent hypertriglyceridaemic agent in both rats and man (Nikkila, 1969).

Although adipose tissue is normally a major site of lipogenesis in the rat, it has been suggested that the lack of fructokinase in adipose tissue causes a shift in the major site of lipogenesis from adipose tissue to liver in fructose-fed animals (Chevalier et al., 1972). Because of this, we also examined the in vivo and in vitro incorporation of acetate-1-^{14}C and glucose-U-^{14}C into various liver lipid classes in chow-fed and fructose-fed test and control progeny.

METHODS

In vivo and in vitro studies were carried out in four groups of animals at 10 or 30 weeks of age: (1) chow-fed control progeny (C); (2) chow-fed test progeny (T); (3) fructose-supplemented control progeny (FC); (4) fructose-supplemented test progeny (FT).

Adipose tissue studies

Animals used for in vitro studies were killed by decapitation 2 or 24 hours after the removal of chow and fructose. Blood was collected for triglyceride, cholesterol and glucose determinations and 100–200 mg pieces of epididymal adipose tissue were excised and incubated in 3 ml Krebs-Ringer bicarbonate buffer, 20 mg albumin, 1.5 mg glucose and 1.5 μCi glucose-U-^{14}C. Insulin (1500 μU) was added to some flasks. After 2 hours tissues were removed and the lipid was extracted (Goldrick and McLoughlin, 1970). The radioactivity in total lipid and in glyceride-fatty acid (following saponification) was determined and that in glyceride-glycerol was calculated from these values. The number and size of adipocytes were determined as previously described (Dole and Meinertz, 1960) and the results were expressed as nmoles glucose incorporated per adipocyte. In vivo studies were carried out in the absorptive state in 10-week-old progeny and lipids were extracted and radioassayed half an hour after the injection of 5 μCi of glucose-U-^{14}C into a tail vein.

Liver studies

The liver studies were carried out in vivo and in vitro in 10-week-old progeny in the absorptive state. Liver slices were incubated in Krebs-Ringer bicarbonate buffer for 2 hours in the presence of 0.45 μCi acetate-1-^{14}C or glucose-U-^{14}C. Studies were again carried out with or without the addition of insulin (0.1 U). Liver lipids were extracted in chloroform:methanol and separated by thin-layer chromatography. The radioactivity in the various lipids was determined and the results were expressed per unit weight of tissue.

RESULTS AND DISCUSSION

Body weights, epididymal fat pad weight, liver weight, the numbers of adipocytes and hepatocytes and the size of adipocytes were reduced in test progeny (Table 1). Plasma triglyceride,

TABLE 1 Body, fat pad and liver weight and the number and size of adipocytes and hepatocytes in 10-week-old progeny of protein-deficient dams (test progeny)[1]

Group[2]	Body weight (g)	Epididymal fat pad weight (mg)	Adipocyte size (μg triolein)	Adipocyte number ($\times 10^6$)	Liver weight (g)	Hepatocyte size (g protein /g DNA)	Hepatocyte numbers $\times 10^8$ (mg DNA/ 6.2 pg)
C	230	3186	0.247	10.00	9.81	112	40 13
T	173***	1878***	0.173***	6.93+	6.89*	110	27.76***
FC	211†	3215	0.257	9.48	9.06	109	38.76
FT	162***†	1925***	0.186***	6.56***	6.16**	113	25.12***

[1] Mean ± 1 S.E.M. for groups of eight rats.
[2] C = chow-fed control progeny.
  T = chow-fed test progeny.
  FC = fructose-fed control progeny.
  FT = fructose-fed test progeny.

Significant differences between control and test progeny:* P<0.05, ** P<0.01, *** P<0.001.
Significant differences due to fructose feeding:† P<0.05.

cholesterol and glucose concentrations were similar in Groups C and T. Fructose feeding had no effect on plasma cholesterol or blood glucose concentrations in either of these groups but significantly elevated the triglyceride concentration. The elevation was of the same order in both test and control progeny but was more evident in older progeny of these two groups.

Adipose tissue studies

Whether tested in the absorptive or fasted state, and in either the presence or absence of insulin, incorporation per adipocyte of glucose-U-^{14}C into glyceride-glycerol and glyceride-fatty acid was similar in the two chow-fed groups (C and T) at either age (see Table 2). With one exception (basal incorporation into glycerol in older progeny) fructose feeding increased both basal and insulin-stimulated incorporation into glyceride-glycerol and glyceride-fatty acids, but to a significantly smaller extent in the test progeny than in the control progeny. With the exception of fructose-fed test progeny, there was a rise in baseline glycerol production with age in all groups

TABLE 2   The incorporation of glucose-U-^{14}C, following fructose supplementation of the diet, into glyceride-glycerol and glyceride-fatty acids in epididymal adipose tissue of progeny from control and protein-deficient dams[1]

Group[2]	10-week-old progeny		30-week-old progeny	
	Glycerol	Fatty acids	Glycerol	Fatty acids
		(nmoles glucose/10[6] cells/2 hours)		

Absorptive state

Group[2]	Glycerol	Fatty acids	Glycerol	Fatty acids
C + Insulin	334	309	352	12.2
C — Insulin	216	63	307	6.0
T + Insulin	333	317	375	11.3
T — Insulin	209	64	312	6.1
FC + Insulin	823††	2833††	581††	49.0††
FC — Insulin	351††	496††	432††	39.1††
FT + Insulin	620††**	1231††**	479††**	27.2††*
FT — Insulin	328††	227††**	327*	22.0††*

24 hour-fasted state

Group[2]	Glycerol	Fatty acids	Glycerol	Fatty acids
C + Insulin	501	81	347	3.2
C — Insulin	385	18	281	1.6
T + Insulin	471	78	298	2.9
T — Insulin	352	12	247	1.7
FC + Insulin	891††	1102††	592††	12.5††
FC — Insulin	554††	108††	421††	8.1††
FT + Insulin	683††**	410††**	501††*	6.2††**
FT — Insulin	501††	60††**	329†**	4.3††*

[1] Mean for groups of eight rats.
[2] C = control progeny; no fructose supplement.
   T = progeny of protein-deficient dams; no fructose supplement.
   FT = fructose-fed test progeny.
   FC = fructose-fed control progeny.
Significant differences between control and test progeny: ** P<0.001, * P<0.01.
Significant differences between fructose-supplemented and chow-fed groups: †† P<0.001, † P<0.01.

when tested in the absorptive state but a fall when tested in the fasted state. There was also a marked reduction with age in both basal and insulin-stimulated fatty acid synthesis in all groups. Following fructose feeding, the adipocytes from test progeny were less responsive to insulin than those of controls. The decreased responsiveness cannot be related to cell size as the chow-fed control and test progeny show similar degrees of responsiveness despite comparable differences in cell size. In vivo studies carried out on 10-week-old progeny confirmed the in vitro findings.

Liver studies

When expressed as a function of unit tissue weight, no differences were apparent between control and test groups in the incorporation of acetate or glucose into phospholipid, triglyceride-fatty acid, triglyceride-glycerol, free fatty acid, cholesterol or cholesterol-ester. Fructose feeding increased lipogenesis per hepatocyte but to the same extent in control and test progeny. Similarly, there was no difference in insulin responsiveness between control and test animals.

   We conclude that intrauterine protein deprivation affects liver lipogenesis only in so far as it reduces the total number of hepatocytes, whereas in adipose tissue both the total number of adipocytes and the ability of each individual adipocyte to respond to a lipogenic stimulus is reduced. This decreased ability to respond may be the result of an effect of early malnutrition on the capacity of the cell membrane transport system or the lipogenic enzyme system or both to respond to a stimulus in later life.

REFERENCES

Chevalier, M., Wiley, J. and Leveille, G. (1972): Effect of dietary fructose on fatty acid synthesis in adipose tissue and liver of the rat. J. Nutr., 102, 337.

Dole, V. and Meinertz, H. (1960): Microdetermination of long chain fatty acids in plasma and tissues. J. biol. Chem., 235, 2595.

Girolamo, M. and Mendlinger, S. (1969): Variations in glucose metabolism, lipogenesis, glyceride synthesis and insulin responsiveness of enlarging adipose cells in three mammalian species. Diabetes, 18, 353.

Goldrick, R. B. and McLoughlin, G. (1970): Lipolysis and lipogenesis from glucose in human fat cells of different sizes. Effects of insulin epinephrine and theophylline. J. clin. Invest., 49, 1213.

McLeod, K. I., Goldrick, R. B. and Whyte, H. M. (1972a): The effect of maternal malnutrition on the progeny in the rat. Studies on growth, body composition and organ cellularity in first and second generation progeny. Aust. J. exp. Biol. med. Sci., 50, 435.

McLeod, K. I., Goldrick, R. B. and Whyte, H. M. (1972b): The effect of maternal malnutrition on the progeny in the rat. Studies on nitrogen balance. Aust. J. exp. Biol. med. Sci., 50, 731.

Nikkila, E. (1969): Control of plasma and liver triglyceride kinetics by carbohydrate metabolism and insulin. Advanc. Lipid Res., 7, 63.

Vintent, J. and Gliemann, J. (1970): Glucose metabolism and insulin sensitivity of single fat cells. Diabetologia, 6, 651.

Zinder, O. and Shapiro, B. (1971): Effect of cell size on epinephrine- and A.C.T.H.-induced fatty acid release from isolated fat cells. J. Lipid Res., 12, 91.

Developmental changes in intrauterine growth: retardation of rat interscapular brown adipose tissue

C. Tordet-Caridroit and **A.-M. Cogneville**

Centre de Recherches Biologiques Néonatales, Unité 29 INSERM, Paris, France

The function of brown adipose tissue, although not studied for a long period of time, has recently been the subject of further work (Barnard and Skala, 1970; Smalley, 1970) which has led to the recognition of its importance in chemical, non-shivering thermogenesis in newborn homeothermic mammals. Clinical observations of 'small for date' children exhibiting the 'cold syndrome' established that interscapular brown adipose tissue (ISBAT) cells were totally or partially depleted of fat (Aherne and Hull, 1966). We have created an experimental model which enables us to obtain small for date rats. We report below the variations in weight and chemical composition (water, total lipid, protein and DNA content) of ISBAT in hypotrophic animals and controls during the perinatal period.

MATERIAL AND METHODS

Sherman strain rats were used and the litters were normalised to 6 animals. Using the method developed by Wigglesworth (1964), animals were obtained whose intrauterine growth had been repressed by restriction of the blood flow to one of the uterine horns (IUGR rats). The animals of the other horn were used as controls. An animal was considered as being hypotrophic (IUGR) if the reduction of body weight was greater or equal to 20% when compared to the control of the same age (Tordet-Caridroit et al., 1969). The ISBAT was removed and weighed. Owing to the low weight of the ISBAT, tissue of very young animals was pooled. Dry weights were obtained by lyophilization to a constant weight. Total lipid was extracted according to the procedure of Folch et al. (1957), the determination being obtained by gravimetry. Protein was determined according to Lowry et al. (1951). DNA was extracted using the Schmidt-Thannhauser technique (Volkin and Cohn, 1957) and quantitatively determined using the procedure of Burton (1956). All results are expressed in mg per 100 mg of fresh tissue. The qualitative determination of lipids was carried out using thin-layer chromatography. A histological study of the ISBAT was performed using hematoxylin-PAS staining. The results were analysed by standard statistical procedures (Student-Fischer t test).

RESULTS

Tissue weight

For 21-day-old foetuses (obtained by caesarean section on the day before birth) and rats aged up to 10 days, the weight of the ISBAT in IUGR rats is less than that of the controls of the same age (Fig. 1). The reduction is 70% for the foetuses and 30% for the 10-day-old rats. The ratio of ISBAT wet weight to body weight has frequently been used as an indication of the physiological importance of this tissue. As seen in Figure 1, this ratio when compared to the controls is lower for the IUGR animals up to the age of 2 days. After 3 days, hypotrophic animals have an identical ratio to that of controls.

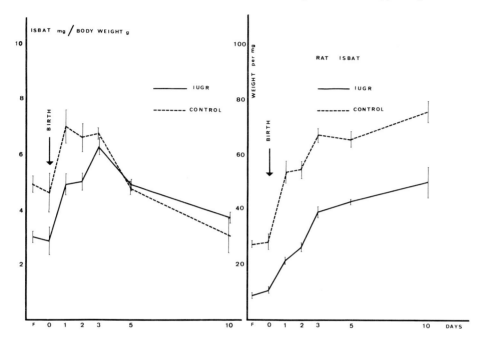

Fig. 1. Left: Interscapular brown adipose tissue as a percentage of body weight in control and IUGR rats during early development. Means ± S.E.M. (vertical bars).

Right: Fresh weight of interscapular brown adipose tissue (ISBAT) in control and hypotrophic (IUGR) rats during early development. Means ± S.E.M. (vertical bars).

Gross composition

As shown in Table 1, an inverse relationship between water and lipid content exists in both hypotrophic and control rats. The percentage of total lipid increases during the first 3 days, decreases between the 3rd and 5th day and subsequently remains constant, while the percentage water content varies in the opposite direction. Although the water content of ISBAT is greater in the IUGR than in control animals, the differences are only highly significant ($P < 0.01$) during the first 24 hours of life. Lipid concentration is lower in the IUGR animals. The observed differences are highly significant ($P < 0.01$) at 24 hours after birth, gradually diminish up to the 3rd day and become significant ($P < 0.05$) once more on the 5th postnatal day.

Qualitative determination of the lipid extracts enabled us to conclude that the greater part consisted of triglyceride (80–90%), the remainder being mainly phospholipid. The results were identical for both categories of animals.

Total protein varied between 8 and 12% during the period studied and was found to be slightly lower in hypotrophic animals than in controls, the observed difference only becoming significant ($P < 0.05$) on the 3rd postnatal day. The determination of total mitochondrial protein (Fig. 2) shows that the level is lower in the IUGR animals ($P < 0.01$). These results would seem to imply that the ISBAT cells of the IUGR animals have less mitochondria than those of the control animals.

In both categories of animals, the DNA concentration was found to decrease with age. From the day before birth up to 24 hours after birth IUGR animals had a greater concentration of DNA compared to the controls. The ratios lipid/DNA, dry weight/DNA, protein/DNA, and fresh weight/DNA were always lower in the IUGR animals up to the 3rd day of postnatal life.

TABLE 1   Water, lipids, proteins and DNA concentrations (mg/100 mg fresh tissue ± S.E.) of developing interscapular brown adipose tissue in control and intrauterine growth-repressed (IUGR) rats.

Age (days)	No. of animals	H$_2$O		Lipid		Protein		DNA	
		IUGR	Control	IUGR	Control	IUGR	Control	IUGR	Control
21 (foetus)	26	86.66 ± 3.93**	79.93 ± 0.26	7.41 ± 1.06**	8.99 ± 0.67	8.31 ± 0.99	9.12 ± 0.50	0.29 ± 0.08*	0.22 ± 0.08
0 (birth)	28	74.07 ± 4.69	73.89 ± 5.23	10.65 ± 4.62**	12.87 ± 4.62	10.57 ± 4.62	12.63 ± 5.77	0.28 ± 0.05*	0.18 ± 0.12
1	20	69.49 ± 9.25**	61.11 ± 7.73	21.38 ± 9.57**	30.31 ± 9.61	9.69 ± 0.83	9.37 ± 1.10	0.28 ± 0.08*	0.21 ± 0.07
2	14	51.74 ± 9.23	48.68 ± 8.80	37.25 ± 9.86	41.72 ±10.22	8.37 ± 1.13	8.40 ± 0.81	0.20 ± 0.05	0.21 ± 0.09
3	17	46.32 ± 6.62	45.98 ± 6.77	43.42 ± 4.65	42.11 ± 7.42	7.81 ± 1.20*	8.63 ± 1.43	0.22 ± 0.10	0.19 ± 0.07
5	17	61.45 ± 7.81	58.68 ± 5.49	26.14 ± 7.86*	30.51 ± 6.51	10.38 ± 1.66	10.44 ± 1.33	0.16 ± 0.07	0.17 ± 0.10
10	8	56.34 ± 4.50	54.15 ± 7.91	29.74 ± 7.43	32.42 ± 6.94	9.25 ± 2.52	9.59 ± 2.74	0.16 ± 0.07	0.14 ± 0.05

** Highly significant (P<0.01).
 * Significant (P<0.05).

This implies that the number and size of the ISBAT cells in these animals are diminished. Histological studies corroborate the biochemical results.

DISCUSSION

Since at birth rats possess very limited reserves of lipid (these reserves being essentially BAT), it is clear that lipogenesis and fat deposition begin very quickly after birth. IUGR animals have a significantly lower ISBAT total weight and a lower lipid content. Furthermore, as seen from histological and biochemical studies, the number and size of ISBAT cells are lower in the IUGR animals. The decrease of mitochondrial protein could imply a smaller number of ISBAT mitochondria.

The carcass lipids are lower in the IUGR animals up to the age of 5 days. At 2 days after birth 5% of the carcass is lipid in the IUGR animals, as against 8% in the controls (P<0.05).

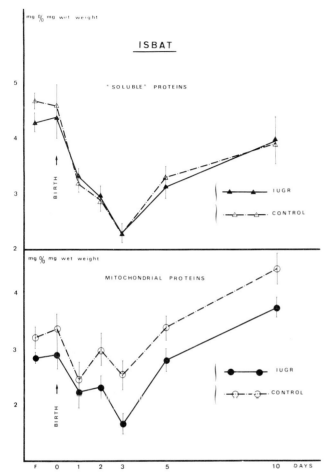

Fig. 2. 'Soluble' and mitochondrial protein concentrations of developing interscapular brown adipose tissue (ISBAT) in control and hypotrophic (IUGR) rats. Means ± S.E.M. (vertical bars).

Although smaller, a difference is still found at 5 days: 7.2% versus 9% (P <0.01) (unpublished data). At 10 days, IUGR and control animals show no differences. In line with this latter observation, it has been found that hepatic glycogen is lower at birth in IUGR animals (Chanez et al., 1971).

Taken together, our results allow us to assume that IUGR animals have a lack of energy reserves available for use in autonomous life and in starting up thermogenesis immediately after birth. This deficiency could play a role in the high mortality level observed during the first 48 hours of life. Furthermore, the decreased lipid content present in the ISBAT of the IUGR animals can be linked to a delay in the mobilisation of fatty acids in the serum of these animals, as described in a previous work (Chanez et al., 1972).

REFERENCES

Aherne, W. and Hull, D. (1966): Brown adipose tissue and heat production in the newborn infant. J. Path. Bact., 91, 223.

Barnard, T. and Skala, J. (1970): The development of brown adipose tissue. In: Brown Adipose Tissue, p. 33. Editor: O. Lindberg. American Elsevier Inc., New York.

Burton, K. (1956): Study of the conditions and mechanism of the diphenylamine reaction for the colorimetric estimation of deoxyribonucleic acid. Biochem. J., 62, 315.

Chanez, C. and Tordet-Caridroit, C. (1972): Glucose, acides gras libres et glycerol du plasma. Arch. franç. Pédiat., 29, 593.

Chanez, C., Tordet-Caridroit, C. and Roux, J. M. (1971): Studies on experimental hypotrophy in the rat. II. Development of some liver enzymes of gluconeogenesis. Biol. Neonat. (Basel), 17, 58.

Folch, J., Lees, M. and Sloan Stanley, G. H. (1957): A simple method for the isolation and purification of total lipids from animal tissues. J. biol. Chem., 226, 497.

Lowry, G., Rosebrough, N., Farr, H. and Randal, R. J. (1951): Protein measurement with the folin phenol reagent. J. biol. Chem., 193, 265.

Smalley, R. (1970): Changes in composition and metabolism during adipose tissue development. In: Brown Adipose Tissue, p. 73. Editor: O. Lindberg. American Elsevier Inc., New York.

Tordet-Caridroit, C., Roux, J. and Chanez, C. (1969): Etude du développement post-natal du rat né dysmature. C.R. Soc. Biol. (Paris), 163, 1321.

Volkin, E. and Cohn, W. (1957): Estimation of nucleic acids. In: Methods of Chemical Analysis, Vol. 1, p. 287. Editor: D. Glick. Interscience Publishers Inc., New York.

Wigglesworth, J. S. (1964): Experimental growth retardation in the foetal rat. J. Path. Bact., 88, 1.

Effects of milk intake on fatty acid distribution during the early postnatal period in the mouse

J. Winand and **J. Christophe**

Laboratory of Biochemistry and Nutrition, University of Brussels Medical School, Brussels, Belgium

The number and size of fat cells are determined early in life. Aubert et al. (1971) and Lemonnier et al. (1973) observed that adult mice originating from small litters had an increased number of enlarged adipose cells and high plasma insulin levels. These parameters were directly related to the amount of milk taken before weaning. Since insulin enhances lipogenesis and fatty acid desaturation it is possible that the caloric intake of the newborn mouse influences not only adipose growth but also fatty acid distribution. In the present study the effect of milk intake on fatty acid distribution during the first 6 weeks after birth has been examined.

In order to achieve marked variations in milk intake, 3-day-old mice of both sexes were randomly distributed to foster mothers in small litters (4 animals), normal litters (8 animals), and large litters (12 animals). At weaning (21 days) mice were allowed free access to the same standard mouse chow. The animals were killed at 3-day intervals up to the time of weaning and the last groups were sacrificed at 6 weeks of age.

The carcass and liver were pooled in each group. After saponification, the fatty acids were extracted, weighed, methylated and analysed by gas-liquid chromatography (Winand, 1970). Milk fat was analysed similarly, after collection from the stomachs of 7-, 9- and 13-day-old mice, suckling for 1 hour after a 3 hour fast. No distinction was made between triglycerides, phospholipids and other lipids. However, it is clear that the fatty acids extracted from the carcass were mostly representative of fat depots while those from the liver were largely present as phospholipids and those from milk fat mostly present as triglycerides.

The wet weight of carcass and liver was adversely influenced by litter size. The differences were already significant on the sixth postnatal day and were greatest at weaning. They did not increase any further in 42-day-old mice, i.e. 3 weeks post-weaning. Slight hepatic hypertrophy was observed at that time in animals previously fed in small groups.

Carcass fatty acids accumulated for 2 weeks in all groups. At the end of this period, they represented only 5% of the wet weight in poorly fed animals as compared to 7% in controls, and as much as 12% in well fed animals. These relative fatty acid contents decreased thereafter in all groups, being 4%, 5% and 7% at 21 days of age in underfed, control and well fed animals, respectively. When the experiment ended at the age of 42 days, the relative carcass lipid content of mice from large litters was still 1% lower than in animals formerly suckling in small litters, and the absolute lipid content was therefore significantly lower than in these previously privileged animals.

In contrast to the data on the carcass, the relative lipid content of the liver was not influenced by milk availability. A slight and parallel decrease in total fatty acid content was observed in all groups.

A comparison of the fatty acid pattern of lipids accumulated in carcass and liver is of interest. If we consider first the case of control animals suckling in litters of 8, it is clear that the carcass was accumulating medium chain fatty acids much more easily than the liver, lauric and myristic fatty acids accounting together for approximately 15% of total fatty acids in the carcass of 2-week-old mice. Oleic acid was also well represented in the carcass but the difference with liver was less obvious than in the case of medium chain fatty acids. At 3 weeks post-weaning the proportions of lauric and myristic acids decreased markedly in the carcass and were almost as low as those observed in the liver, while the proportion of oleic acid remained elevated. The fatty acids of medium chain length accumulating in the carcass were of dietary origin: lauric and myristic acids were well represented in milk fat, and their relative content was even higher after 13 days of suckling than after 3 days. The accumulation of 18:1 in the carcass could be easily accounted for by milk fat until day 12. However the proportion of oleic acid in milk fat decreased during the second week of suckling and the young obviously had to rely on endogenous synthesis and monodesaturation of stearic acid in order to maintain their relatively high content of carcass oleic acid.

When underfed (litters of 12) and overfed (litters of 4) animals were compared to the control animals (litters of 8), there was no obvious difference in fatty acid composition. This indicates that milk fatty acids were essentially stored without preliminary modification in all instances during the first 2 weeks after birth, the only limiting factor being milk availability. During the third postnatal week, endogenous synthesis of oleic acid was taking place and this process was moderate in underfed animals and very active in overfed animals. A slight relative excess of oleic acid was even observed in the carcass of animals bred in small litters during the second and third postnatal weeks.

These results may be correlated with considerable metabolic changes occurring at birth and during the weaning period (Greengard, 1971). During the first days of extrauterine life, lipogenic enzymes are depressed by the new diet, rich in fat and protein but poor in carbohydrate. During the third week of postnatal life, the transition from milk to a high carbohydrate chow is accompanied by higher plasma insulin levels, enhanced glucose utilization, increased lipogenesis and an important remodeling of accumulated lipids. From our experiments and from the data of Aubert (1971) and Lemonnier (1973), it is apparent that the size and fatty acid distribution of fat depots were directly related to milk intake during the pre-weaning period. The quantitative and qualitative changes in the fatty acids stored in post-weaning mice depended in part on the influence exerted by the amount of milk taken up before weaning on the rate of insulin secretion.

REFERENCES

Aubert, R., Suquet, J. P. and Lemonnier, D. (1971): Effets à long terme de trois niveaux d'ingesta lactés sur les lipides corporels, la taille et le nombre de cellules adipeuses de la souris. C. R. Acad. Sci. (Paris), 273, 2636.

Greengard, O. (1971): Enzymic differentiation in mammalian tissues. Essays in Biochem., 7, 159.

Lemonnier, D., Suquet, J. P., Aubert, R. and Rosselin, G. (1973): Long term effect of mouse neonate food intake on adult body composition, insulin and glucose serum levels. Hormone metab. Res., 5, 223.

Winand, J. (1970): Aspects Qualitatifs et Quantitatifs du Métabolisme Lipidique de la Souris Normale et de la Souris Congénitalement Obèse. Arscia, Brussels.

Cholesterol storage in adipose tissue*

A. Angel, J. Farkas, J. Rosenthal and **F. Smigura**

Department of Medicine, University of Toronto, Toronto, Canada

It is now well established that white adipose tissue contains one of the largest pools of exchange-able cholesterol in man as well as lower animals (Vague and Garrigues, 1955; Angel and Farkas, 1970; Farkas et al., 1973). While the concentration of cholesterol in relation to total cell lipid is low (0.6–1.6 mg/g), the concentration per unit DNA or per unit protein exceeds that of most other tissues including liver, muscle and kidney (Fig. 1) (Farkas et al., 1973). The purpose of this communication is to emphasize the physiological significance of cholesterol metabolism in

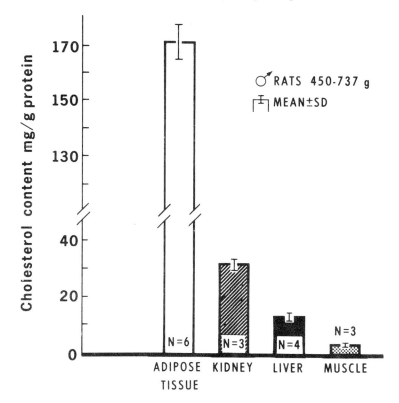

Fig. 1. The cholesterol content of rat epididymal adipose tissue as compared to that of kidney, liver and muscle. It is apparent that adipose tissue has the highest level.

* This work was supported by grants from the Medical Research Council of Canada and the Ontario Heart Foundation.

adipose tissue by (1) describing the anatomical compartmentation of cholesterol in fat tissue, (2) indicating how dietary and nutritional manipulations regulate cholesterol storage in fat tissue, (3) describing how adipose cell growth and obesity affects the adipose tissue cholesterol storage pool, and (4) demonstrating the cholesterogenic properties of fat tissue.

COMPARTMENTATION OF CHOLESTEROL IN ADIPOSE TISSUE

Structural compartmentation of cholesterol in fat tissue exists at both the cellular and subcellular level. In epididymal fat of young rats (6–7 weeks, 150–160 g) two-thirds of fat tissue cholesterol was in collagenase-derived adipocytes and one-third in stromal-vascular elements. In older animals (22 weeks, 450–480 g) 90% of tissue cholesterol was found in adipocytes and the remainder in stromal-vascular elements (Fig. 2). Age-related differences in subcellular cholesterol

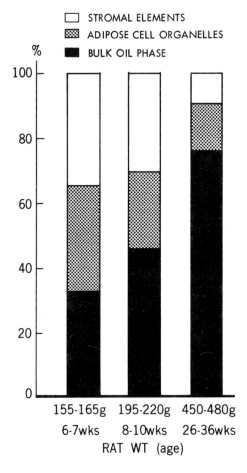

Fig. 2.   Effect of age on cholesterol distribution in adipose tissue. Isolated fat cells were prepared by collagenase digestion of rat epididymal fat. The isolated cells were homogenized in 0.25 M sucrose and centrifuged at 300 *g* for 30 seconds. The upper oil phase, which represents the bulk lipid storage pool, and lower aqueous phase containing membranous organelles, enzymes and soluble cellular constituents were separated and analyzed for cholesterol. The stromal vascular fraction was similarly analyzed. The results are expressed as proportions of total adipose tissue cholesterol.

distribution are also observed. In young rats, 30% of the fat tissue cholesterol was associated with the lipid storage phase (i.e. floating oil layer following homogenization and brief centrifugation) while in tissue of older rats (450–480 g) the proportion increased to 75% (Fig. 2). Additionally, chemical analysis of purified plasma membrane fraction, mitochondria and microsomal membranes from fat cells of young and old animals showed identical cholesterol: phospholipid molar ratios. Thus, cholesterol accumulates in fat tissue in relation to animal age (and hence cell size) and this accumulation occurs primarily in the lipid storage vacuole of the fat cells rather than in membranous organelles.

Almost all the cholesterol in fat tissue is in free form and it exists in a dynamic state of turnover. Following intravenous administration of [4-^{14}C] cholesterol to rats, equilibration of serum and adipose tissue cholesterol specific activity occurred at 7.5 days (Fig. 3). This rate of equilibration resembles that previously reported for muscle and kidney, and is unlike that in liver, red blood cells and small intestine, which equilibrate rapidly (Avigan et al., 1962). The t 1/2 disappearance of adipose tissue cholesterol was found to be 27 days (Fig. 3). These kinetic data demonstrate the slow turnover of fat tissue cholesterol and are consistent with its role as a slowly exchangeable storage pool.

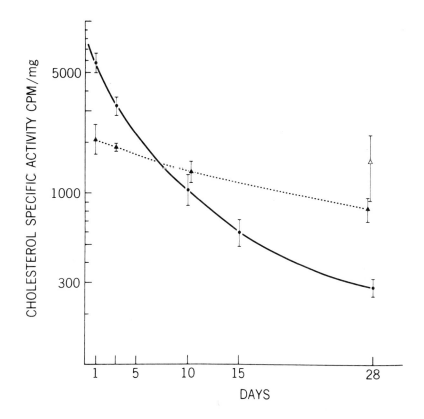

Fig. 3. Serum and adipose tissue cholesterol specific radioactivity time curve. [4-^{14}C] cholesterol (5 μCi) was dispersed in 5 ml of fresh rat serum and was administered via the femoral vein to male rats (411–529 g). Serum and epididymal fat (250–500 mg) biopsies were taken at the times shown. On the 28th day, adductor muscle (300–500 mg) and perirenal fat were removed. Nonsaponifiable fractions of lipid extracts were analyzed for cholesterol content and radioactivity. ●——● = serum, ▲ - - - ▲ = adipose tissue, △ = muscle. Each point is the mean ± S.E.M. (n = 5 or 6).

DIETARY AND NUTRITIONAL CONTROL OF CHOLESTEROL ACCUMULATION IN ADIPOSE TISSUE

Rats raised on a high carbohydrate, cholesterol-free diet were shown to have higher plasma and adipose tissue cholesterol levels than those fed Purina chow (Table 1). In this study, rats were fed semi-synthetic diets containing 0–5% cholesterol. Cholesterol accretion occurred in the absence of a dietary source. In fact, the serum and adipose tissue cholesterol levels were higher in rats on a cholesterol-free diet than in those maintained on 0.05–0.1% cholesterol. Thus endogenous synthesis of cholesterol, when unopposed by dietary feedback suppression, resulted in an expansion of storage pools exceeding that which would occur had small amounts of dietary cholesterol been present. However, with diets containing 0.5% cholesterol or more, a progressive rise in adipocyte cholesterol occurred.

Accumulation of cholesterol in fat tissue may be prevented by early nutritional deprivation. Thus, rats alternately fasted 4 days and fed 4 days for 2 months, then allowed free access to food for 2 months had 40% less cholesterol in fat tissue than did age-matched controls. Mobilization of adipose tissue cholesterol occurred during brief periods of starvation. This was demonstrated by analysis of the cholesterol/DNA ratios in adipocytes isolated from one epididymal fat pad and compared to the opposite pad after 4–6 days' starvation (Table 2). A fall in adipocyte cholesterol/DNA ratio was consistently observed. However, the proportion of cholesterol mobilized (29%) was significantly less than the proportion of triglyceride mobilized (47%).

These studies showed that the adipose cholesterol pool is labile and may be reduced by early nutritional deprivation as well as by acute starvation.

TABLE 1   Effect of dietary cholesterol on cholesterol storage in adipocytes

Dietary cholesterol level (w/w) (%)	Adipocyte cholesterol (mg/mg DNA)*
0.00	20.0 ± 0.8
0.05	15.3 ± 1.2
0.10	15.5 ± 0.7
0.50	22.2 ± 1.6
5.00	26.7 ± 2.0

*Mean ± S.E.M., n = 6.

Groups of rats weighing 200 g were fed semi-synthetic diets consisting of 20.4% vitamin-free casein, 'alphacel' cellulose 16.0%, sucrose 56.7%, salt mixture 3.9%, and corn oil 3%, to which cholesterol was added in the amounts shown. The animals were fed this diet for 2 months before removal of epididymal fat for isolation of adipocytes. The cholesterol content of fat cells at the beginning of the experiment was 4.9 ± 0.1 mg/mg DNA.

TABLE 2   Effects of acute starvation on adipocyte cholesterol and triglyceride content

	Cholesterol content (mg/mg DNA)*	Triglyceride content (g/mg DNA)*
Control	38.5 ± 4.4	47.8 ± 4.3
6 days' starvation	27.5 ± 3.7	25.1 ± 3.1

* Mean ± S.E.M., n = 6.

EFFECTS OF GROWTH AND OBESITY ON ADIPOSE TISSUE CHOLESTEROL LEVELS

The cholesterol content per unit DNA in intact adipose tissue fragments and isolated fat cells increased progressively with rat age (7–36 weeks) (Table 3). This accumulation is not simply a function of increasing fat cell size since cholesterol accretion exceeded that of triglyceride. This was shown by comparing the cholesterol/triglyceride ratio in isolated fat cells obtained from young (7 weeks, 100–120 g) rats to those obtained from older animals (26–36 weeks, 500–700 g). The values obtained were 0.66 ± 0.44 and 1.41 ± 0.07 mg cholesterol per g cell lipid (mean ± S.E.M., n = 4) respectively (Farkas et al., 1973).

Obesity has profound effects on a number of aspects of cholesterol metabolism in man. In obesity cholesterol production and excretion are increased (Miettinen, 1971). Total body cholesterol content is also increased (Nestel et al., 1969). To determine whether adipose tissue itself was responsible for these findings a study was carried out using the ob/ob mouse as the experimental model. Here, accumulation of cholesterol in fat tissue was related to fat cell size and number. The cholesterol content per unit tissue lipid in adipose tissue of obese animals was identical to that of littermate controls; therefore, the size of the adipose pool was strictly a function of organ mass (Table 4). These studies indicated that expansion of the slowly exchangeable pool(s) (pools B and C) of cholesterol, as noted in obesity (Nestel et al., 1969), is indeed a function of an expanded adipose organ.

TABLE 3   Effect of age on adipose tissue cholesterol content

Rat age (weeks)	Rat weight (g)	Cholesterol content (mg/mg DNA)*
6	120–130	1.74 ± 0.1
7–10	200	2.90 ± 0.2
10–11	260–310	4.48 ± 0.5
26–36	510–580	7.69 ± 0.4

*   Mean ± S.E.M., n = 6.

TABLE 4   Cholesterol content of adipose tissue in relation to triglyceride stores

Source	Cholesterol content (mg/g triglyceride)*	P
ob/ob	0.88 ± 0.80	> 0.4 (n.s.)
Littermate controls	0.82 ± 0.12	

* Mean ± S.D., n = 4.

Epididymal fat pads were extracted with chloroform:methanol 2:1 and saponified and analyzed for cholesterol (Farkas et al., 1973).

CHOLESTEROL SYNTHESIS IN ADIPOSE TISSUE

Finally, mention should be made of recent studies from this laboratory (Angel and Farkas, 1971; Angel et al., 1972; Rosenthal et al., 1974) on the role of adipose tissue as a cholesterogenic organ. These studies were prompted by the observation (see above) that cholesterol

TABLE 5   Lipid synthesis in isolated rat adipocytes

Substrate	Total lipids*	Cholesterol and sterol intermediates	Squalene and hydrocarbons
		(mmoles/g lipid/4 hours)	
Acetate-1-¹⁴C (1 mM)	16,200 ± 1,400	3.0 ± 0.4	2.1 ± 0.2
Glucose-U-¹⁴C (16 mM)	29,900 ± 2,100	2.1 ± 0.7	5.1 ± 3.5
Mevalonate-2-¹⁴C (0.24 mM)	18 ± 0.6	0.47 ± 0.04	5.3 ± 0.37
Leucine-U-¹⁴C (0.1 mM)	245 ± 24	0.228**	0.43 ± 0.02

* Mean ± S.D., n = 4.
** Average of duplicates.

accumulation in fat tissue occurred in the absence of a dietary source. Glucose, acetate, mevalonate, leucine and ^3H_2O were all readily converted to cholesterol in rat epididymal fat incubated in vitro (Table 5). The rate of cholesterol synthesis from glucose in the total adipose organ was 1/8th that of liver while the rate of leucine conversion to sterols was 1/2 that of liver. In obese hyperglycemic mice the rate of sterol synthesis in fat tissue was markedly increased compared to littermate controls (Angel et al., 1972). These studies indicate that a portion of the adipose cholesterol pool may be derived from in situ synthesis.

CONCLUSION

The present report demonstrates that cholesterol synthesis, storage and turnover are important physiological properties of adipose tissue. While expansion of total body cholesterol in obesity is probably due almost entirely to enlargement of the adipose pool the significance of this finding with respect to diseases of cholesterol metabolism such as cholelithiasis, hypercholesterolemia or atherosclerosis remains to be studied.

REFERENCES

Angel, A. and Farkas, J. (1970): Structural and chemical compartments in adipose cells. In: Adipose Tissue, p. 152. Editors: B. Jeanrenaud and D. Hepp. G. Thieme Verlag, Stuttgart.
Angel, A. and Farkas, J. (1971): Cholesterol synthesis in white adipose tissue. Circulation, 44, Suppl. 2, 2, 1.
Angel, A., Farkas, J. and Rosenthal, J. (1972): Regulation of cholesterol synthesis in adipose tissue. Clin. Res., 20, 540.
Avigan, J., Steinberg, D. and Berman, M. (1962): Distribution of labelled cholesterol in animal tissues. J. Lipid Res., 3, 216.
Farkas, J., Angel, A. and Avigan, M. I. (1973): Studies on the compartmentation of lipid in adipose cells. II. Cholesterol accumulation and distribution in adipose tissue components. J. Lipid Res., 14, 344.
Miettinen, T. A. (1971): Cholesterol production in obesity. Circulation, 54, 842.
Nestel, P. J., Whyte, H. M. and Goodman, W. S. (1969): Distribution and turnover of cholesterol in humans. J. clin. Invest., 48, 982.
Rosenthal, J., Angel, A. and Farkas, J. (1974): Metabolic fate of leucine: A significant cholesterol precursor in adipose tissue and muscle. Amer. J. Physiol., in press.
Vague, J. and Garrigues, J. C. (1955): Recherches sur la composition du tissu adipeux humain et notamment sa teneur en steroïdes. Ann. Endocr. (Paris), 16, 805.

DISCUSSION

Lefebvre: Have you any idea of the mechanism involved in the adipose tissue cholesterol mobilization which you have observed during fasting?

Angel: The mechanism of cholesterol uptake or release in adipose tissue is entirely unknown. While we have been able to show rapid uptake of cholesterol in vitro using isolated human fat cells we have not been able to demonstrate release of cholesterol from prelabelled cells. I have been working on the assumption that movement of cholesterol between the fat cells and the capillary lumen must be determined by a carrier system such as a sterol carrier protein or its equivalent, e.g. an LDL apo-lipoprotein. On the other hand, LCAT may be involved and should be considered as a possible mechanism determining directional flow of cholesterol out of the fat cell.

Lefebvre: Could you also comment on the curious finding of the relatively high cholesterol content of the animals fed a cholesterol-free diet?

Angel: It was also a surprise for us to find cholesterol accumulation in fat tissue in rats maintained on a cholesterol-free diet. We expected it to be lower than in tissues from cholesterol-fed rats. It was obvious that the cholesterol must have originated from endogenous sources by de novo synthesis. The question is which tissues are responsible — liver, gut, kidney, adipose tissue, etc. This led us to study the cholesterogenic capability of adipose tissue resulting in the observations reported here, namely that fat tissue is able to synthesize cholesterol from all substrates tested. What remains to be done is to establish the quantitative significance of the reaction. The most interesting finding was that the fat tissue cholesterol level was lower in rats receiving lower dietary cholesterol levels (0.05%-0.1%) compared to the cholesterol-free group. I interpret this to mean that small amounts of dietary cholesterol exert sufficient feedback inhibition on endogenous synthesis to lower storage levels as compared to the situation of unrestrained cholesterogenesis seen on cholesterol-free diets.

Chlouverakis: Dr. Angel, you suggested that cholesterol synthesis is under hormonal control and showed us experiments in which cholesterol synthesis was enhanced by insulin. However, in these experiments glucose was used as a precursor and thus the effect of insulin might be entirely due to increase of glucose availability inside the cell. What happens when a non-insulin-dependent precursor, for instance mevalonate, is used?

Angel: We have been able to demonstrate that conversion of glucose to cholesterol was enhanced by addition of insulin to the incubation medium. But this may, as you suggest, simply represent enhancement of glucose transport. It is of interest that conversion of leucine (in the presence of glucose) to cholesterol was also enhanced by insulin addition. This may also reflect enhancement of transport, but enhancement of transport is a legitimate hormone-mediated effect. We have not studied effects of insulin on mevalonate incorporation into fat tissue cholesterol. I would hasten to add, however, that lack of effects of insulin on mevalonate incorporation would not necessarily mean an absence of insulin effect on the integrated reaction starting from the primary substrates such as glucose or leucine, since the substrate you mention bypasses many control points in the biosynthetic sequence.

Chlouverakis: Your data show that the rate of cholesterol synthesis is higher in young animals but that cholesterol concentration of the tissue is higher in older animals. These findings suggest to me that the rate of turnover of intracellular cholesterol decreases with age.

Angel: With regard to turnover of cholesterol in adipose tissue, we have not examined the effects of age so that it is not possible to make a definitive comment. However, I would agree with you that a reduction in synthetic rate with age together with an increase in fat cell cholesterol content may be a function of slower turnover rate. Other age-related mechanisms could also be operative e.g. age-related change in adipose lipid composition resulting in increased solubility and greater storage of cholesterol.

Yen: You showed evidence that cholesterol synthesis in the tissue of ob/ob mice was higher than normal, but the relative amount of cholesterol was normal. Does that mean that these mice have a high cholesterol mobilization rate?

Angel: If the adipose tissue cholesterol level remains constant in spite of an apparent increase in the rate of cholesterol synthesis, then a net increase in cholesterol mobilization must have occurred. Since total body cholesterol turnover is also increased in obesity it is possible, indeed likely, that adipose tissue cholesterol turnover is also increased. Precisely how the fat cell maintains a relatively constant cholesterol/ triglyceride ratio in the ob/ob as compared to its littermate in the face of increased synthesis and turnover has yet to be established.

The effect on regulation of adipose tissue differentiation and evolution of excessive doses of essential fatty acids

J. Raulin, C. Loriette, M. Launay, D. Lapous, M. F. Goureau-Counis, R. Counis and **J. P. Carreau**

E. R. 91 du CNRS, U. 56 de l'INSERM, Hôpital de Bicêtre, Le Kremlin, Bicêtre, France

The rate of adipose cell formation in relation to the physiological state has been studied extensively. During the last few years there have been numerous experiments concerning the problem of increasing or reducing weight in relation to nutritional treatments. In this field, Tepperman (1968) proposed the 'working hypertrophy of enzymes', claiming that enzymic induction occurs through the intervention of multiple factors, for example, refeeding after fasting periods, feeding a balanced diet after giving a low protein one and taking proper meals instead of eating by nibbling. All these experimental conditions produced the same change in adipose tissue – overdevelopment either by adipocyte repletion and enlargement or by differentiation and evolution of new adipose cells.

These enzymic adaptations can also be produced in rat epididymal adipose tissue by feeding a large supply of essential fatty acids with no change in the total amount of lipid calories (Raulin and Launay, 1967; Raulin et al., 1967; Raulin, 1969a, b). Our observations, which indicated variations in content and in specific radioactivity of nucleic acids, have shown fluctuations in the number of adipose cells in relation to the intake of linoleic acid (Launay et al., 1968, 1969, 1972). These results were first obtained by administering to the weaning animal a 20% fat diet containing either lard (L) with 7% linoleic acid or sunflower oil (SO) with 65% linoleic acid. After 3 weeks on experimental diets the epididymal adipose cells were definitely more numerous, but smaller, in SO than in L animals under ad lib but identical food intakes.

Although it was highly probable that the linoleic acid proportion in the exogenous lipid might be the factor controlling the rate of cell formation, the possible effect of unsaponifiable compounds should also have been investigated. The activating effect of SO on the synthesis of DNA has been traced to its high content of essential fatty acid. This conclusion was made from the observation that, after in vivo injection of labelled precursors, the specific activity of DNA (which was already increased when SO was administered instead of L) was further increased when the unsaponifiable fraction of prepared fats was removed (Launay et al., 1972). Conversely, the RNA specific activity was always lower after feeding excessive proportions of linoleic acid.

The bulk of cytoplasmic or nuclear RNA was extracted at $0°C$ or at $65°C$ from the isolated epididymal adipose cells in order to determine whether this low specific activity in SO-fed animals was related to faster RNA utilization. In order to localize the site of this eventual increased activity, 42-day-old rats were intravenously injected with either $[6-^{14}C]$ orotic acid or $[^{32}P]$ phosphate, and prepared RNAs were fractionated by ultracentrifugation in a sucrose gradient. In the cell nuclei from rats fed SO the synthesis of rapidly labelled heterodispersed RNAs seemed accelerated. At the ribosomal level these RNAs seemed to be used faster. This observation was interpreted as indicative of enzymic inductions (Launay and Raulin, 1972) and further experiments are now in progress to test this hypothesis (Launay, 1973).

The reason why SO adipocytes were smaller in size than L adipocytes seemed to be their rapid mobilization of lipids in association with an intensive de novo fatty acid biosynthesis. That is to say that in SO animals we observed a lack of exogenous lipid accumulation, in contrast to what happened in L animals. In fact, all the lipogenic enzyme activities (supply of reduced coenzymes via G-6-PD, 6-GPD, NADP-MD; acetyl-CoA availability via ATP-CL; malonyl-CoA production via CBX; insulin stimulated fatty acid biosynthesis from $[U-^{14}C]$ glucose or $[2-^{14}C]$

pyruvate) were higher in SO than in L animals. These determinations were carried out in (1) fetal brown fat collected from females on experimental diets and (2) epididymal adipose tissue of young rats (Loriette et al., 1969, 1970, 1971, 1973; Loriette and Raulin, 1971, 1972). This lipogenic activity was about as high in SO-fed animals as after feeding a carbohydrate diet. This can be related to the fact that linoleic acid, being an essential fatty acid, was not able to regulate the endogenous synthesis of lipids by feedback, since it never appeared as the end product of any biochemical pathway in animals (Loriette and Raulin, 1972).

Moreover, all the lipolytic enzyme activities under noradreline, ACTH or glucagon stimulation (adenylate cyclase lipase) were definitely increased with the proportion of L in the fats given. The basal and fluoride-stimulated adenylate cyclase and phosphodiesterase activities were identical whether or not fats were high in essential fatty acids (Carreau et al., 1971, 1972; Counis, 1973; Counis and Raulin, 1970).

Thus, the membrane phospholipid configuration (Raulin et al., 1971), which became pluridimensional in SO animals by feeding lipids rich in linoleic acid but more linear in L subcellular fractions rich in oleic acid, actually had an influence upon the reactivity of adipose cells or on their ghosts, at least under lipogenic or lipolytic hormonal stimulation. The relative unmasking of their hormonal receptors can be suspected when the plasma membrane phospholipids were enriched in essential fatty acids.

In order to study further the metabolic interrelationships produced by excessive doses of essential fatty acids, the experiments were extended to the nuclear chromatin level. In an earlier analysis (Goureau and Raulin, 1970), we observed that high linoleic acid proportions influenced the phospholipid fatty acid composition close to the nucleoprotein of the rat liver. Obviously, every variation in the phospholipid composition served to produce differences in the area surrounding the chromatin. One molecule of our extracted chromatin contained about 400 more unsaturated bonds in SO than in L animals and this higher number of unsaturated bonds could increase the effective strength of the Van der Waals forces and lead to stabilization of the chromatin. This would explain (Goureau et al., 1972) the higher temperature of denaturation (T_m) obtained in the SO group with the more unsaturated chromatin (84.1°C) and the lower T_m obtained in the L group with the less unsaturated chromatin (83.1°C).

Similar chromatin samples were used as templates for the in vitro reaction using deoxyribonucleotide triphosphates and [³H]dTMP in the presence or absence of a purified regenerating rat liver DNA polymerase (gift of A. M. De Recondo) (De Recondo et al., 1973). The rate of [³H]dTMP incorporation in the acid nonsoluble material was always higher (Goureau-Counis, 1973a, b) when L chromatin was chosen as a template than when a chromatin with phospholipids rich in essential fatty acids was used (whatever the in vitro DNA concentration or the incubation time). Whether or not phospholipids are real constituents of chromosomes or only contaminants from the nuclear membrane is unknown. Variation of their fatty acid composition and, in particular, the linoleic acid proportion appeared to influence the T_m of the DNA in the whole chromatin and its role during the in vitro enzymic polymerization of deoxyribonucleotides.

Finally, the evolution and the conservation state (Desnoyers et al., 1971) of adipose cells, their subcellular organelles (especially mitochondria) and the annular cytoplasmic fraction appeared in better condition and seemed to persist longer when they were enriched in linoleic acid. Their active metabolism led to a net decrease in lipid storage and to an increase of structural and enzymic proteins – at least in the rat epididymal adipose tissue – as indicated by electron microscopic examination.

REFERENCES

Carreau, J. P., Counis, R. and Raulin, J. (1971): Adénylcyclase foetale et néonatale du tissu adipeux. Relation avec la proportion d'acides gras essentiels des lipides exogènes. J. Physiol. (Paris), 63, 184A.

Carreau, J. P., Loriette, C., Counis, R. and Ketevi, P. (1972): Relative non-masking of noradrenaline receptors by linoleic acid-rich phospholipids in fat cell membranes. I. Lipase activity. Biochim. biophys. Acta (Amst.), 280, 440.

Counis, R. (1973): Eventual un-masking of noradrenaline receptors by linoleic acid-rich phospholipids in fat cell membranes. 2. Adenylcyclase and phosphodiesterase activities. Biochim. biophys. Acta (Amst.), 306, 391.

Counis, R. and Raulin, J. (1970): Activité adénylcyclase au cours du développement foetal et néonatal du tissu adipeux brun. Rappel et critique des procédés de mesure. Bull. Soc. Chim. biol. (Paris), 52, 1393.

De Recondo, A. M., Lepesant, J. A., Fichot, O., Grasset, L., Rossignol, J. M. and Cazillis, M. (1973): Synthetic template specificity of deoxyribonucleic acid polymerase from regenerating rat liver. J. biol. Chem., 248, 131.

Desnoyers, F., Vodovar, N., Lapous, D. and Raulin, J. (1971): Influence du degré d'insaturation des acides gras du régime sur la morphologie du dépot adipeux mésentérique et épididymaire. C. R. Acad. Sci. (Paris), 272, 2836.

Goureau, M. F., Cernohorsky, I., Chapman, D. and Raulin, J. (1972): Phospholipid configuration and thermal denaturation of DNA in rat liver nucleus chromatin. Physiol. Chem. Phys., 4, 399.

Goureau, M. F. and Raulin, J. (1970): Répercussion du degré d'insaturation des lipides exogènes sur les phospholipides associés à la chromatine du noyau hépatique. Bull. Soc. Chim. biol. (Paris), 52, 941.

Goureau-Counis, M. F. (1973a): Configuration des phospholipides et modulation de l'effet template de la chromatine (Activité DNA polymérase). C.R. Acad. Sci. (Paris), 276, 403.

Goureau-Counis, M. F. (1973b): Polymérisation des desoxynucleotides et configuration des phospholipides de la chromatine. J. Physiol. (Paris), 67, 204A.

Launay, M. (1973): RNA cytoplasmiques particuliers et surcharge lipidique de la cellule adipeuse. J. Physiol (Paris), 67, 203A.

Launay, M., Dauvillier, P. and Raulin, J. (1969): Développement du tissu adipeux. Activité spécifique des acides nucléiques et rôle des acides gras polyinsaturés. Bull. Soc. Chim. biol. (Paris), 51, 95.

Launay, M. and Raulin, J. (1972): Distribution and specific radioactivities of RNAs in isolated adipose cells with relation to exogenous lipids. Lipids, 7, 221.

Launay, M., Richard, M., Alavoine, R. and Raulin, J. (1972): Excess of linoleic acid and incorporation of radioactive precursors in DNA and RNA of adipose cells. Nutr. Rep. Int., 5, 339.

Launay, M., Vodovar, N. and Raulin, J. (1968): Développement du tissu adipeux. Nombre et taille des cellules en fonction de la valeur énergétique et de l'insaturation des lipides du régime. Bull. Soc. Chim. biol. (Paris), 50, 439.

Loriette, C., Jomain, M., Macaire, I. and Raulin, J., (1969): Lipides exogènes et activités enzymatiques du tissu adipeux, G-6-PDH, 6-PGDH, NADP-MDH. C.R. Acad. Sci. (Paris), 269, 1581.

Loriette, C., Jomain, M., Macaire, I. and Raulin, J. (1970): Modification de l'activité de l'enzyme malique de l'ATP-citrate lyase et des déshydrogénases du cycle des pentoses du tissu adipeux par administration de lipides insaturés. J. Physiol. (Paris), 62, 183.

Loriette, C., Jomain-Baum, M., Macaire, I. and Raulin, J. (1971): Lipogénèse de novo et insaturation des lipides exogenes dans le tissu adipeux du rat. Europ. J. clin. biol. Res., 16, 366.

Loriette, C., Lapous, D. and Cresteil, T. (1973): Lipogénèse, lipolyse au cours de la différenciation du tissu adipeux brun. J. Physiol. (Paris), 67, 206A.

Loriette, C. and Raulin, J. (1971): Activité Acétyl CoA carboxylase du tissu adipeux et proportion d'acide linoléique des lipides exogènes. J. Physiol. (Paris), 63, 250 A.

Loriette, C. and Raulin, J. (1972): A few comments on the CBX activity and the regulation of lipogenesis by polyunsaturated fatty acids in the liver and adipose tissue. Biochimie, 54, 1467.

Raulin, J. (1969a): Vitesse de synthèse des acides nucléiques et développement du tissu adipeux fonctions de la nature des lipides alimentaires. Rev. franç. Corps gras, 16, 767.

Raulin, J. (1969b): In: Physiopathology of Adipose Tissue, Discussion, pp. 76–77. Editor: J. Vague. Excerpta Medica, Amsterdam.

Raulin, J., Lapous, D., Dauvillier, P. and Loriette, C. (1971): Remaniements structuraux du tissu adipeux et apport excessif d'acides gras essentiels. Nutr. Metab., 13, 249.

Raulin, J. and Launay, M. (1967): Enrichissement en ARN et ADN du tissu adipeux périépididymaire du rat par administration de lipides insaturés. Nutr. et Dieta (Basel), 9, 208.

Raulin, J., Launay, M., Dauvillier, P. and Lapous, D., (1967): Variabilité de l'activité spécifique du phosphore nucléique et lipidique des fractions subcellulaires du tissu adipeux du rat en fonction des conditions nutritionelles. J. Physiol. (Paris), 59, 285.

Tepperman, J. (1968): Metabolic and Endocrine Physiology, 2nd ed. Medical Year Book Publishers, Chicago.

Metabolism in experimental obesity induced in the rat by varying levels of dietary fat*

N. Zaragoza-Hermans, C. Schindler and J. P. Felber

Département de Biochimie Clinique, Clinique Médicale Universitaire, Lausanne, Switzerland

Fat-rich diets have been shown to induce experimental obesity in the rat (Peckam et al., 1962; Lemonnier, 1967; Schemmel et al., 1969) which is dependent on the presence in the diet of a certain proportion of carbohydrate (Lemonnier, 1967). The present work was undertaken to investigate the metabolic alteration produced by varying levels of fat and carbohydrate in the diet. The results are discussed in relation to the relative degree of obesity observed in the various dietary conditions.

Several groups of rats were fed one of four synthetic diets ad lib for four weeks after weaning (Table 1). In the control, fat-poor diet, carbohydrate was supplied as starch. To prepare the

TABLE 1   Composition (%) of the diets

| | Control diet | | Mixed diets | | | | Fat diet | |
| | | | Low-fat | | High-fat | | | |
	g	calories	g	calories	g	calories	g	calories
Protein	24	28	28	29	32	29.5	35	29.5
Carbohydrate	58	67	39	40	20.5	19	3	2.5
Lipid	2	5	13.5	31	24.5	51.5	35.5	68
Noncaloric	16	—	19.5	—	23	—	26.5	—
Caloric equivalent	355		400		443		486	

other diets, starch was partially (mixed diets) or completely (fat diet) replaced by lard. The composition of each diet was established so as to maintain a constant caloric proportion of protein and non-protein components. Although no difference in the total caloric intake was noted during the experimental period, the three fat-supplemented diets produced a significant increase of the mean body weight (Table 2). The maximal effect was obtained with the low-fat mixed diet. These differences were related to parallel variations of the adipose tissue mass. In fact, the epididymal fat pad was enlarged up to 30% above the control value in rats fed the low-fat mixed diet.

The results of glucose, insulin and glyceride measurements in the serum are presented in Table 3. When compared with the control diet, the mean serum glucose was significantly higher in the rats fed the three fat-supplemented diets. While unaltered in the low-fat mixed diet, the level of serum insulin was lower than the control value in both fat-rich diets. Similarly, a marked decrease of the total liver glycogen was only observed in rats fed both fat-rich diets. As also

* This work was supported by a grant from the Fonds National Suisse de la Recherche Scientifique (request No. 3344.70).

TABLE 2   Effect of varying levels of dietary fat on body and epididymal fat pad weights*

	Control diet	Mixed diets		Fat diet
		Low-fat	High-fat	
Body weight (g)	234 ± 5 (17)	264 ± 7 (17) (P< 0.001)	251 ± 4 (17) (P< 0.01)	248 ± 5 (15) (P< 0.05)
Fat pad weight (g)	1.21 ± 0.06 (14)	1.60 ± 0.09 (14) (P< 0.005)	1.44 ± 0.66 (16) (P< 0.02)	1.37 ± 0.06 (12) (n.s.)

* Mean ± S.E.M. with number in parentheses.

TABLE 3   Effect of varying levels of dietary fat on serum glucose, insulin (IRI) and triglycerides, and on liver glycogen and fat*

	Control diet	Mixed diets		Fat diet
		Low-fat	High-fat	
Serum:				
Glucose (mg/100 ml)	149 ± 2	162 ± 4***	159 ± 1	163 ± 2
Insulin (μU/ml)	99 ± 5	103 ± 6 (n.s.)	86 ± 4**	72 ± 5
Triglycerides (mg/100 ml)	102 ± 4	219 ± 20	153 ± 11	310 ± 31
Liver:				
Glycogen (mg/liver)	1020 ± 63	1028 ± 40 (n.s.)	673 ± 37	593 ± 34
Fat (mg glycerol/liver)	24 ± 2	40 ± 3	69 ± 5	96 ± 9

* Mean ± S.E.M. of 15–17 values.
** P< 0.05 when compared with the control diet.
*** P< 0.01 when compared with the control diet.
When not indicated P< 0.001.

shown in Table 3, a three- or fourfold increase in the serum glyceride level and liver fat content was induced by the fat diet, when compared with the control diet. Similar effects of a fat diet on liver fat and glycogen, as well as on serum glucose and insulin, have been previously reported (Zaragoza and Felber, 1970). With the mixed diets, while the variation of total liver fat was parallel to the change in proportion of dietary fat, the response of serum glyceride was unexpected. In rats fed the low-fat mixed diet, the level of serum glyceride was much higher than in rats fed the fat-rich mixed diet. Thus, the low-fat mixed diet appeared to be related to high levels of serum glucose, insulin and glyceride.

The effect of varying levels of dietary fat and carbohydrate on glucose metabolism in diaphragm in vitro is presented in Table 4. Lipogenesis in skeletal muscle is known to be low. Nevertheless, a highly significant and progressive decrease of glucose conversion into fatty acid was shown to be produced by increasing the proportion of dietary fat. While unaltered in both fat-rich diets, glycogen synthesis was increased by 50% in the low-fat mixed diet. In the experiments reported in Table 4, no significant difference in glucose oxidation was found between the control and the low-fat mixed diet. In other experiments, not presented here, a 30% increase of glucose conversion into CO_2 was observed in diaphragm from rats fed the low-fat mixed diet. In contrast, glucose oxidation was markedly decreased by both fat-rich diets. From these observations, it may be concluded that glucose metabolism in muscle appears to be unaltered or even stimulated in rats fed the low-fat mixed diet, but is impaired in both fat-rich diets.

TABLE 4   Effect of varying levels of dietary fat on glucose metabolism in diaphragm in vitro*

	Control diet	Mixed diets		Fat diet
		Low-fat	High-fat	
$^{14}CO_2$	932 ± 49	1063 ± 110	684 ± 40	488 ± 22
^{14}C-Glycogen	1544 ± 8	2006 ± 24	1555 ± 20	1524 ± 15
^{14}C-Fatty	22.3 ± 0.7	9.34 ± 0.21	3.74 ± 0.15	2.76 ± 0.13
^{14}C-Glyceride glycerol	75.0 ± 1.8	104 ± 3	104 ± 5	115 ± 3

* Results are expressed as nmoles glucose/g per 2 hours (mean ± S.E.M. of 6–8 values).
Krebs-Ringer bicarbonate buffer, 2% albumin, 10 mM[U^{14}C] glucose.

The fat diet-induced impairment of carbohydrate catabolism in diaphragm has been investigated previously (Bringolf et al., 1972). The inhibition of glucose oxidation was shown to be completely reversed in the presence of 2-bromostearate (Bringolf et al., 1972), a compound which is known as a specific inhibitor of long-chain fatty acid oxidation. This finding indicates that the fat diet-induced alteration of carbohydrate oxidation in diaphragm is probably the result of an accelerated oxidation of fatty acid. The mechanisms involved in this process have been demonstrated by Randle et al. (1966) in starvation and alloxan diabetes.

In both experimental conditions, i.e. starvation and alloxan diabetes, Denton and Randle (1967) showed an increase in muscle glyceride. Similarly, the level of glyceride in diaphragm was fourfold higher in rats fed the fat diet than in those fed the control diet (Bringolf et al., 1972). Muscle glyceride was also measured in rats fed both mixed diets. From preliminary results, not presented in this work, no parallelism appears between the variation of muscle glyceride content and the proportion of dietary fat. In fact, the glyceride level in diaphragm was equivalent in rats fed the high-fat mixed diet or the fat diet, whereas it was decreased below the control value in rats fed the low-fat mixed diet. The latter finding is consistent with the high rate of glucose oxidation in diaphragm observed in vitro under this dietary condition. A low muscle glyceride content appears to be in opposition to the high level of circulating glyceride (Table 3). This could be brought about by a relative diminution of the muscle capacity to take up and store exogenous glyceride in rats fed the low-fat mixed diet. The validity of this interpretation will be tested in further work on the activity of tissue lipoprotein lipase under these same dietary conditions.

The metabolism in epididymal adipose tissue was investigated in vitro in the presence of [U^{14}C] glucose and insulin (Table 5). Lipolysis was significantly ($P < 0.005$) reduced below the control value in adipose tissue of rats fed the low-fat mixed diet but slightly increased in both fat-rich diets. A marked, progressive decrease in total glucose carbon utilization was produced by increasing the proportion of dietary fat (Table 5). The lactate plus pyruvate fraction was the only one which was increased, and, in the tissue of rats fed the fat diet, up to 36% of glucose carbon was released into the medium as lactate and pyruvate. A concomitant progressive increase of the lactate:pyruvate ratio was also observed. The significance of this observation with regard to the regulation of glucose metabolism in adipose tissue is discussed elsewhere (Zaragoza-Hermans, 1973). An inverse relationship appeared between the rates of glucose oxidation and conversion into fatty acid and the proportion of fat in the diet (Table 5). A change from 2 to 13.5% (w/w) of dietary fat produced a 59% decrease of lipogenesis from glucose. This metabolic pathway was highly reduced in the tissue of rats fed the fat diet. Similar effects of this carbohydrate-poor, fat-rich diet on adipose tissue metabolism have been previously reported and discussed (Zaragoza-Hermans and Felber, 1972; Zaragoza-Hermans, 1973).

Conversion of glucose into glyceride in epididymal adipose tissue in vitro was estimated by Goldrick et al. (1972) under different nutritional conditions in relation to the rate of fat deposition in vivo in the same tissue. Lipogenesis from glucose could account for the accumulation of glyceride fatty acid only in young glucose-supplemented animals (Goldrick et al., 1972). In the present experiments, no relation appeared between the relative development of the epididymal fat (Table 2) and the different rates of fatty acid synthesis in vitro from glucose (Table 5).

TABLE 5   Effect of varying levels of dietary fat on glucose metabolism in epididymal adipose tissue in vitro*

	Control diet	Mixed diets		Fat diet
		Low-fat	High-fat	
$^{14}CO_2$	9.40 ± 0.30	4.88 ± 0.20	3.62 ± 0.16	2.42 ± 0.11
^{14}C-Fatty acid	10.25 ± 0.44	4.24 ± 0.24	2.61 ± 0.18	1.23 ± 0.09
^{14}C-Glyceride glycerol	1.05 ± 0.03	0.79 ± 0.04	0.86 ± 0.02	0.67 ± 0.005
Lactate (+ pyruvate) release	1.85 ± 0.09	2.30 ± 0.13	3.66 ± 0.30	4.83 ± 0.38
Glycerol release	0.92 ± 0.04	0.75 ± 0.02	1.11 ± 0.02	1.14 ± 0.04
Glucose recovered**	21.62	11.06	8.92	6.73
Lactate/pyruvate	6.2 ± 0.3	12.5 ± 1.0	16.4 ± 0.8	21.9 ± 0.8

 * Results are expressed as nmoles glucose converted (or other compound released)/mg per 2 hours (mean ± S.E.M. of 5–6 values).
** Total of [U ^{14}C] glucose converted into CO_2, fatty acid and glyceride glycerol, plus half the number of nmoles lactate (+ pyruvate) released into the medium.
Krebs-Ringer bicarbonate buffer, 2% albumin, 10 mM [U ^{14}C] glucose, 1000 μU/ml insulin.

Therefore, when a minimal amount of dietary fat is supplied, the deposition of fat is not controlled by the lipogenic capacity of the tissue. It is more probably regulated by the ability of the tissue to take up and esterify exogenous fatty acid which is supplied in the glyceride fraction of the blood. Then, intensive fat storage is expected to result from high levels of circulating neutral fat. However, the process is possibly limited by the activity of adipose tissue lipoprotein lipase. This enzymatic activity is increased by a high-fat diet (De Gasquet et al., unpublished data) and is dependent on a complex regulatory system in which insulin appears to play a central role (Austin and Nestel, 1968; Patten, 1970). In rats fed the low-fat mixed diet, the concomitance of high levels of insulin and glyceride in the serum may promote the accumulation of fat. In addition, the low rate of lipolysis observed in vitro and the significant lipogenic capacity of the tissue are probably also dependent on insulin and most likely cooperate in the development of the adipose mass. On the basis of these cooperative effects, insulin is proposed as one of the major factors involved in the induction of nutritional obesity.

REFERENCES

Austin, W. and Nestel, P. J. (1968): The effect of glucose and insulin in vitro on the uptake of triglyceride and on lipoprotein lipase activity in fat pads from normal fed rats. Biochim. biophys. Acta (Amst.), 164, 59.
Bringolf, M., Zaragoza, N., Rivier, D. and Felber, J. P. (1972): Studies on the metabolic effects induced in the rat by a high-fat diet. Inhibition of pyruvate metabolism in diaphragm in vitro and its relation to the oxidation of fatty acids. Europ. J. Biochem, 26, 360.
Denton, R. M. and Randle, P. J. (1967): Concentrations of glycerides and phospholipids in rat heart and gastrocnemius muscles. Effects of alloxan diabetes and perfusion. Biochem. J., 104, 416.
Goldrick, R. B., Hoffmann, C. C. and Reardon, M. (1972): Studies on lipid and carbohydrate metabolism in the rat. Effects of diets on the metabolism and the responses in vitro to insulin of epididymal adipose tissue and hemidiaphragms. Aust. J. exp. Biol. med. Sci., 50, 289.
Lemonnier, D. (1967): Aspects physiologiques de l'obésité expérimentale produite par des régimes hyperlipidiques. Cah. Nutr. Diét., 3, 29.
Patten, R. L. (1970): The reciprocal regulation of lipoprotein lipase activity and hormone-sensitive lipase activity in rat adipocytes. J. biol. Chem., 245, 5577.

Peckam, S. C., Entemman, C. and Carroll, H. W. (1962): The influence of a hypercaloric diet on gross body and adipose tissue composition in the rat. J. Nutr., 77, 187.

Randle, P. J., Garland, P. B., Hales, C. N., Newsholme, E. A. and Denton, R. M. (1966): Interactions of metabolism and the physiological role of insulin. Recent. Progr. Horm. Res., 22, 1.

Schemmel, R., Mickelsen, O. and Tolgay, Z. (1969): Dietary obesity in rats: influence of diet, weight, age and sex on body composition. Amer. J. Physiol., 216, 373.

Zaragoza, N. and Felber, J. P. (1970): Studies on the metabolic effects induced in the rat by a high-fat diet. I. carbohydrate metabolism in vivo. Horm. metab. Res., 2, 323.

Zaragoza-Hermans, N. (1973): Studies on the metabolic effects induced in the rat by a high-fat diet. Estimation of glucose carbon utilization through various metabolic pathways in epididymal adipose tissue. Europ. J. Biochem., 38, 170.

Zaragoza-Hermans, N. and Felber, J. P. (1972): Studies on the metabolic effects induced in the rat by a high fat diet. (U-^{14}C) glucose metabolism in epididymal adipose tissue. Europ. J. Biochem., 25, 89.

The regulation of adipose tissue activity

B. Shapiro

Department of Biochemistry, Hebrew University-Hadassah Medical School, Jerusalem, Israel

The maintenance of a nearly constant adipose tissue mass (ATM) in non-growing organisms requires a control mechanism which keeps caloric intake equivalent to output. Any hypothesis on the nature of this control mechanism must take into account the following observations. Intake and output are not in permanent balance and long periods of negative balance alternate with periods of positive balance (Edholm et al., 1970). ATM thus fluctuates around a certain level ('ideal weight'). When larger deviations in this level are induced by diet, subsequent free access to food will bring about a return to the original level. However this readjustment was only evident when a deficit of about 2000 kcal had been built up (Thompson et al., 1961). The common experience is that, following weight reduction by diet, most subjects return to their previous weight when the dieting effort is relaxed. Most subjects can easily keep a certain weight but will not be able to maintain a reduced weight without special effort. This rule seems to be true whether the subject is of normal weight, obese or lean. In the obese the ideal weight is shifted to higher values.

The same principle has also been established in animal experiments showing that, following weight gain by forced feeding, return to normal weight was attained by spontaneous lowering of food intake (Kennedy, 1950). Here too the response was not immediate but started several days after return to voluntary food intake. The preservation of ATM must therefore be based on continuous correction of errors in the energy balance and requires that the organism will take food in excess of its immediate caloric requirement when ATM is too low or, on the other hand, take less than the daily demand to remove the excess stored.

This consideration and similar arguments (Kennedy, 1952; Mayer and Thomas, 1967) have led to the conclusion that the size of adipose tissue stores should be a regulatory factor on energy balance, i.e. food intake and caloric output. Some systemic factor related in its potency to the size of ATM was assumed to influence the physiological regulation of food intake (Kennedy, 1952). The existence of such a mechanism of body weight regulation, named a 'ponderostat' (Cabanac et al., 1971), has been postulated in view of the effect of excess body weight on olfactory and gustatory sensations.

No information is available concerning the nature of the 'signal' which can transmit the state of body energy reserves to the systems of energy regulation or how such a signal can modify the regulation. Neither is it known why the ATM which is maintained differs between various individuals. In some ('obese') the regulation is affected at a high ATM and in others ('lean') at a mass below that found in the general population.

In order to arrive at a better formulation of the problem, one may assume that the signal is related to the total activity of the adipose tissue in the body, named 'active energy reserves', and that it is the latter quantity (Q) which is regulated and not the mass per se (Shapiro and Zinder, 1972). Q is proportional, but not equal, to ATM, i.e. $Q = K(ATM)$ where K is an 'activity factor' or the activity per unit mass. Q increases with the size and energy demands of the organism. This increase can be brought about by increase in ATM. In the young organism this is brought about mainly by the formation of new cells. Later, when the formation of new cells decreases or ceases completely, the increase in ATM is based on the expansion of existing cells. According to findings in our laboratory (Zinder et al., 1967; Zinder and Shapiro, 1971), the activity of adipose tissue cells (in deposition, formation and release of free fatty acid (FFA) does not rise in proportion to the mass.

The increase in activity was found to depend on the increased surface area of the larger cells. In older subjects, in whom adipose tissue is made up of larger cells, the relative surface area (surface area per unit mass) is smaller and thus K will be smaller than in younger subjects. The total activity regulated (Q = K (ATM)) has therefore to be maintained by an increase in ATM. With increasing age, one thus obtains the state of 'non-obese obesity' (increased proportion of adipose tissue per body weight).

A similar case of decreased K was found in a strain of obese mice (Shapiro and Zinder, 1972) compared to mice of normal weight from the same colony. Here, too, the obesity was based on the production of larger cells. When activity was expressed per surface area of the cells, it was found to be nearly equal in the obese and the normal mice. The increase in ATM in this case will also be to compensate for the lower K. Some evidence is also available for the compensatory regulation of ATM when a deficit is created by surgical removal of tissue. Thus, improved viability of adipose tissue grafts was reported following such an intervention (Liebelt et al., 1968), and an increased growth of the contralateral epididymal fat pad was observed when one pad was removed (Shapiro and Zinder, 1972). In the latter case growth was attained by increase in cell number (maturation?).

The signal reflecting the state of adipose tissue activity and affecting the energy balance is not known. The most logical candidate for this function would be the FFA formed by adipose tissue and circulating in the blood. A certain rate of FFA mobilization may be required to keep energy intake and output in balance so that an excess would replace part of the caloric intake, while a deficit would call for larger intake. However, FFA in the plasma is found to be elevated in most obese individuals. This does not invalidate the thesis since (1) one would expect the level of FFA

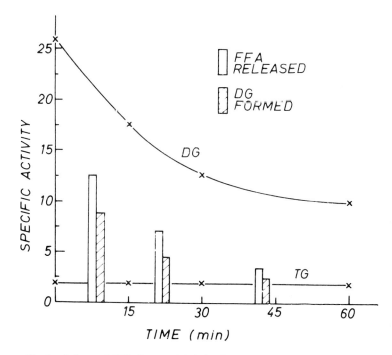

Fig. 1. Release of FFA from prelabeled cells and changes in cell glycerides. Epididymal fat cells in 2 ml bicarbonate Ringer containing 4% serum albumin and 6 μmole glucose with 1 μmole sodium palmitate-^{14}C (2×10^4 d.p.m.) were incubated in an atmosphere of 5% CO_2 and 95% O_2. The cells were then washed 4 times with 3 ml warm albumin-Ringer solution and resuspended in 2 ml of this medium, with the addition of 4 μg epinephrine and incubated for the periods stated. Specific activity values are expressed as d.p.m./ μmole FA in all fractions.

to be 'normal' when body weight is at the 'ideal' point and in many obese subjects the body weight may be above this value; and (2) the relevance of FFA levels to energy balance may be in the capacity of the organism to keep elevated FFA levels for long periods of caloric deficit. This capacity seems to be defective in obese individuals.

If this is so, how can one explain the lack of FFA reserves, necessary for prolonged response in the obese individual, in the face of increased ATM? It may be speculated that this is due to the existence of an 'active pool' of glycerides, as postulated by a number of authors using different approaches (Stein and Stein, 1961; Kerpel et al., 1961; Angel, 1970).

It could be shown that the specific activity of FFA released from adipose cells, prelabelled by incubation with precursors of triglycerides, was much higher than that of the bulk triglycerides, when tested shortly after labelling (Zinder and Shapiro, 1973) (Fig. 1). The diglyceride fraction of adipose cells, on the other hand, had specific activities exceeding that of the released FFA (Figs. 2 and 3). Fatty acids from the diglyceride fraction make up part of the FFA released on epinephrine addition. The rest of the FFA was derived from a pool of triglycerides of high specific activity, which also replenished the diglyceride pool. This high-activity triglyceride pool equilibrated with the bulk triglyceride in about one hour in the presence of epinephrine.

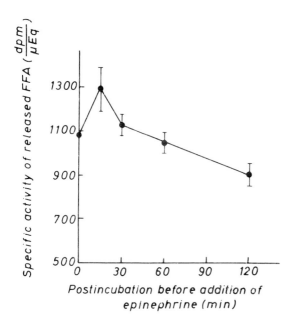

Fig. 2.   Specific activity of FFA after preincubation without epinephrine. Cells were prepared as in Figure 1 and incubated for the periods stated without epinephrine. 4 µg epinephrine was then added and the specific activity of FFA determined.

In the absence of epinephrine, equilibration was very slow. Thus, epinephrine, in addition to promoting FFA release, accelerates the conversion of triglyceride from the inert to the active pool. It would be of interest to compare the size of the active pool (probably by measuring specific activity and size of the diglyceride pool) and rapidity of mixing between obese and normal subjects. This would supply an additional parameter to those previously used to estimate adipose tissue 'activity' and one which may be of relevance to the regulation of energy balance.

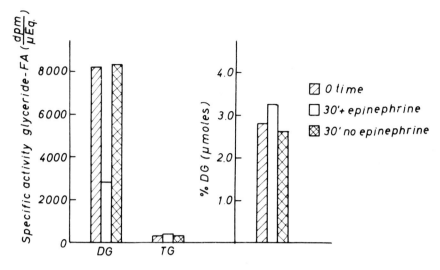

Fig. 3. Specific activity of diglyceride (DG) after incubation with and without epinephrine. Cells prepared as in Figure 1 and then incubated for 30 minutes in the absence and presence of 4 μg epinephrine.

REFERENCES

Angel, A. (1970): Studies on the compartmentation of lipid in adipose cells. I. Subcellular distribution, composition and transport of newly synthesized lipid: Liposomes. J. Lipid Res., 11, 420.

Cabanac, M., Duclane, R. and Spector, P. H. (1971): Sensory feedback in regulation of body weight: is there a ponderostat? Nature (Lond.), 229, 125.

Edholm, O. G., Adam, J. M., Healey, M. J. R., Wolff, H. S., Goldsmith, R. and Best, T. W. (1970): Food intake and energy expenditure of army recruits. Brit. J. Nutr., 24, 1091.

Liebelt, R. A., Vismara, L. and Liebelt, A. G. (1968): Autoregulation of adipose tissue mass in the mouse. Proc. Soc. exp. Biol. (N.Y.), 127, 458.

Kennedy, G. C. (1950): The hypothalamic control of food intake in rats. Proc. roy. Soc. B., 137, 535.

Kennedy, G. C. (1952): The role of depot fat in the hypothalamic control of food intake in the rat. Proc. roy. Soc. B., 140, 578.

Kerpel, S., Shafrir, E. and Shapiro, B. (1961): Mechanism of fatty acid assimilation in adipose tissue. Biochim. biophys. Acta (Amst.), 46, 495.

Mayer, J. and Thomas, D. W. (1967): Regulation of food intake and obesity. Science, 156, 328.

Shapiro, B. and Zinder, O. (1972): Adipose tissue activity as regulator of energy balance. Israel J. med. Sci., 8, 317.

Stein, Y. and Stein, O. (1961): Metabolic activity of rat epididymal fat pad labelled selectively by an in vivo incubation technique. Biochim. biophys. Acta (Amst.), 54, 555.

Thompson, A. M., Billewitz, W. Z. and Passmore, R. (1961): The relation between caloric intake and body weight in man. Lancet, 1, 1027.

Zinder, O., Arad, R. and Shapiro, B. (1967): Effect of cell size on the metabolism of isolated fat cells. Israel J. med. Sci., 3, 787.

Zinder, O. and Shapiro, B. (1973): Compartmentation of glycerides in adipose tissue cells. I. The mechanism of free fatty acid release. J. biol. Chem., in press.

Zinder, O. and Shapiro, B. (1971): Effect of cell size on epinephrine- and ACTH-induced fatty acid release from isolated fat cells. J. Lipid Res., 12, 91.

Lipolytic mechanisms

Lipoprotein lipase activity in adipose cells from genetically and nutritionally obese rats and mice

P. de Gasquet and **E. Péquignot**

Unité de Recherches Diététiques, INSERM U1, Hôpital Bichat, Paris, France

Among the factors contributing to increase in the fat stores in animals, adipose tissue lipoprotein lipase (LPL) might be of importance since hydrolysis of lipoprotein triglycerides is a prerequisite for the entry of esterified fatty acids into the adipocyte (see review by Robinson, 1970). This report deals with the activity of adipose tissue LPL in several types of obesity. Since there may be considerable variation of adipocyte size and/or number in obese animals, expression of enzyme activity per g fresh weight (or protein) or per whole tissue is of little physiological meaning. We therefore determined the cellularity of the various adipose tissues used in order to express the LPL activities in obese and lean animals in terms of fat cell number.

EXPERIMENTAL

The animals (both sexes) used in this study were:

a. 3-5-month-old genetically obese hyperglycaemic mice (ob/ob) and their lean littermates fed on laboratory chow;

b. 3-month-old genetically obese Zucker rats (fa/fa) and their lean littermates fed on laboratory chow;

c. 7-month-old Swiss mice rendered obese by a high-fat diet containing 40% lard (w/w) and lean controls fed on a low-fat diet (the diets were as described by Lemonnier, 1972);

d. 12-month-old genetically obese Zucker rats (fa/fa) and their lean littermates. Obese and lean rats were fed either a low-fat diet for 12 months or a low-fat diet for 5 months and a high-fat diet for the following 7 months. The diets were as described by Lemonnier (1972).

The animals were killed in the fed state and the perigenital and perirenal adipose tissue was excised. LPL was usually assayed in homogenates of the fresh tissue (De Gasquet and Péquignot, 1972). In the experiment with Swiss mice the enzyme activity was assayed in homogenates of acetone-ether-dried powders of perigenital adipose tissue. Adipocyte number and size were determined by a histological technique (Lemonnier, 1972).

RESULTS

Genetically obese hyperglycaemic mice (ob/ob)

LPL activity per two fat pads in obese mice was 5-11-fold that in lean controls (Table 1). However, the increase in adipocyte number in obese animals of the same age was only 1-5-fold. Accordingly, the estimated activity of LPL per adipocyte was 2-4-fold that in obese mice.

Young, genetically obese Zucker rats (fa/fa)

The total enzyme activity in adipose tissue was increased 4-9-fold in genetically obese rats, whereas the adipocyte number in the corresponding adipose tissue from obese rats of the same

TABLE 1   Increase of lipoprotein lipase (LPL) activity per adipocyte in genetically obese mice and rats

Adipose tissue	Sex of the animals	Increase in obese animal values over lean animal values (-fold):			Estimated LPL activity per 10⁶ adipocytes	
		in tissue wt	in LPL activity per two fat pads	in adipocyte number per two fat pads	Lean	Obese
*3–5-month-old genetically obese hyperglycaemic mice (ob/ob)**						
Perigenital	Male	4.9	4.4	1.0	13	58
	Female	9.1	9.4	4.1	19	44
Perirenal	Male	8.9	8.2	4.6	22	40
	Female	11.1	10.9	4.8	17	38
3-month-old genetically obese Zucker rats (fa/fa)						
Perigenital	Male	4.8	4.1	0.7**	16	94
	Female	7.2	8.8	2.3	16	108
Perirenal	Male	6.4	7.7	1.2	14	87
	Female	8.3	8.5	1.6	17	89

* LPL activities are taken from the data of De Gasquet and Péquignot (1972).
** There are less adipocytes in epididymal adipose tissue from young obese Zucker rats than from lean rats (Lemonnier, 1971; Johnson et al., 1972).
LPL activity per 10⁶ adipocytes was estimated by dividing the enzyme activity (μmoles of fatty acids released/hour at 37° C) per two fat pads by the adipocyte number (in millions) per two fat pads. Enzyme activities and cellularity were separately determined on animals of the same age. There were 5–13 animals per group.

age was increased to a much lesser extent (Table 1). Consequently the estimated LPL activity per adipocyte was 5–7-fold that in obese rats.

7-month-old Swiss mice made obese with a high-fat diet

In nutritionally obese mice adipocyte hypertrophy alone accounted for the weight increase in epididymal adipose tissue, whereas in parametrial adipose tissue cell size and number were both increased (Lemonnier, 1972). The LPL activity per two fat pads was higher in mice fed on the high-fat diet (Fig. 1a). The enzyme activity per adipocyte in obese males and females was respectively 4- and 2-fold that of the corresponding controls (Fig. 1b). The enzyme activity per unit of adipocyte surface area was higher (P < 0.01) in obese male mice (Fig. 1c).

1-year-old genetically obese Zucker rats (fa/fa)

The combined effects of genetic obesity and the high-fat diet on perigenital and perirenal adipose tissue LPL activity were studied. The LPL activities per g fresh weight were quite similar in the 16 groups under study and the wide spectrum of activities per two fat pads (Fig. 2a) reflected the weights of the various fat depots. However, the LPL activities per adipocyte were distributed at only two levels, indicated in Figure 2b by the horizontal dotted lines. The basal level corresponded to the genetically lean rats, whereas the high level corresponded to the genetically obese rats. No significant effect of the diet, fat depot location or sex of the animals could be detected. The LPL activity per unit of adipocyte surface area was similar in all groups (Fig. 2c).

LPL activity per
two fat - pads

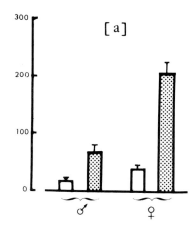

LPL activity per
million adipocytes

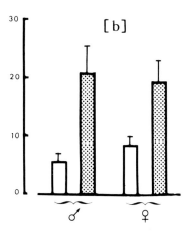

LPL activity per unit of
adipocyte surface area

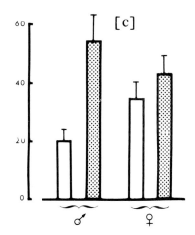

Fig. 1. Lipoprotein lipase (LPL) activity in perigenital adipose tissue from nutritionally obese Swiss mice. The mice were fed on either a low-fat diet (white columns) or a high-fat diet (stippled columns) from weaning until the age of 32 weeks. LPL activity (μmoles of fatty acids released/hour at 37°C) was assayed in homogenates of acetone-ether-dried powders of epididymal or parametrial adipose tissue and was expressed per two fat pads (a), per 10^6 adipocytes (b) and per 10^3 cm^2 of adipocyte surface area (c). The vertical lines represent one S.E.M. (5-7 animals per group).

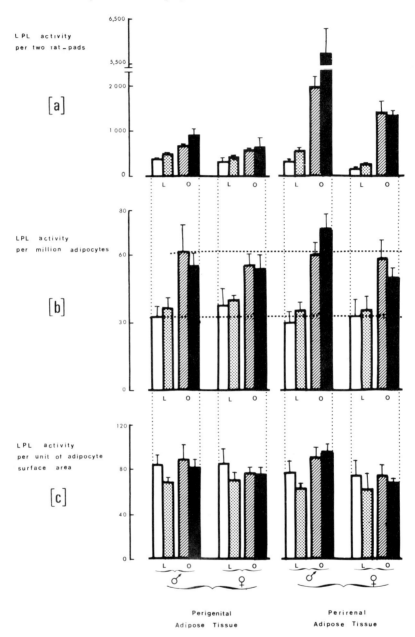

Fig. 2. Lipoprotein lipase (LPL) activity in perigenital and perirenal adipose tissue from 1-year-old genetically lean and obese Zucker rats fed on either a low-fat diet or a high-fat diet. LPL activities (μmoles
of fatty acids released/hour at 37° C) were determined in fresh tissue homogenates and were expressed per
two fat pads (a), per 10^6 adipocytes (b) and per 10^3 cm^2 of adipocyte surface area (c). White columns =
lean (Fa/-) rats, low-fat diet; stippled columns = lean (Fa/-) rats, high-fat diet; shaded columns = obese
(fa/fa) rats, low-fat diet; black columns = obese (fa/fa) rats, high-fat diet. L = lean (Fa/-) rats. O = genetically obese (fa/fa) rats. The horizontal dotted lines are explained in the text. The vertical lines represent
one S.E.M. (6-8 animals per group). The results pertaining to the rats on the low-fat diet are derived from
the data of De Gasquet et al. (1973).

DISCUSSION AND CONCLUSION

In genetically obese mice (ob/ob) and rats (fa/fa) LPL activity per adipocyte was 2-5-fold that in lean controls. There was no effect of the sex of the animals or the adipose tissue site.

The increase in LPL activity per adipocyte was also demonstrated in Swiss mice made obese with a high-fat diet given ad lib from weaning.

However, the high-fat diet administered to 5-month-old lean rats (Fa/-) failed to increase the activity of LPL per adipocyte though it produced a significant, still moderate, increase in the fat depot weights.

The discrepancy between the effect of the high-fat diet on LPL in mice (Fig. 1) and the lack of effect of this diet on LPL in genetically lean rats (Fig. 2) could come from a species difference. More likely it could result from the fact that whereas the mice had access to the high-fat diet from weaning, the high-fat diet was offered to the rats only when they were 5 months old, i.e. after the completion of the active phase of development of adipose tissue (Hirsch and Han, 1969). Supporting the latter hypothesis is the usual observation that dietary manipulations are less effective in changing a number of enzyme activities in adipose tissue of older animals (Griglio, 1970). Besides, the difference in LPL activity between genetically obese (fa/fa) and lean (Fa/-) rats was higher in younger animals (Table 1 and Fig. 2), suggesting that the increase in LPL activity occurs early in the onset of obesity.

Insulin and glucose are necessary to induce high LPL activity in adipose tissue from normal animals. In this study the increase in LPL activity per adipocyte in obese animals was observed in hyperinsulinaemic-hyperglycaemic mice (ob/ob), in hyperinsulinaemic-normoglycaemic rats (fa/fa) and in normoinsulinaemic-normoglycaemic mice (Swiss mice). However, nothing is known about the insulinaemia and the glycaemia of young animals becoming obese, precisely when the LPL activity starts increasing. Further investigation is needed in order to elucidate the events occurring at the very onset of either genetically or nutritionally induced obesity.

REFERENCES

De Gasquet, P. and Péquignot, E. (1972): Lipoprotein lipase activities in adipose tissues, heart and diaphragm of the genetically obese mouse (ob/ob). Biochem. J., 127, 445.
De Gasquet, P., Péquignot, E., Lemonnier, D. and Alexiu, A. (1973): Adipose tissue lipoprotein lipase activity and cellularity in the genetically obese Zucker rat (fa/fa). Biochem. J., 132, 633.
Griglio, S. (1970): Modification de quelques activités enzymatiques du foie et du tissu adipeux épididymaire chez le rat rendu obèse par un regime hyperlipidique. Thesis, Paris.
Hirsch, J. and Han, P. W. (1969): Cellularity of rat adipose tissue: effects of growth, starvation and obesity. J. Lipid Res., 10, 77.
Johnson, P. R., Zucker, L. M., Cruce, J. A. F. and Hirsch, J. (1971): The cellularity of adipose depots in the genetically obese Zucker rat. J. Lipid Res., 12, 706.
Lemonnier, D. (1971): Sex difference in the number of adipose cells from genetically obese rats. Nature (Lond.), 231, 50.
Lemonnier, D. (1972): Effect of age, sex and site on the cellularity of the adipose tissue in mice and rats rendered obese by a high fat diet. J. clin. Invest., 51, 2907.
Robinson, D. S. (1970): The function of the plasma triglycerides in fatty acid transport. Comprehens. Biochem., 18, 51.

Lipoprotein lipase activity release by human adipose tissue in vitro: effect of adipose cell size*

B. Guy-Grand and **B. Bigorie**

Clinique Médicale de l'Hôtel Dieu, UER Broussais, University of Paris VI, Paris, France

Recent reports have emphasized the role of adipose cell size as an important determinant of its in vitro metabolism in rats or man. Basal and stimulated lipolysis (Zinder and Shapiro, 1971), glucose oxidation and incorporation into triglyceride (TG) (Smith, 1971; Björntorp and Karlsson, 1970) and palmitate esterification (Zinder and Shapiro, 1967) increase with cell size when these activities are expressed on a per cell basis. Insulin sensitivity of adipose tissue is reduced with increasing cell size (Salans et al., 1968). On the other hand, such activities as palmitate esterification or free fatty acid release have been reported to be constant per cell surface area unit (Zinder and Shapiro, 1967, 1971). Although some of these effects are dependent on other factors than cell size (Salans and Dougherty, 1971), they seem relevant to the problem of the importance of the anatomical and physiological characteristics of adipose tissue as factors of its overall in vivo metabolism or mass regulation, particularly in obesity.

Uptake of fatty acids from plasma TG, one of the major sources in man for adipose tissue TG synthesis (Björntorp et al., 1971a), is likely to be an important physiological parameter which is probably determined by the activity of the enzyme lipoprotein lipase (Robinson and Wing, 1970). Since the measurement of post heparin lipolytic activity (PHLA) in plasma does not reflect the lipoprotein lipase release from adipose tissue (La Rosa et al., 1972), this was studied directly in human adipose tissue samples. An attempt was made to correlate this release with cell size or surface area and with some metabolic factors. The results of such an investigation may be relevant to the problem of whether the active form of lipoprotein lipase (Robinson and Wing, 1970) is determined by cell number or tissue mass or both, and is an indirect approach for TG fatty acid uptake estimates.

MATERIAL AND METHODS

Adipose tissue samples were taken from the upper part of the buttock by percutaneous needle biopsy (Hirsch and Goldrick, 1964). Lipoprotein lipase assay was performed according to the method described by Persson et al. (1966) using Ediol as substrate. Since we found that the lipoprotein lipase activity (LPLA) eluted from the tissue was slightly dependent on its incubated weight, 160-180 mg of tissue from every subject were run in duplicate. The LPLA of the eluates was assayed in triplicate by a colorimetric method (Björntorp, 1965) and, as the fatty acid release from Ediol was linear up to 45 minutes, the activities were expressed as μmoles fatty acid released per 45 minutes.

No activity was found in elution medium which had never been in contact with adipose tissue, when heparin was omitted, or when NaCl 0.5 M was added.

An aliquot of adipose tissue was taken for cell diameter measurement according to the method described by Sjöström et al. (1971) and another for lipid extraction with chloroform-methanol 2:1. Mean cell volume was calculated by the formula of Goldrick (1967) and mean cell surface

* Supported by crédits de recherche, University of Paris VI and a grant from the Caisse Nationale d'Assurance Maladie des Travailleurs Salariés.

area by the formula of Zinder and Shapiro (1971). The results were calculated in three ways: per cell, per mm² surface area and per nl TG.

Forty outpatients attending for obesity were included in the study: 15 men (mean age 36.4, range 15-68) and 25 females (mean age 36.5, range 16-75). The biopsy was taken after an overnight fast and they received a 50 g oral glucose load. Fasting TG, fasting and post-loading blood glucose and immunoreactive insulin (IRI) were determined. Plasma PHLA, when obtained, was measured some days later after an intravenous injection of 0.1 mg/kg heparin. Body fat was obtained by calculating total body water.

Five subjects were of normal weight and had normal glucose tolerance tests (GTT) and fasting TG (< 150 mg/100 ml). Four were 'plump' (> 10% < 20% ideal body weight) with normal GTT and fasting TG. Thirty one were obese (> 20% ideal body weight) and 18 of these had normal GTT and fasting TG. All the subjects had been of stable weight for at least 2 months and were instructed to eat at least 250 g carbohydrate daily for 3 days before testing. In addition, a careful dietary history was taken in order to estimate the caloric intake during the period preceding the tests.

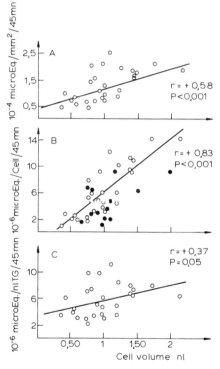

Fig. 1. Effect of cell volume on the lipoprotein lipase activity (LPLA) eluted from human adipose tissue. Open circles = patients without diabetes or hypertriglyceridemia. Black circles = obese diabetics (with or without hyperlipemia). A: LPLA expressed per unit cell surface area; regression equation: $y = 0.87x + 0.40$, $r = + 0.58$, P < 0.001. B : LPLA expressed per unit cell surface area; regression equation for non-diabetic, non-hypertriglyceridemic subjects only: $y = 8.24x - 2.04$ $r = + 0.84$, P < 0.001. C : LPLA expressed per nl triglyceride; regression equation: $y = 2.31x + 3.56$, $r = 0.37$, P = 0.05.

RESULTS

Influence of cell size

It is apparent from Figures 1A and B that a positive correlation was found between LPLA eluted from adipose tissue and cell volume when the data were expressed on a per cell basis or per unit cell surface area. When expressed per unit TG volume, the correlation coefficient was only +0.37 with limited significance (Fig. 1C).

Adipose tissue from obese and normal weight subjects appeared to release LPLA in a similar manner on a per cell basis, since the regression equations for LPLA eluted per cell in terms of cell volume were $y = 8.34x—2.15$ and $y = 8.24x—2.04$ for obese only and for the whole group respectively. Figure 1B also shows that most of the representative points for diabetic and hypertriglyceridemic subjects were located under the mean regression line. However, no significant correlation was found for these patients between cell size and LPLA. When the subjects were divided into three groups according to their cell volume (Table 1), the previous correla-

TABLE 1　Effect of cell volume on lipoprotein lipase activity eluted from adipose tissue in non-diabetic, non-hypertriglyceridemic subjects and in obese diabetics with and without hypertriglyceridemia (mean ± S.E.M.)

Cell volume (nl)	Lipoprotein lipase activity (10^{-6} μmol/nl triglyceride/45 min)		Lipoprotein lipase activity (10^{-6} μmol/cell/45 min)	
	Non-diabetics	Diabetics	Non-diabetics	Diabetics
> 1.20	7.48 ± 0.32	5.41 ± 0.71*	11.31 ± 0.72	9.21 ± 0.13
< 1.20 } > 0.75	5.74 ± 0.74	4.49 ± 0.82	5.36 ± 0.61**	4.12 ± 0.83**
< 0.75	4.13 ± 0.65**	2.39	2.04 ± 0.25***	1.53

　* Significantly different to non-diabetics.
　** Significantly different to largest cells.
*** Significantly different to medium cells.

tions between cell size and LPLA per cell and, to a much lesser extent, per nl TG, were found again in non-diabetic non-hypertriglyceridemic subjects. In each size group, the LPLA was lower in diabetic and hypertriglyceridemic patients, although these differences were not statistically significant. However, a significant difference appeared between LPLA in diabetics and hypertriglyceridemics with a cell size of more than 1.20 nl when compared to those with a cell size of less than 1.20 nl, suggesting that in such patients some correlation between LPLA and cell size may persist.

Correlations between LPLA, plasma TG and plasma PHLA

No significant correlation was found between LPLA per cell or per nl TG and fasting plasma TG ($r = + 0.10$) or plasma PHLA ($r = + 0.17$) in the whole population under study. This lack of correlation persisted when the different pathological groups were studied separately. By multiplying LPLA per cell by the total number of fat cells calculated from fat body mass, the overall lipoprotein lipase release capabilities of a given subject could be roughly estimated. Such estimates did not correlate significantly with fasting TG either in the whole group or in the different pathological groups.

Blood glucose

As shown in Table 2, a weak but significant negative correlation appeared between LPLA per nl TG and the mean blood glucose during the GTT (all subjects). When the subjects were grouped by cell size this negative correlation was strengthened in the largest cell size group. No significant correlation was found between LPLA and fasting glycemia in the whole group or in any of the different pathological groups. LPLA per cell was not correlated with fasting blood glucose or mean of glycemia during the GTT.

TABLE 2   Correlation between lipoprotein lipase activity and some metabolic parameters

	No.	Lipoprotein lipase activity ($10^{-6}\mu$mol/nl triglyceride/45 min)		Lipoprotein lipase activity (10^{-6} μmol/cell/45 min)	
		r	P	*r*	P
Cell volume*	27	+ 0.37	0.05	+ 0.83	<0.001
Fasting triglycerides	40	+ 0.10	n.s.	+ 0.10	n.s.
Post heparin lipolytic activity	34	+ 0.17	n.s.	+ 0.10	n.s.
Mean glucose all subjects	40	— 0.28	<0.05	— 0.10	n.s.
Cell size >1.20 nl	7	— 0.89	<0.001		
Cell size 0.75–1.20 nl	25	— 0.34	0.05		
Cell size <0.75 nl	5	— 0.14	n.s.		
Fasting blood glucose	40	+ 0.22	n.s.	+ 0.10	n.s.
Fasting IRI*	25	+ 0.22	n.s.	+ 0.50	<0.001

* Diabetic and hypertriglyceridemic subjects excluded.

Insulin

In non-diabetic, non-hypertriglyceridemic subjects fasting insulin, but not the mean of insulin during the GTT, was correlated with LPLA per cell (r = + 0.50, P < 0.001). Since a positive correlation between cell volume and fasting insulin was also found (r = + 0.51, P < 0.001), partial regression analysis was performed. The partial correlation coefficient for LPLA and fasting insulin when the influence of cell size was adjusted for was + 0.17 (not significant), whereas the partial coefficient between LPLA per cell and cell size when the influence of insulin was adjusted for remained significant (+ 0.72 P < 0.001).

DISCUSSION

The specificity and reliability of the enzyme activity assay in this study has been extensively discussed by others (Persson et al., 1966; Persson and Hood, 1970) and the convincing evidence supporting its ability to measure lipoprotein lipase presented earlier by these authors was further documented in our laboratory (Guy-Grand and Bigorie, unpublished data). Since the physiology of lipoprotein lipase remains partly unknown (Robinson and Wing, 1970) a measure of the enzyme release from adipose tissue may be of more physiological significance than the determination of the whole tissue or isolated cell content generally used in most of the work on this topic.

Our results demonstrate a positive correlation between lipoprotein lipase release by human adipose tissue and cell size in subjects without diabetes. These data give further evidence of the intracellular origin of the enzymatic activities measured and are in agreement with previous observations that heparin plays a role in the mechanism involved in the release of lipoprotein lipase (Persson et al., 1966; Patten and Hollenberg, 1969).

On the other hand, these findings are at variance with those reported by Nestel et al. (1969) in normal growing rats, in which the enlargement of cells was associated with a decrease in enzyme content, but they are in agreement with recent work (De Gasquet et al., 1973) demonstrating an increase in LPLA per adipocyte in both genetically and nutritionally obese rats and mice. A specific effect of obesity by itself appears unlikely since we found no difference between the regression lines for obese and normal subjects, provided that diabetics and hyperlipemics were excluded. Thus, enlarged fat cells are able to release more LPLA, as they increase most of their lipogenic capabilities.

In contrast to the data in rats (De Gasquet et al., 1973), we did not find constant activities per unit cell surface area, as has been found for lipolysis (Zinder and Shapiro, 1971), a metabolic activity known to depend on membrane receptors. On the other hand, the activities per unit volume were poorly correlated with cell size, suggesting that the increase in LPLA per cell with increasing cell size could be mainly due to an increase in the cell content of some regulating factor. Obviously more details on the metabolic and hormonal factors regulating lipoprotein lipase in adipose tissue are needed before a hypothesis can be formulated. However these data may fit with the observations that free fatty acid (Nikkilä and Pykälistö, 1968) or cyclic AMP (Robinson and Wing, 1970) might be involved.

An apparent correlation was found between fasting IRI and LPLA per cell. However, fasting IRI was also correlated with cell size, in agreement with recent reports (Björntorp et al., 1971b; Stern et al., 1972). Partial regression analysis showed that, when cell size influence was adjusted for, LPLA was no longer correlated to fasting IRI. This finding contrasts with the well known promoting role of insulin in lipoprotein lipase synthesis (Robinson and Wing, 1970). However, plasma IRI levels may not reflect the activity of insulin at the cellular level, particularly in obese subjects. At least it is suggested that elevated plasma IRI, as present in our patients, is not involved in the demonstrated size effect. A maximal effect of insulin could be reached when its concentration is adapted to cell size.

Conversely, in diabetic obese subjects characterized by an inability to maintain normal glucose metabolism and insulin resistance, LPLA tended to be less for a given cell size. Further, LPLA correlated negatively with mean blood glucose during the GTT irrespective of the actual IRI levels. A relative lack of insulin may be suspected (Bagdade et al., 1969) which blunts the cell size effect on lipoprotein lipase release.

The lack of correlation between LPLA and plasma PHLA is in accordance with the recent report (La Rosa et al., 1972) that the latter is a poor index for adipose tissue lipoprotein lipase. In contrast to previous reports (Persson et al., 1966; Persson et al., 1972), fasting TG was not correlated to LPLA, even if an attempt to evaluate the overall capabilities for LPLA release was made. However, the TG values were not widely distributed among the patients studied. This lack of correlation does not exclude, of course, the physiological significance of the enzymatic activities measured. Plasma TG are obviously dependent on several mechanisms and very low activities were found in Type I hyperlipoproteinemia (Guy-Grand and Bigorie, unpublished data).

ACKNOWLEDGEMENTS

Grateful thanks are due to Drs Tchobroutsky and Assan for the insulin assay, to Mme Rebuffe for fat body mass determination and to Messrs G. Bonhomme and A. Legall for their technical assistance.

REFERENCES

Bagdade, J. D., Porte, D. and Bierman, E. L. (1969): The interaction of diabetes and obesity on the regulation of fat mobilization in man. Diabetes, 18, 759.

Björntorp, P. (1965): The effect of nicotinic acid on adipose tissue metabolism. Metabolism, 14, 836.

Björntorp, P., Berchtold, P., Holm, J. and Larsson, B. (1971a): The glucose uptake of human adipose tissue in obesity. Europ. J. clin. Invest., 1, 480.

Björntorp, P., Berchtold, P. and Tibbling, G. (1971b): Insulin secretion in relation to adipose tissue in men. Diabetes, 20, 65.

Björntorp, P. and Karlsson, M. (1970): Triglyceride synthesis in human subcutaneous adipose tissue cells of different size. Europ. J. clin. Invest., 1, 112.

De Gasquet, P., Péquignot, E., Lemonnier, D. and Alexiu, A. (1973): Adipose tissue lipoprotein lipase activity and cellularity in the genetically obese Zucker rat (fa/fa). Biochem. J., 132, 633.

Goldrick, R. G. (1967): Morphological changes in the adipocyte during fat deposition and mobilisation. Amer. J. Physiol., 212, 777.

Hirsch, J. and Goldrick, R. B. (1964): Serial studies on the metabolism of human adipose tissue. I. Lipogenesis and free fatty acid uptake and release in small aspirated samples of subcutaneous fat. J. clin. Invest., 43, 1776.

La Rosa, J. C., Levy, R. I., Windmueller, H. G. and Fredrickson, D. S. (1972): Comparison of the triglyceride lipase of liver, adipose tissue and post heparin plasma. J. Lipid Res., 13, 356.

Nestel, P. J., Austin, W. and Foxman, C. (1969): Lipoprotein lipase content and triglyceride fatty acid uptake in adipose tissue of rats of differing body weights. J. Lipid Res., 10, 383.

Nikkilä, E. A. and Pykälistö, O. (1968): Induction of adipose tissue lipoprotein lipase by nicotinic acid. Biochim. biophys. Acta (Amst.), 152, 421.

Patten, R. L. and Hollenberg, (1969): The mechanism of heparin stimulation of rat adipocyte lipoprotein lipase. J. Lipid Res., 10, 374.

Persson, B., Björntorp, P. and Hood, B. (1966): Lipoprotein lipase activity in human adipose tissue. I. Conditions for release and relationship to triglycerides in serum. Metabolism, 15, 730.

Persson, B. and Hood, B. (1970): Characterization of lipoprotein lipase activity eluted from human adipose tissue. Atherosclerosis, 12, 241.

Persson, B., Schröder, G. and Hood, B. (1972): Lipoprotein lipase activity in human adipose tissue: assay methods. Relations to serum triglycerides level in a normolipemic population. The effect of ethyl chlorophenoxy isobutyrate. Atherosclerosis, 16, 37.

Robinson, D. S. and Wing, D. R. (1970): Regulation of adipose tissue clearing factor lipase. In: Adipose Tissue: Regulation and Metabolic Functions, p. 41. Editors: B. Jeanrenaud and D. Hepp. G. Thieme, Stuttgart.

Salans, L. B. and Dougherty, J. W. (1971): The effect of insulin upon glucose metabolism by adipose cells of different size. Influence of cell lipid and protein content, age and nutritional state. J. clin. Invest., 50, 1399.

Salans, L. B., Knittle, J. L. and Hirsch, J. (1968): The role of adipose cell size and adipose tissue sensitivity in the carbohydrate intolerance of human obesity. J. clin. Invest., 47, 153.

Sjöström, L., Björntorp, P. and Vrana, J. (1971): Microscopic fat cell size measurements of frozen cut adipose tissue in comparison with automatic determination of osmium-fixed fat cells. J. Lipid Res., 12, 521.

Smith, U. (1971): Effect of cell size on lipid synthesis by human adipose tissue in vitro. J. Lipid Res., 12, 65.

Stern, J., Batchelor, B., Hollander, N., Cohen, C. and Hirsch, J. (1972): Adipose cell size and immunoreactive insulin levels in obese and normal weight adults. Lancet, 2, 948.

Zinder, O. and Shapiro, B. (1967): Effect of cell size on the metabolism of isolated fat cells. Israel J. med. Sci., 3, 787.

Zinder, O. and Shapiro, B. (1971): Effect of cell size on epinephrine and ACTH induced fatty acid release from isolated fat cells. J. Lipid Res., 12, 91.

Lipolytic activities in human tissues*

H. Giudicelli, N. Pastré, M. Charbonnier and **J. Boyer**

Laboratoire de la Clinique Endocrinologique, Hôpital de la Conception, Marseilles, France

Human tissues are known to contain several lipase activities, among which are a 'hormone-sensitive' triglyceride lipase (TGL) and a monoglyceride lipase (MGL) activity. The former has been reported to have a pH optimum in the neutral range and to be controlled in vivo through a cyclic AMP-dependent kinase system (Huttunen et al., 1970). Another tissue triglyceride lipase activity has also been described (lipoprotein lipase, LPL) which appears to be active only in the presence of a plasma lipoprotein cofactor, to have an optimum activity in the alkaline pH range and to be released into the blood stream by exogenous heparin.

In this communication, we present an attempt to evaluate the comparative levels of these various lipolytic activities in adipose tissue and liver extracts from a single human subject, and to find out whether or not TGL and LPL tissue activities may be differentiated on the basis of their behaviour after simple treatments of the crude extracts. Specimens of omental adipose tissue and liver were obtained from an adult male (49 years old, 1.70 m, 73 kg) under general anaesthesia. They were immediately homogenized in 50 mM phosphate buffer containing 20% glycerol (pH 7.4) at 4° C (Giudicelli and Boyer, 1973). The first enzyme assays were performed 30 minutes after biopsy. In all enzyme assays, emulsification of the radioactive substrates was achieved by sonication (Boyer et al., 1970). All assays were performed with constant agitation of the incubation mixtures. The release of labeled fatty acids was determined after isolation of the acids by using a liquid-liquid partition system previously described (Boyer and Giudicelli, 1970). In all cases, lipase activities refer to initial reaction rates calculated from the amounts of labeled fatty acids released 0, 10 and 20 minutes after the addition of enzyme. Under these conditions, assays are reproducible within 10%. One enzyme unit corresponds to the release of one μmole of acid per minute. Protein concentrations were determined by the method of Lowry et al. (1951).

RESULTS

The levels of TGL activity were 20-30% those of MGL activity both in human fat and liver (Table 1). The specific activities of TGL or MGL were nearly identical in both adipose tissue and liver, but in both the relative specific activity of MGL was 3-4 times that of TGL. These observations are in agreement with the accepted theory that in tissue from animal sources TGL is rate-limiting in the process of triglyceride catabolism.

Due to its greater mass, and from a rough estimation of its protein content, adipose tissue appeared to have a total lipolytic capacity (TGL + MGL) potentially 2-3 times that of liver (Table 2).

Tables 3 and 4 show that TGL and LPL activities behaved identically when adipose tissue and liver homogenates were either centrifuged at 12,000 × g for 30 minutes or treated with 90% acetone at –10° C. The recoveries and specific activities followed the same pattern. In adipose tissue, 34-35% of both activities were recovered in the 12,000 × g supernatant fraction, and 15-17% in the 90% acetone-precipitated extract. Both specific activities increased by 26-34%. In

* This work was supported by grants from INSERM (Contrat No. 712 132) and from la Fondation pour la Recherche Médicale Française.

TABLE 1   Triglyceride (TGL) and monoglyceride lipase (MGL) activities in human adipose tissue and liver

12,000 g supernatant fraction	Lipase activity (mU/g tissue)		Protein conc. (mg/g tissue)	Specific activity (mU/mg protein)	
	TGL	MGL		TGL	MGL
Adipose tissue	4.6	13	6.6	0.7	2.0
Liver	86	360	160	0.5	2.2

The tissue homogenates were centrifuged at $12,000 \times g$ for 30 minutes at 4° C and the supernatants used as sources of enzyme. TGL activity was assayed in 50 mM sodium phosphate buffer containing 1 mM [^{3}H] triolein and 0.5% bovine serum albumin to give a final volume of 5 ml at 37° C, pH 7.0. MGL activity was assayed in 0.1 M Tris-HCl buffer containing 2.5 mM [^{3}H] ethyl oleate and 0.5% bovine serum albumin to give a final volume of 5 ml at 37° C, pH 7.4. Justification of the use of [^{3}H] ethyl oleate as substrate is given elsewhere (Charbonnier et al., 1973; Arnaud et al., 1973).

TABLE 2   Total lipolytic activity of human adipose tissue and liver

Tissue	Total protein (g)	Total TGL activity (units)	Total MGL activity (units)
Adipose tissue	400	280	800
Liver	220	110	480

TABLE 3   Triglyceride lipase (TGL) and lipoprotein lipase (LPL) activities in various preparations of human adipose tissue

Tissue preparation	TGL activity (pH 7)			LPL activity (pH 8.5)		
	mU/g tissue	Yield (%)	mU/mg protein	mU/g tissue	Yield (%)	mU/mg protein
Homogenate	13.2	100	—	9.6	100	—
12,000 g supernatant fraction	4.6	35	0.70	3.3	34	0.50
90% acetone-precipitated extract	2.0	15	0.95	1.6	17	0.76

Cold acetone was added to aliquot fractions of the homogenates to a concentration of 90% (v/v) at −10° C. The preparation was mixed continuously for 15 minutes. The insoluble material was collected by centrifugation, dissolved in phosphate buffer (pH 7.4) and immediately assayed. LPL activity was assayed in 0.1 M Tris-HCL buffer containing 1 mM [^{3}H] triolein, 0.5% bovine serum albumin, 0.1 M $(NH_4)_2SO_4$, 10 units sodium heparin and 0.5 ml human fresh serum to give a final volume of 5 ml at 37° C, pH 8.5.

liver, 32-34% of both activities were recovered in the $12,000 \times g$ supernatant fraction, and 30-34% in the 90% acetone-precipitated extract. Both specific activities remained practically unchanged. This similarity is better illustrated in Table 5 where it can be seen that the ratio

TABLE 4   Triglyceride lipase (TGL) and lipoprotein lipase (LPL) activity in various preparations of human liver

Tissue preparation	TGL activity (pH 7)			LPL activity (pH 8.5)		
	mU/g tissue	Yield (%)	mU/mg protein	mU/g tissue	Yield (%)	mU/mg protein
Homogenate	250	100	—	270	100	—
12,000 g supernatant fraction	86	34	0.5	87	32	0.5
90% acetone-precipitated extract	74	30	0.5	91	34	0.6

TABLE 5   Ratio of lipoprotein lipase (LPL) to triglyceride lipase (TGL) activity in human adipose tissue and liver

Tissue preparation	Ratio of LPL to TGL activity (mU/g tissue)	
	Adipose tissue	Liver
Homogenate	0.7	1.1
12,000 g supernatant fraction	0.7	1.0
90% acetone-precipitated extract	0.8	1.2

of LPL to TGL activity remained virtually constant in all preparations from both tissues. As a consequence, it cannot be excluded that human 'hormone-sensitive' and lipoprotein lipase activities may derive from a single intracellular enzymatic entity. Serial comparisons of specific activity and catalytic properties at various stages of purification are needed to further substantiate this hypothesis.

ACKNOWLEDGEMENT

The expert collaboration of O. Nobili is gratefully acknowledged.

REFERENCES

Arnaud, J., Charbonnier, M. and Boyer, J. (1973): Further characterization of fatty acyl monoester lipase from human adipose tissue. Biochim. biophys. Acta (Amst.), 316, 162.

Boyer, J. and Giudicelli, H. (1970): Hydrolyse de la [³H] trioléine par des extraits de tissu adipeux humain. Biochim. biophys. Acta (Amst.), 202, 219.

Boyer, J., Le Petit, J. and Giudicelli, H. (1970): L'activité lipolytique du tissu adipeux. II. L'activité triglycéride-lipase du tissu adipeux humain. Biochim. biophys. Acta (Amst.), 210, 411.

Charbonnier, M., Arnaud, J. and Boyer, J. (1973): Partial purification and characterization of a fatty acyl monoester lipase from human adipose tissue. Biochim. biophys. Acta (Amst.), 296, 471.

Giudicelli, H. and Boyer, J. (1973): Effects of glycerol on the human adipose tissue triglyceride lipase activity. J. Lipid Res., 14, 592.

Huttunen, J. K., Steinberg, D. and Mayer, S. E. (1970): ATP-dependent and cyclic AMP-dependent activation of rat adipose tissue lipase by protein kinase from rabbit skeletal muscle. Proc. nat. Acad. Sci. (Wash.), 67, 290.

Lowry, O. H., Rosebrough, N. J., Farr, A. L. and Randall, R. J. (1951): Protein measurement with the Folin phenol reagent. J. biol. Chem., 193, 265.

Mechanisms regulating the mobilization of free fatty acids from human and rat adipose tissue*

D. Steinberg, J. C. Khoo and S. E. Mayer

Division of Metabolic Disease and Division of Pharmacology, Department of Medicine, School of Medicine, University of California, San Diego, La Jolla, Calif., U.S.A.

Regulation of the adipose tissue mass, the central theme of this Meeting, reduces at the level of the adipocyte primarily to the balance between rates of deposition of carbon atoms as triglycerides and rates of their mobilization as free fatty acids (FFA) and glycerol. In situ oxidation of fatty acids by adipose tissue will affect the balance also but this (although perhaps incorrectly in some instances) is usually assumed to be quantitatively minor relative to the rates of deposition of fatty acids taken up from the plasma and the rates of FFA mobilization. De novo synthesis of fatty acids from nonlipid precursors is included in 'deposition of carbons as triglycerides' and is certainly a factor as well. In adult life the number of adipocytes appears to be fixed and the mass contributed by adipocyte stores of triglycerides is overwhelmingly preponderant. Thus, it is formally correct to consider the input-output balance to be the determinant of adipose tissue mass and to seek disturbances in that balance to explain 'fatness' or 'leanness'.

An intimate understanding of the processes regulating FFA mobilization is thus relevant to our central theme, and it is certainly relevant to the short-term provision of oxidizable substrate, as in exercise or in forced fight or flight. The relevance of mechanisms involved in acute regulation of fat-mobilization to long-term adjustments of adipose tissue mass is less clear. Dr. Shapiro has dealt with this problem in his paper (p. 40) and we shall not say more here.

We would like to review quickly the studies showing that hormone sensitive lipase is controlled by cyclic AMP-dependent protein kinase and that the activation involves covalent modification. We shall present recent evidence that the human enzyme is similarly regulated. Finally, we shall present recent studies on two other systems in adipocytes regulated by cyclic AMP-dependent protein kinase – glycogen synthase and phosphorylase – and close with some evidence suggesting that the ability of insulin to counteract catecholamine effects on these three enzyme systems is not adequately accounted for by an effect at the level of cyclic AMP.

HORMONE-SENSITIVE LIPASE OF RAT ADIPOSE TISSUE AND ITS ACTIVATION

Mobilization of triglyceride fatty acids from adipose tissue requires prior hydrolysis of the glyceryl ester bonds. Many hormones accelerate the process and the rate-limiting step is, at least under most conditions, the hydrolysis of the first ester bond (Vaughan et al., 1964; Strand et al., 1964). Thus, hormonal treatment of intact adipose tissue increases activity in homogenates (made subsequent to hormone exposure) toward triglycerides but not toward lower glycerides. The name 'hormone-sensitive lipase' was introduced to refer to this regulated triglyceride lipase activity without necessarily implying that it represented a single enzyme protein. Only recently has it been possible to obtain any significant purification and indeed purity in the classical sense has still not yet been achieved. What has been purified is a lipid-rich particle of enormous size, relatively homogeneous by several criteria but aggregating perniciously on concentration (Huttunen et al., 1970a). Some electron micrographs show fairly uniform spherical particles ranging in diameter from 180-220 Å; others show some particles in this same size range but, in

* This work was supported by USPHS Grant HL-12373 and the Weight Watchers Foundation, Inc.

addition, many more huge aggregates 1,000-10,000 Å in diameter. The purified particles contain high levels of activity against diglycerides and monoglycerides and many of the properties of these lower glyceridases differ markedly from those of the triglyceride lipase, suggesting that we may in fact be dealing with some sort of multi-enzyme complex peculiar to adipose tissue (Heller and Steinberg, 1972).

That cyclic AMP was involved somehow in lipase activation was well established by a number of lines of evidence nicely reviewed by Butcher (1966). Using the partially purified enzyme (particles) it was shown that its triglyceride lipase activity (triolein-^{14}C as substrate) was enhanced by about 50% on incubation with cyclic AMP, ATP, Mg^{++} and protein kinase prepared from rabbit skeletal muscle (Huttunen et al., 1970b). All four additions were required. In cruder fractions the addition of protein kinase is not required, as in the studies of Rizack (1964) and of Tsai et al. (1970), presumably because the levels of endogenous protein kinase in such fractions is adequate (Corbin and Krebs, 1969). This was shown by Corbin et al. (1970) who inhibited the endogenous enzyme by adding protein kinase inhibitor after which added skeletal muscle protein kinase stimulated activation.

Evidence that the protein kinase activation in cell-free preparations does indeed reflect the process stimulated by hormones in the intact cell was shown by comparing the residual activatability of hormone-sensitive lipase isolated from cells previously treated with lipolytic hormones. If in intact cells the hormones trigger the same kind of activation demonstrated in cell-free fractions with the protein kinase system, enzyme from hormone-treated cells should already be largely in the activated form and therefore show a smaller response to cyclic AMP-dependent kinase. This was in fact shown to be the case both with fat pads (Huttunen et al., 1970b; Huttunen and Steinberg, 1971) and with isolated adipocytes (Khoo et al., 1972a, 1973). Finally, it should be noted that monoglyceride and diglyceride lipase activities, either in crude 100,000 × g supernatant fractions or in more purified fractions, were not altered by the protein kinase activation system. The 100-fold purified enzyme particle still contains considerable activity against mono- and diglycerides but this is not enhanced under conditions that yield 40 to 60% of activation of the triglyceride lipase (Heller and Steinberg, 1972).

PHOSPHORYLATION IN RELATION TO ACTIVATION

Using the 100-fold purified rat enzyme it was shown that activation in the presence of [γ-^{32}P]ATP was accompanied by transfer of radioactivity to the enzyme protein (Huttunen et al., 1970c; Huttunen and Steinberg, 1971). Essentially all the label was protein-bound and its properties were compatible with those of serine-bound phosphate but the nature of the binding site has not been further characterized. From 2 to 4 moles of phosphate were bound per 10^{-6} g protein. When the time course of activation and of phosphorylation were followed simultaneously there was a close parallelism between the two.

Indirect evidence suggests that the protein kinase acts directly in the phosphorylation of the lipase, i.e. that there is no intermediate enzyme playing a role analogous to that of phosphorylase kinase in the activation of muscle phosphorylase (Krebs, 1972). Thus, it appears that the overall process of hormone-stimulated lipolysis in rat adipocytes can now be characterized as shown in Figure 1. By analogy with current nomenclature the activated form of lipase should be termed 'lipase a' and the nonactivated from 'lipase b'.

Since the purification procedure has not thus far resolved lipase a from lipase b it is not possible to say how much activity the latter displays in the assays currently used. Clearly the tissue must contain a system for converting lipase a back to lipase b, presumably by dephosphorylation, but attempts to demonstrate reversible deactivation-activation have thus far been unsuccessful. The presence in rat adipose tissue of an unusual ATP-dependent system that irreversibly inactivates lipase may contribute to the difficulty (Tsai and Vaughan, 1972).

HORMONE-SENSITIVE LIPASE IN HUMAN ADIPOSE TISSUE

Recently Dr. Khoo has shown that hormone-sensitive lipase in human adipose tissue (surgical biopsies) is activated by a mechanism parallel to that operating in rat adipose tissue (Khoo et al.,

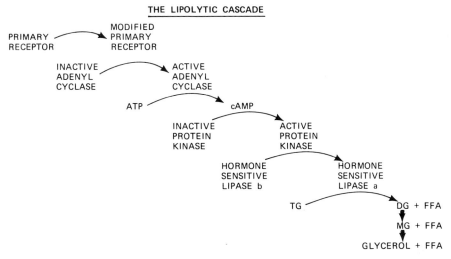

Fig. 1. Scheme summarizing current concepts of the mechanisms involved in hormone stimulation of free fatty acid (FFA) release from rat and human adipose tissue.

1972b; Khoo and Steinberg, 1973). The human enzyme, like that of the rat, precipitates at pH 5.2, has a density less than 1.12 and shows a similar particle size in electron micrographs. Activation of the partially purified enzyme requires the same general cofactor supplementation. Again, there is no evidence that an intermediate enzyme is involved and, again, the enzyme isolated from catecholamine-treated tissue shows reduced activatability, indicating that conversion to the a form in the intact cell is probably the same process as that stimulated by protein kinase in the partially purified preparations.

LIPOPROTEIN LIPASE IN HUMAN ADIPOSE TISSUE

In many situations the activities of hormone-sensitive lipase and lipoprotein lipase in adipose tissue shift in a reciprocal fashion (Robinson and Wing, 1970; Patten, 1970). This has led to speculation that lipoprotein lipase might also be controlled by cyclic AMP, i.e. it might be deactivated when cyclic AMP concentrations rise. This is an attractive hypothesis since it would represent a push-pull control analogous to that regulating glycogen synthesis and breakdown. However, so far our attempts to demonstrate a cyclic AMP-dependent protein kinase deactivation of lipoprotein lipase have been unsuccessful. As shown in Figure 2, treatment of lipoprotein lipase extracted from acetone powders of human adipose tissue under the same activation conditions that convert lipase b to lipase a was without effect. We considered the possibility that the enzyme extracted from acetone powders might already be largely in the postulated deactivated form. We, therefore, studied both hormone-sensitive lipase and lipoprotein lipase activity in the pH 5.2 isoelectric precipitate. When activity was assayed in the absence of added serum, to minimize the contribution of lipoprotein lipase, treatment with cyclic AMP-dependent protein kinase yielded 67% activation (Fig. 3). This reflects activation of hormone-sensitive lipase. Addition of serum in the assay, to bring out the contribution of lipoprotein lipase, increased basal lipase activity markedly. However, now cyclic AMP-dependent protein kinase effected little or no percentage change in activity (Fig. 3). These findings suggest that protein kinase does not control lipoprotein lipase activity directly. This does not rule out the possibility that cyclic AMP may be involved in some other way, for example, by regulating enzyme biosynthesis directly or indirectly.

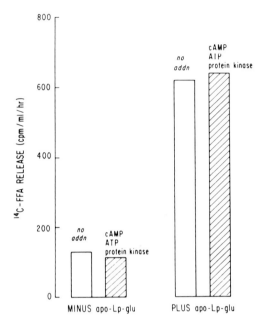

Fig. 2.   Lack of effect of cyclic AMP-dependent protein kinase in lipoprotein lipase activity in acetone-powder extracts of human adipose tissue. Lipase activity was assayed against ^{14}C-triolein emulsions at pH 8.2 (1 hour at 30°) either in the absence (left pair of bars) or in the presence (right pair of bars) of the apolipoprotein activator isolated from plasma very low density lipoprotein (10 μg/ml) (Brown and Baginsky, 1972). Prior to assay, the enzyme was incubated for 10 minutes either with no cofactor added (open bars) or with addition of cyclic AMP (10^{-3} M), ATP (5×10^{-4} M), Mg^{++} (5×10^{-5} M) and skeletal muscle protein kinase (shaded bars).

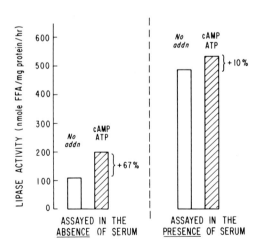

Fig. 3.   Activation of lipase in a pH 5.2 isoelectric precipitate prepared from the 100,000 × *g* supernatant fraction of human adipose tissue. Activation as described in legend to Figure 2. Assayed at pH 7.0 in the absence of added serum (left), lipoprotein lipase contributes little to the total lipase activity; assayed in the presence of added serum (right), lipoprotein lipase activity predominates.

REGULATION OF HORMONE-SENSITIVE LIPASE, GLYCOGEN SYNTHASE AND PHOSPHORYLASE IN RAT ADIPOCYTES

The so-called lipolytic hormones not only regulate hormone-sensitive lipase in adipose tissue but also phosphorylase and glycogen synthase (Vaughan and Steinberg, 1965; Jungas, 1966; Moskowitz and Fain, 1969; Allen and Ashmore, 1972). We have asked whether these three systems, all presumably linked to cyclic AMP, respond strictly according to changes in cyclic AMP concentration or whether additional, differential controls are exercised beyond the level of cyclic AMP. The functional connection, if any, between regulation of lipolysis and regulation of glycogenolysis remains unknown and it is difficult to know a priori whether or not to expect differential control.

When rat adipocytes were incubated for 5 minutes at 37° C with graded concentrations of epinephrine, there was a very close parallelism between the dose response curves for (a) stimulation of glycerol release, (b) stimulation of phosphorylase activity (both total activity and the ratio of activities in the absence and in the presence of 5′-AMP), (c) inhibition of glycogen synthase activity (ratio in the absence and in the presence of glucose-6-phosphate), (d) accumulation of cyclic AMP (cells plus medium) (Khoo et al., 1973). These results are compatible with the proposition that regulation of all three systems is tightly linked through cyclic AMP.

EFFECTS OF INSULIN

When insulin (0.4 nM) was added together with epinephrine (0.5 μM) the lipolytic effect of the latter was almost totally blocked and the inhibitory effect on glycogen synthase was completely reversed (Table 1). However, the effect on cyclic AMP accumulation was very small. The mean values were not significantly different although by paired data analysis there was a small but statistically significant effect. One must ask whether this difference is really large enough (even if statistically significant) to account for the insulin reversal of epinephrine effects on these

TABLE 1   Effect of insulin and epinephrine on adipocyte phosphorylase, glycogen synthase, glycerol, and cyclic AMP accumulation in the presence of 10 mM glucose (mean ± S.E. of four experiments)

Addition	Phosphorylase activity ratio (—AMP:+AMP)	Glycogen synthase activity ratio (—glucose-6-P: +glucose-6-P)	Glycerol (μmoles/hr/mg protein)	Cyclic AMP (pmoles/mg protein)
None	0.19 ± 0.02	0.23 ± 0.03	0.158 ± 0.038	10.0 ± 1.9
Epinephrine (0.5 μM)	0.50 ± 0.06 [a]	0.13 ± 0.01[a]	1.099 ± 0.145[c]	26.8 ± 5.8 [a] ⎤
Epinephrine (0.5 μM) + Insulin (0.4 nM)	0.24 ± 0.04	0.22 ± 0.03	0.220 ± 0.021	20.1 ± 3.4 [a] ⎦ d
Insulin (0.4 nM)	0.16 ± 0.02	0.35 ± 0.01[b]	0.102 ± 0.058	8.7 ± 1.7

[a]P <0.05,   [b]P <0.01,   [c]P <0.001. Significance of differences from results in the absence of any additions ('None').   [d] Group means are not significantly different (P >0.2). However, when the data from individual experiments (using a single preparation of adipocytes) are subjected to paired comparisons, P <0.001.

systems. Speaking against this is the observation that, in a series of 5 minute incubations with epinephrine alone, much higher rates of lipolysis were associated with levels of cyclic AMP equal to those reached in the presence of epinephrine plus insulin in the experiments shown in Table 1. When the epinephrine-insulin antagonism was studied in the same way using adipocytes from fasted rats the results were as shown in Figure 4. In cells from fasted animals the glycogen synthase was already predominantly in the b form and thus no epinephrine effect could be

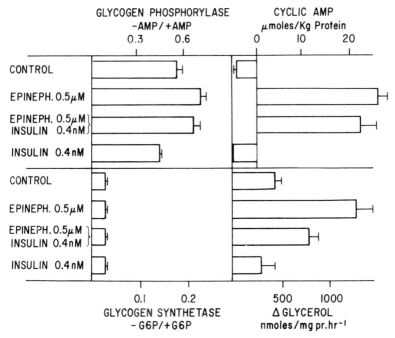

Fig. 4.   Effects of epinephrine and of insulin on adipocytes from rats fasted 48-72 hours. Cells were incubated for 5 minutes with the additions indicated and assays were carried out as previously described (Khoo et al., 1973).

demonstrated. The phosphorylase activity ratio was higher than in cells from fed animals but increased further on exposure to epinephrine. The epinephrine effect was partially reversed when insulin was present. Basal glycerol release was higher than in fed animals, as expected, but epinephrine induced a substantial increase in lipolytic rate. This lipolytic effect of epinephrine was significantly reduced (by about 50%) by insulin but with no statistically significant effect on cyclic AMP accumulation.

A similar independence of the antilipolytic effect of insulin from its ability to suppress cyclic AMP accumulation or effects on adenyl cyclase activity has been reported by several laboratories (Vaughan and Murad, 1969; Cryer et al., 1969; Allen and Clark, 1970; Jarett et al., 1972). Now, it has been shown unequivocally that *under some circumstances* insulin does very definitely reduce epinephrine-induced accumulation of cyclic AMP. This was first shown by Butcher et al. (1966, 1968), but it is important to note that the effect was seen only in the presence of caffeine, a phosphodiesterase inhibitor. Under these conditions extremely high cyclic AMP levels were reached in the presence of the lipolytic hormone. Insulin certainly reduced cyclic AMP levels reached under these conditions but there was no reduction in glycerol release rate, presumably because even the lower levels of cyclic AMP were adequate to maximally activate the lipase.

When caffeine was omitted insulin did not reduce cyclic AMP levels but did inhibit lipolysis, results similar to those reported here! In a recent paper Soderlinge et al. (1973) reported insulin suppression of epinephrine-induced cyclic AMP accumulation in whole *fat pads*. However, a very high level of epinephrine was used (11 μM) and the cyclic AMP concentration reached in the adipocytes themselves is not known. We suggest that there may be an important difference between experimental conditions associated with very high cellular concentrations of cyclic AMP and those associated with lower concentrations. In view of the reports that insulin increases phosphodiesterase activity (Senft et al., 1968; Loten and Sneyd, 1970; Vaughan, 1972) one might suggest that that effect only becomes readily apparent when the intracellular levels of cyclic AMP reach high values, values that may approach or exceed those required for maximal stimulation of lipolysis. This might be the case if much of the cyclic AMP is actually bound to protein kinase and/or other cell proteins and thus is unavailable to phosphodiesterase when total levels are low. The issue of whether or not insulin has a direct effect on isolated membrane-bound adenylate cyclase remains controversial (Illiano and Cuatrecasas, 1972) and judgment must be deferred until the reasons for the disparity in results are elucidated.

We cannot rule out a compartmentalization of cyclic AMP as proposed by Kuo and De Renzo (1969) and by Jarett et al. (1972) such that insulin *might* reduce levels in a subcellular compartment (in close relation to protein kinase) without a significant effect on total cell concentration. We must consider also that the insulin molecule might directly affect lipase activity or the ability of protein kinase to convert lipase b to lipase a. Our preliminary studies in which insulin was added directly during kinase activation of crude or purified rat lipase fail to show such an effect (Aquino and Steinberg, unpublished data). The remaining alternatives are that insulin exposure generates some 'message' that reduces the effectiveness of a given level of protein kinase free catalytic unit on the hormone-sensitive lipase (and on the phosphorylase-glycogen synthase systems) or that such a 'message' reduces the level of catalytic unit generated *at a given cell concentration of cyclic AMP*.

In closing it should be pointed out that insulin can affect other cyclic AMP-regulated systems without necessarily affecting cyclic AMP accumulation. Thus, Goldberg et al. (1967) demonstrated insulin-stimulated increases in glycogen synthase *a* in skeletal muscle in vivo that were actually associated with a small *increase* in tissue cyclic AMP concentrations. Craig et al. (1969) found that insulin antagonized the effects of epinephrine on glycogen synthase in rat diaphragm in vitro without affecting total cyclic AMP concentrations reached. Whether the insulin effects discussed above can be accounted for by action at the level of cyclic AMP alone thus remains uncertain. We feel that additional studies are in order before the issue is considered closed.

REFERENCES

Allen, D. O. and Ashmore, J. (1972): Assay of glycogen phosphorylase in isolated fat cells of the rat. Biochem. Pharmacol., 21, 1441.

Allen, D. O. and Clark, J. B. (1970): Effect of various antilipolytic compounds on adenyl cyclase and phosphodiesterase activity in isolated fat cells. Advanc. Enzyme Regulation, 9, 99.

Brown, W. V. and Baginsky, M. L. (1972): Inhibition of lipoprotein lipase by an apoprotein of human very low density lipoprotein. Biochem. biophys. Res. Commun., 46, 375.

Butcher, R. W. (1966): Cyclic 3´,5´-AMP and the lipolytic effect of hormones on adipose tissue. Pharmacol. Rev., 18, 237.

Butcher, R. W., Baird, C. E. and Sutherland, E. W. (1968): Effect of lipolytic and antilipolytic substances on adenosine 3´,5´-monophosphate levels on isolated fat cells. J. biol. Chem., 243, 1705.

Butcher, R. W., Sneyd, J. G. T., Park, C. R. and Sutherland, E. W. (1966): Effect of insulin on adenosine 3´,5´-monophosphate in rat epididymal fat pad. J. biol. Chem., 241, 1651.

Corbin, J. D. and Krebs, E. G. (1969): A cyclic AMP-stimulated protein kinase in adipose tissue. Biochem. biophys. Res. Commun., 36, 328.

Corbin, J. D., Reiman, E. M., Walsh, D. A. and Krebs, E. G. (1970): Activation of adipose tissue lipase by skeletal muscle cyclic adenosine 3´,5´-monophosphate-stimulated protein kinase. J. biol. Chem., 245, 4849.

Craig, J. W., Rall, T. W. and Larner, J. (1969): The influence of insulin and epinephrine on adenosine 3´,5´-phosphate and glycogen transferase in muscle. Biochim. biophys. Acta (Amst.), 177, 213.

Cryer, P. E., Jarett, L. and Kipnis, D. M. (1969): Nucleotide inhibition of adenyl cyclase activity in fat cell membrane. Biochim. biophys. Acta (Amst.), 177, 586.

Goldberg, N. D., Villar-Palasi, C., Sasko, H. and Larner, J. (1967): Effects of insulin treatment on muscle 3´,5´-cyclic adenylate levels in vivo and in vitro. Biochim. biophys. Acta (Amst.), 148, 665.

Heller, R. A. and Steinberg, D. (1972): Partial activatable triglyceride lipase from rat adipose tissue. Biochim. biophys. Acta (Amst.), 270, 65.

Huttunen, J.K., Aquino, A. A. and Steinberg, D. (1970a): A purified triglyceride lipase, lipoprotein in nature, from rat adipose tissue. Biochim. biophys. Acta (Amst.), 224, 295.

Huttunen, J. K. and Steinberg, D. (1971): Activation and phosphorylation of purified adipose tissue hormone-sensitive lipase by cyclic AMP-dependent protein kinase. Biochim. biophys. Acta (Amst.), 239, 411.

Huttunen, J. K., Steinberg, D. and Mayer, S. E. (1970b): ATP-dependent and cyclic AMP-dependent activation of rat adipose tissue lipase by protein kinase from rabbit skeletal muscle. Proc. nat. Acad. Sci. (Wash.), 67, 290.

Huttunen, J. K., Steinberg, D. and Mayer, S. E. (1970c): Protein kinase activation and phosphorylation of a purified hormone-sensitive lipase. Biochem. biophys. Res. Commun., 41, 1350.

Illiano, G. and Cuatrecasas, P. (1972): Modulation of adenylate cyclase activity in liver and fat cell membranes by insulin. Science, 175, 906.

Jarett, L., Steiner, A., Smith, R. M. and Kipnis, D. M. (1972): The involvement of cyclic AMP in the hormone regulation of protein synthesis in rat adipocytes. Endocrinology, 90, 1277.

Jungas, R. (1966): Role of cyclic $3',5'$-AMP in the response of adipose tissue to insulin. Proc. nat. Acad. Sci. (Wash.), 56, 757.

Khoo, J. C., Fong, W. W. and Steinberg, D. (1972b): Activation of hormone-sensitive lipase from human adipose tissue by cyclic AMP-dependent protein kinase. Biochem. biophys. Res. Commun., 49, 407.

Khoo, J. C., Jarett, L., Mayer, S. E. and Steinberg, D. (1972a): Subcellular distribution of and epinephrine-induced changes in hormone-sensitive lipase, phosphorylase and phosphorylase kinase in rat adipocytes. J. biol. Chem., 247, 4812.

Khoo, J. C. and Steinberg, D. (1973): The mechanism of activation of hormone-sensitive lipase in human adipose tissue. Clin. Res., 21, 629.

Khoo, J. C., Steinberg, D., Thompson, B. and Mayer, S. E. (1973): Hormonal regulation of adipocyte enzymes. The effects of epinephrine and insulin on the control of lipase, phosphorylase kinase, phosphorylase and glycogen synthase. J. biol. Chem., 248, 3823.

Krebs, E. G. (1972): Protein kinases. Curr. Top cell. Regulation, 5, 99.

Kuo, J. F. and De Renzo, C. E. (1969): A comparison of the effects of lipolytic and antilipolytic agents on adenosine $3',5'$-monophosphate levels in adipose cells as determined by prior labelling with adenine-8-^{14}C. J. biol. Chem., 244, 2252.

Loten, E. G. and Sneyd, J. G. T. (1970): An effect of insulin on adipose tissue adenosine $3',5'$-cyclic monophosphate phosphodiesterase. Biochem. J., 120, 187.

Moskowitz, J. and Fain, J. N. (1969): Hormonal regulation of lipolysis and phosphorylase activity in human fat cells. J. clin. Invest., 48, 1802.

Patten, R. L. (1970): The reciprocal regulation of lipoprotein lipase activity and hormone-sensitive lipase activity in rat adipocytes. J. biol. Chem., 245, 5577.

Rizack, M., (1964): Activation of an epinephrine-sensitive lipolytic from adipose tissue by adenosine $3',5'$-phosphate. J. biol. Chem., 239, 392.

Robinson, D. S. and Wing, D. R. (1970): Regulation of adipose tissue clearing factor lipase activity. In: Adipose Tissue, p. 41. Editors: B. Jeanrenaud and D. Hepp. Academic Press, New York.

Senft, G., Shultz, G., Munske, K. and Hoffman, M. (1968): Influence of insulin on cyclic $3',5'$-AMP phosphodiesterase activity in liver, skeletal muscle, adipose tissue and kidney. Diabetologia, 4, 322.

Soderling, T. R., Corbin, J. D. and Park, C. R. (1973): Regulation of adenosine $3',5'$-monophosphate-dependent protein kinase II. J. biol. Chem., 248, 1822.

Strand, O., Vaughan, M. and Steinberg, D. (1964): Rat adipose tissues: hormone-sensitive lipase activity against triglycerides compared with activity against lower glycerides. J. Lipid Res., 5, 554.

Tsai, S-C., Belfrage, P. and Vaughan, M. (1970): Activation of hormone-sensitive lipase in extracts of adipose tissue. J. Lipid Res., 11, 466.

Tsai, S-C. and Vaughan, M. (1972): Inactivation of hormone-sensitive lipase and monoglyceride lipase activities in adipose tissue. J. biol. Chem., 239, 409.

Vaughan, M. (1972): The role of insulin in regulation of cyclic AMP metabolism. In: Insulin Action, Chapter XI, p. 297. Editor: I. B. Fritz. Academic Press, New York.

Vaughan, M., Berger, J. E. and Steinberg, D. (1964): Hormone-sensitive lipase and monoglyceride lipase activities in adipose tissue. J. biol. Chem., 239, 409.

Vaughan, M. and Murad, F. (1969): Adenyl cyclase activity in particles from fat cells. Biochemistry, 8, 3092.

Vaughan, M. and Steinberg, D. (1965): Glyceride biosynthesis, glyceride breakdown in adipose tissue: mechanisms and regulation. In: Handbook of Physiology – Section 5: Adipose Tissue, p. 239. Editors: A. S. Renold and G. F. Cahill, Jr. American Physiological Society, Washington, D. C.

DISCUSSION

Angel: I was most interested in your finding that the antilipolytic effect of insulin was associated with a statistically significant reduction in cyclic AMP accumulation. We have been able to confirm this observation and have also shown that the lipolytic effect of insulin (enhancement of norepinephrine-stimulated lipolysis) is associated with a rise in adipocyte cyclic AMP levels. These observations are particularly important because initial rates of cyclic AMP accumulation and glycerol release were measured. I wonder if you would explain the mechanism of the cyclic AMP-lowering effect of insulin and why you think that this effect is an unlikely explanation of the antilipolytic effect of insulin.

Steinberg: We doubt that the antilipolytic effect of insulin is adequately accounted for by its effect on cyclic AMP levels but, of course, it is still an open question, requiring further experimentation. Let me point out that while some workers have found clear effects of insulin on cyclic AMP levels, the experiments have been characterized by very high initial levels (because of the inclusion of a phosphodiesterase inhibitor in the medium or the use of supramaximal levels of hormone) and often the decrease in cyclic AMP level has *not* been accompanied by a decrease in lipolytic rate. This was the case in the first studies by Butcher et al. (1966, 1968) when caffeine was included in the medium; when it was omitted, insulin suppressed lipolysis without affecting cyclic AMP levels, as in our own studies. There is no doubt that insulin can affect cyclic AMP levels under some circumstances but we suggest that this effect is not necessarily linked to the antilipolytic activity. Perhaps the reported effects of insulin on phosphodiesterase activity only become apparent at high levels of cyclic AMP. I would also point out that an insulin effect on membrane adenylate cyclase is at best elusive. Some have observed it; others are unable to find an effect. Finally, other workers have reported the same kind of dissociation we report here – Jaret et al. (1972) and Allen and Clark (1970) among them. Similar dissociation has been reported with regard to insulin-epinephrine antagonism in regulation of glycogen synthase (Craig et al., 1969). I submit that the anti-cyclic AMP hypothesis is simple and attractive but the case is not proved and the book is not closed. We need more experiments.

Triglyceride storage disease:
a study of an affected family*

C. Gilbert, D. J. Galton and **J. Kaye**

Diabetes Research Laboratory, St. Bartholomew's Hospital, London, U. K.

We wish to report on a patient (D.D.) who appears to have a defect in the breakdown of triglyceride in adipose tissue associated with obesity, and although the exact nature of this metabolic defect has not yet been established it appears to be transmitted on a familial basis.

METHODS

Clinical details of the propositus and daughter have been previously described (Gilbert et al., 1973).

Six samples of adipose tissue were taken from D. D. from the anterior abdominal wall under a variety of conditions (fed, overnight fast, fast for 14 days, inpatient, outpatient) and the results were meaned. Three biopsies were performed on her daughter (N. D.) as an inpatient under fed and fasted regimes. Control tissue was obtained from 7 obese inpatients under fed conditions (mean weight 102 kg (range 62-170 kg), mean height 142 cm, mean age 48 years). Adipose tissue from all sources was transferred to the laboratory in 0.9% saline for metabolic studies.

Tissue (about 100 mg) was incubated for 1 hour with or without isoprenaline (10^{-5} M) in an Earle's bicarbonate buffer (pH 7.4) containing NaCl (120 mM), KCl (5.4 mM), $CaCl_2$ (1.8 mM), $MgCl_2, 7H_2O$ (0.8 mM), $NaH_2PO_4.2H_2O$ (1.1 mM), $NaHCO_3$ (26 mM) and 1% bovine serum albumin (crystalline, Fraction V). Release of glycerol into the medium was assayed enzymatically by the method of Garland and Randle (1962).

For the assay of cyclic AMP incubated tissue was extracted in ice-cold HCl (6% v/v) and the fat removed with ether. The aqueous extract was neutralized with 50 μl of KOH and stored at $-20°$ C. Cyclic AMP was measured using the competitive protein binding method of Brown et al. (1971).

Tissue extracts were also incubated in a phosphate buffer with ^{32}ATP (2.25×10^{-5} M) and histones (6 mg/ml). The extent of histone phosphorylation in the absence and presence of cyclic AMP (5×10^{-7} M) was taken as a measure of protein kinase activity.

RESULTS

Dose-response studies show that isoprenaline maximally stimulates release of glycerol and augments levels of cyclic AMP in human adipose tissue at concentrations of 10^{-6} to 10^{-5} M. Table 1 presents the effect of isoprenaline (10^{-5} M) on release of glycerol from adipose tissue of members of the family and includes values for obese controls. Despite similar release of glycerol by all subjects in the basal state (Student's $t = 0.26$, P not significant), stimulation with isoprenaline did not significantly increase the output of glycerol by the mother (Student's $t = 1.13$, P not significant). Her daughter and eldest sister showed a similar tendency to a reduced output of glycerol, whereas a non-identical twin sister (non-obese) and a brother (obese) released glycerol

* Financial support for this project came from the Servier laboratories and the Joint Research Board of St. Bartholomew's Hospital, London.

TABLE 1   Glycerol release from adipose tissue

Subject	Age (years)	% of average body weight increase	Glycerol release (nmole/mg/hour)*	
			Basal	Increment due to isoprenaline
D.D. (mother)	60	30	0.096 ± 0.03(6)	0.06 ± 0.02(6)
N.D. (daughter)	29	35	0.128 ± 0.001(3)	0.111 ± 0.05(3)
A.F. (sister)	64	29	0.195(2)	0.07(2)
G.T. (non-identical twin)	60	0.9	0.25(2)	0.165(2)
R.D. (brother)	48	15	0.097(2)	0.183(2)
Obese controls	48 (mean)	—	0.10 ± 0.036(7)	0.247 ± 0.07(7)

* Figures in parentheses indicate number of observations.

similar to control values. The mean increment due to isoprenaline was significantly different from that in controls both in D. D. (Student's $t = 2.32$, $P < 0.05$) and in the obese members of her family (Student's $t = 3.63$, $P < 0.01$). The mean response to isoprenaline (without subtraction of basal values) in obese members of the family was significantly different from corresponding values of obese controls (Student's $t = 2.4$, $P < 0.05$).

Although the output of glycerol was depressed in the mother and daughter, isoprenaline increased the tissue levels of cyclic AMP to a similar extent as in obese controls (Gilbert et al., 1973). The next step after production of cyclic AMP is a kinase activation of triglyceride lipase, and Table 2 shows that tissue extracts from the mother contained a protein kinase that can be stimulated by cyclic AMP.

TABLE 2   Activity of protein kinase in human adipose tissue of propositus and controls

Subject	Additions	Incorporation of ^{32}P from ATP-γ-^{32}P (pmoles/mg protein)*			
		0	10 minutes	20 minutes	40 minutes
D.D.	—cyclic AMP	3.0 ± 1.4(3)	2.4(2)	3.9 ± 0.96(3)	5.4 ± 1.53(3)
	+cyclic AMP	2.3 ± 0.76(3)	6.2 ± 0.76(3)	9.4 ± 0.83(3)	12.7 ± 1.53(3)
Obese controls (3 patients)	—cyclic AMP	4.0 ± 0.76(3)	5.7 ± 0.63(3)	6.6 ± 0.83(3)	9.7 ± 1.56(3)
	+cyclic AMP	3.5 ± 0.9(3)	11.5 ± 3.6(3)	14.8 ± 5.4(3)	20.5 ± 7.1(3)

* Results are expressed as means of ^{32}P incorporated into histones ± S.E.M. Figures in parentheses indicate number of observations.

DISCUSSION

The mother in this family appears to have a defect in the breakdown of triglyceride to glycerol in adipose tissue when stimulated by isoprenaline. This defect may be on a familial basis since two other members of the family (eldest sister and daughter) appear to have the same trait. The defect may not be on an environmental basis since a non-identical twin sister showed a response to isoprenaline similar to obese controls. An obese brother likewise responded to isoprenaline in a similar way to obese controls. There are other reports in the literature of decreased mobilization of fat in obese patients (Pinter and Pattee, 1969).

The nature of the lipolytic defect in the mother (and eldest sister) is not yet clearly established. It appears to be sited after adenyl cyclase, since the mother responded to isoprenaline by a normal rise in tissue levels of cyclic AMP. The triglyceride lipase is also functional since the basal release of glycerol by the mother and eldest sister is similar to obese controls and the mother reacted to a fast by a rise in plasma fatty acids and loss of weight. The triglyceride lipase may be abnormal in not undergoing activation by the kinase, or the kinase itself may be deficient. If the latter is the case it does not extend to all the cyclic AMP-dependent kinases in the cell, since tissue extracts of the mother could phosphorylate histones and this was stimulated by cyclic AMP.

ACKNOWLEDGEMENTS

The helpful advice of Dr. K.O. Black and Professor J. Landon is gratefully acknowledged.

REFERENCES

Brown, B.L., Albano, J.D.M., Ekins, R.P. and Sgherzi, A.M. (1971): A simple and sensitive assay method for measurement of adenosine 3,5 cyclic monophosphate. Biochem. J., 121, 561.
Corbin, J.D. and Krebs, E.G. (1969): c-AMP-stimulated protein kinase in adipose tissue. Biochem. biophys. Res. Commun., 36, 328.
Garland, P.B. and Randle, P.J. (1962): A rapid enzymatic assay for glycerol. Nature (Lond.), 196, 987.
Gilbert, C., Galton, D.J. and Kaye, J. (1973): Triglyceride storage disease: a disorder of lipolysis in adipose tissue of two patients. Brit. med. J., 1,25.
Pinter, E.J. and Pattee, C.J. (1969): Comparative studies of fat mobilization in lean and obese subjects. In: Physiopathology of Adipose Tissue. p. 126. Editor: J. Vague. Excerpta Medica, Amsterdam.

DISCUSSION

Yen: How long did you keep the plasma fatty acid level elevated in your obese patients by fasting?

Galton: The patient was fasted for 14 days and the plasma fatty acids rose to 1.0 mM by the 7th day and remained elevated for a further 7 days.

Yen: Did that cause any ketosis?

Galton: Yes, but not a severe ketoacidosis requiring termination of the fast.

Roncari: Could you please tell me whether there was a difference in weight between the control group and the patients? Also, what was the nutritional status of the two groups at the time of biopsy?

Galton: The mean weight of controls was 102 kg, whereas the weight of the propositus was 79 kg. The majority of patients were biopsied after an overnight fast. However, the conditions under which the biopsy of the propositus was performed varied (fed, fasted, as inpatient, as outpatient, etc.).

Gurr: Is it possible that at least part of the reduced glycerol release could be explained by an increased ability to reutilize glycerol, for example because of increased activity of glycerol kinase in the patient's tissue? Have you measured glycerol kinase in this patient?

Galton: Yes, this could be so. We have studied human adipose tissue for the conversion of glycerol to CO_2 and neutral lipid and very little is metabolized compared to the amounts released. However we have not tested tissue from the kindred for glycerol utilization.

Catecholamines at the membrane level: an approach to the mode of action

P. Laudat

INSERM, Hôpital Cochin, Paris, France

The interactions of the peptide hormones with their receptors have been studied largely in liver and adrenal. No such study has previously been carried out on adipose tissue and, in particular, on the catecholamines which are the principal lipolytic hormones. Without presuming the existence of other mechanisms of action, it is probable that the primary effect of catecholamines can be located at the level of fat cell membranes, the interaction of the hormone and its receptor being proved by the activation of the adenylate cyclase system. We have, therefore, attempted to study the correlation between the specific binding of epinephrine to its receptors and adenylate cyclase activity.

The fact that high molecular weight complexes such as epinephrine-sepharose are able to activate the adipocyte adenylate cyclase without passing through the plasma membrane implies that a 'receptor site' or 'sites' must exist at the membrane surface where epinephrine binds.

METHODS

For the preparation of fat cell membranes we have used the method previously described (Laudat et al., 1972), which produces a preparation containing an adenylate cyclase system that is very sensitive to lipolytic hormones.

Adenylate cyclase activity was determined using $[a^{-32}P]ATP$ as substrate according to Krishna et al. (1968). Cyclic AMP was isolated by chromatography on neutral alumina (White and Zenser, 1971). 10 μg membrane protein was used for each assay. The [³H]epinephrine (sp. act. 12 Ci/mM) binding study was performed using the following technique. For this experiment epinephrine was at a constant molarity, 10^{-6} M. 80 μg of membrane protein were incubated for 5 minutes at 37° C in 50 mM Tris buffer without albumin at pH 7.4/7.6 in the presence of [³H]epinephrine. The mixture was placed in a sucrose solution (10%) in a plastic tube and centrifuged at 4,500 × g for 10 minutes. The upper phase was discarded and a second wash with sucrose carried out. The membrane pellet was then counted in a scintillation vial containing 10 ml Instagel after overnight solubilization with constant gentle shaking. For the displacement studies cold catecholamine analogues and metabolites at 10^{-4}M were added before the incubation.

RESULTS

Figure 1 shows the approximate values expressed as percentages of the displacement of [³H]epinephrine bound by different analogues or metabolites and the percentage adenylate cyclase stimulation. Figure 2 indicates the approximate values expressed as percentages of the displacement of [³H]epinephrine bound by other analogues or metabolites. From these results it can be seen that the hydroxyl group in the para or meta position plays a key role in the binding reaction. The stereospecific side chain is related more specifically to the stimulation of the adenylate cyclase.

So it appears that two classes of sites are involved in the hormonal activation of the cyclase in

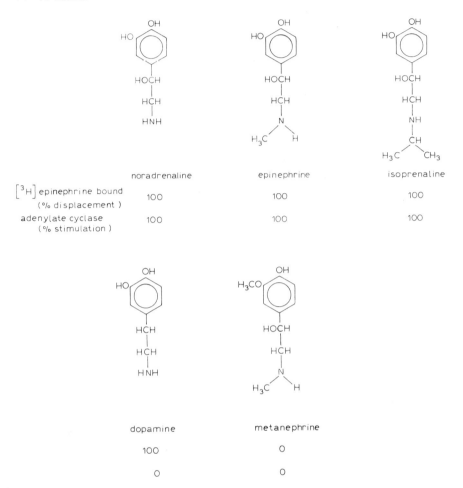

Fig. 1. Approximate values of displacement of [³H] epinephrine bound by different analogues or metabolites and corresponding adenylate cyclase stimulation.

fat cell membranes. One site, R_1 (Fig. 3), is related to the binding of the catechol moiety of the molecule. This site does not seem to be a very specific one, as can be concluded from the displacement results obtained. On the other hand, the stereospecific side chain binding which is required for the cyclase stimulation (R_2) seems to be more specific. It appears that the biological effect of catecholamines on adipose tissue implies the occupancy of both $R_1 + R_2$.

The β-blocking agents such as propranolol, which are known to inhibit the cyclase activation induced by epinephrine, do not completely effect the binding of the hormone. This discrepancy between propranolol action on binding and cyclase is probably due to the high non-specific binding of epinephrine. Either only a small number of sites are responsible for cyclase inactivation or activation, or different affinity sites are present on the fat cell membranes, some being related only to binding and adenylate cyclase. Work is in progress to test this hypothesis.

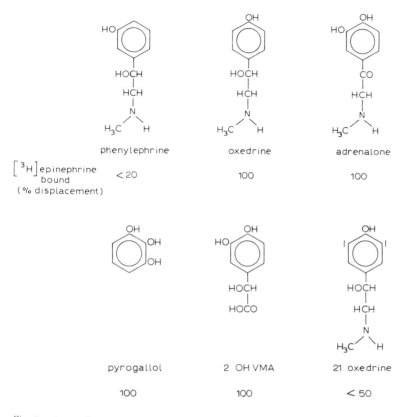

Fig. 2. Approximate values of displacement of [³H] epinephrine bound by various analogues or metabolites.

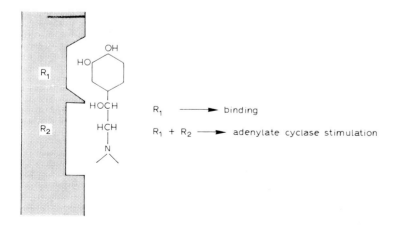

Fig. 3. Hypothetical scheme for catecholamine action at the membrane level.

REFERENCES

Krishna, G., Weiss, B. and Brodie, B.B. (1968): A simple, sensitive method for the assay of adenyl cyclase. J. Pharmacol. exp. Ther., 163, 379.

Laudat, M.H., Pairault, J., Bayer, P., Martin, M. and Laudat, P. (1972): Preparation of fat cell membrane with high sensitivity to lipolytic hormones. Biochem. biophys. Acta (Amst.), 255, 1005.

White, A.A. and Zenser, T.V. (1971): Separation of 3,5 nucleotide monophosphates from other nucleotides on aluminum oxide columns. Ann. Biochem., 41, 372.

DISCUSSION

Krans: Did you also find two types of binding sites with different affinities in human adipose tissue?

Laudat: Binding studies in human fat cell membranes are in progress in our laboratory but we do not have any data at present.

Krans: With regard to the data shown for activation of adenylate cyclase by norepinephrine where concentrations up to 10^{-4} M did not show maximal stimulation, do you think that the two types of receptors are related to two types of adenylates cyclase molecule, and what then is the physiological significance of the low affinity binding sites?

Laudat: I think the two types of receptors for epinephrine are related only to one adenylate cyclase molecule, the activation of the cyclase requiring first the binding of the entire structure of the epinephrine molecule (catechol part and stereospecific side chain). The physiological significance of the low affinity binding sites shown in rat fat cell membranes is unknown.

In vitro lipolytic activity of porcine growth hormone in the rabbit: effects of location, number and size of adipocytes*

A. Vezinhet and **J. Nouguès**

Station de Physiologie Animale, Ecole Nationale Supérieure Agronomique, Montpellier, France

Several pituitary hormones are known to stimulate in vitro lipolysis in the rabbit (Rudman, 1963; Raben, 1965). Although growth hormone produces an effect in vivo, many workers have noted its failure to promote lipolysis from adipocytes or adipose tissue slices in vitro. Fain et al. (1965) have shown that growth hormone and dexamethasone together produce a lipolytic effect on rat adipose cells. Ramachandran et al. (1972) observed that human growth hormone was able to induce glycerol release from and cyclic AMP accumulation in rabbit adipocytes. In our laboratory, we have demonstrated an in vivo lipolytic role of porcine growth hormone (PGH) in the rabbit (Vezinhet et al., 1972). Furthermore, the addition of PGH to an incubation medium containing isolated rabbit adipocytes produces marked lipolysis, as shown by the release of glycerol and free fatty acids (FFA) (Vezinhet, 1973). Shafrir and Wertheimer (1965) have studied the lipolytic and lipogenic activity of adipose tissue with respect to the locations of the deposits, and Björntorp and Karlsson (1970), Smith (1971) and Romsos and Leveille (1972) have shown that lipogenic activity may change as a function of adipocyte size. In man, Sjöström et al. (1972) have studied the variation in the number and size of adipocytes from different anatomical locations, while various workers (e.g. Knittle and Ginsberg-Fellner, 1972) have examined the change in lipolytic activity as a function of adipocyte size. To our knowledge, however, there has been no work on the lipolytic phenomenon with regard to both the location of the adipose tissue and the age of the animals.

MATERIAL AND METHODS

This work was carried out on 30 male New Zealand rabbits separated into 5 identical age groups of 35, 50, 75, 150 and 180 days. After a 24-hour solid fast the animals were sacrificed at the same time (8.30 a.m.) and adipose tissue slices were removed quickly after death. They originated from the perirenal, lesser omentum, interscapular and neck regions. At 35 days tissue from the lesser omentum could not be obtained in sufficient quantity. About 2 g of each tissue type were placed in 6 ml of Krebs-Ringer albumin bicarbonate buffer containing 10 mg of collagenase (Worthington Biochemical Corp., 137 U/mg). The adipocyte dissociation and preparation of the incubation medium were carried out according to Rodbell (1964), as described by Vezinhet (1973) and Vezinhet and Nouguès (1974). For both metabolic and histometric studies, 0.5 ml aliquots of cell suspensions were incubated at 37° C for 3 hours in glucose-free Krebs-Ringer albumin bicarbonate buffer at pH 7.4. The glycerol released was titrated in the medium by Wieland's (1957) method and the FFA by the method of Dole and Meinertz (1960). The total lipids were extracted from the incubation medium according to the procedure of Folch et al. (1957).

The data relating to the number and size of the adipocyte samples used for the metabolic studies were obtained in the following way. 0.5 ml of the cell suspension was fixed in 2% osmium tetroxide solution in phosphate buffer (pH 7.3) for 24 to 48 hours at 37° C. After fixation, the adipocytes contained in 0.5 ml were recovered on a Millipore filter (HAWP

* This research was supported in part by grants from the Délégation Générale à la Recherche Scientifique et Technique, Contrat No. 3129.

04700, 0.45 μm) and were again suspended in 100 ml distilled water. One ml aliquots were removed from the agitated cell suspension and the number of cells present in each aliquot was counted under the microscope. Following the count, all the cells were returned to the medium, so that one sampling did not influence the other. This measurement is accurate since the relative error in the mean value of the number of cells in 1 ml of cell suspension is $< 1\%$. The total cell number was obtained by multiplying the average number of cells contained in 1 ml by the dilution factor.

The suspended adipocytes, fixed in osmium tetroxide, are spherical in shape and the diameter of 300 to 400 cells was measured using a microscope magnification of $200 \times - 500 \times$ according to the adipose tissue site studied. The diameter distribution frequencies stabilize after 200 to 250 measurements.

PGH preparations (NBC 5940, 50 IU and Calbio Chem lot 200753, 1 IU/mg) were added to 2 ml of the medium in 50 μl of solution containing 10 mIU PGH. Distilled water was added to the control media. Each sample was run in triplicate and was followed by an albumin blank. FFA or glycerol results are expressed with respect to the total lipid extracted from the medium, as previously described (Vezinhet, 1973), or as a function of the number of adipocytes or of the unit surface area and volume.

RESULTS AND DISCUSSION

The influence of PGH on FFA and glycerol production per g of lipid (Tables 1 and 2) was always apparent, regardless of the type of adipose tissue and of the age of the animal. The differences between control and PGH-treated means were always significant ($P < 0.01$ or $P < 0.001$).

TABLE 1   Effect of PGH addition (10 mIU per 2 ml incubation medium) on glycerol release from adipocytes isolated from the four adipose deposits removed from rabbits of various ages

Age (days)	Glycerol (μmoles/g lipid/3 hours incubation) (mean ± S.E.)							
	Perirenal		Lesser omentum		Interscapular		Neck	
	Controls	PGH	Controls	PGH	Controls	PGH	Controls	PGH
35	15.95 ± 4.91**	35.35 ± 2.22**			13.15 ± 1.66***	41.35 ± 2.84***	15.75 ± 1.16***	49.95 ± 5.03***
50	4.05 ± 0.53**	39.75 ± 9.03**	7.18 ± 0.81***	52.36 ± 8.86***	11.75 ± 2.40**	50.68 ± 8.83**	23.0 ± 3.59***	84.98 ± 5.28***
75	6.63 ± 0.85***	34.45 ± 3.14***	6.85 ± 0.68***	22.95 ± 3.26***	6.80 ± 0.99***	31.19 ± 3.76***	10.30 ± 1.31***	37.40 ± 3.99***
150	3.25 ± 1.01**	18.73 ± 3.72**	3.18 ± 0.61***	19.58 ± 2.72***	6.58 ± 1.88**	20.23 ±2.61**	10.36 ± 2.86**	24.12 ± 3.00**
180	3.81 ± 0.40***	23.3 ± 3.63***	3.78 ± 0.33***	24.10 ± 3.41***	4.70 ± 0.33***	26.56 ± 3.76***	4.50 ± 0.22***	29.83 ± 1.64***

** P $<$0.01
*** P $<$0.001

We observed that, in controls, lipolysis as a function of age and in the absence of added hormone was significant at 35 days in all types of tissues, but that it rapidly diminished in perirenal and lesser omental internal fats. Lipolysis was most marked in the interscapular and, in particular, in the neck subcutaneous fat, and remained so in the neck up to 150 days.

TABLE 2 Effect of PGH addition (10 mIU per 2 ml incubation medium) on FFA release from adipocytes isolated from the four adipose deposits removed from rabbits of various ages

Age (days)	FFA (μmoles/g lipid/3 hours incubation) (mean $\pm$ S.E.)							
	Perirenal		Lesser omentum		Interscapular		Neck	
	Controls	PGH	Controls	PGH	Controls	PGH	Controls	PGH
35	13.25 ± 4.74***	66.20 ± 6.79***			16.65 ± 4.70***	78.50 ± 6.71***	18.75 ± 3.53***	89.1 ± 5.28***
50	5.43 ± 1.80***	59.45 ± 9.89***	9.35 ± 4.02***	101.30 ± 12.21***	4.70 ± 1.23***	81.55 ± 6.89***	14.10 ± 3.10***	126.0 ±17.71***
75	4.15 ± 0.78***	55.18 ± 9.84***	6.03 ± 1.68***	47.40 ± 5.14***	10.30 ± 3.56***	66.08 ± 6.81***	14.23 ± 6.74***	72.40 ± 9.49***
150	1.96 ± 0.11***	41.05 ± 5.97***	5.88 ± 1.57***	47.13 ± 6.98***	10.10 ± 3.47**	48.20 ± 9.99**	23.85 ± 8.85**	63.50 ± 6.71**
180	2.75 ± 0.23***	47.75 ± 9.06***	2.50 ± 0.24***	45.06 ± 9.03***	2.58 ± 0.15***	43.06 ± 5.21***	2.33 ± 0.18***	66.0 ± 6.53***

** P <0.01
*** P <0.001

In the media containing PGH, the highest values were also encountered in the subcutaneous tissues. The greatest lipolysis seemed to take place at 50 days. By comparing glycerol and FFA release between the perirenal and neck fat, it can be seen that at 35 and 50 days the differences are significant. From 75 to 180 days the differences tend to diminish and then disappear. The observation suggests a precocious specialization of the subcutaneous fat with respect to lipolysis, a finding which agrees with the results of Derry et al. (1972), who observed that in the rabbit the interscapular and neck tissues lose their brown adipose characteristics late in postnatal life, which may explain why they are the site of greater lipolytic activity.

The results reported in Tables 1 and 2 show that at the early ages the ratio of 1 mole of released glycerol for 3 of FFA was not observed, suggesting the reesterification of fatty acids. Although approached in the older animals, this ratio was never attained. A study of adult animals would be necessary to test whether this ratio is ever reached in the rabbit. In an effort to relate the lipolytic activity to histometric characteristics, the results of our study are expressed only in terms of glycerol release, the true lipolysis index (Steinberg and Vaughan, 1965).

Table 3 gives the average values of cell number and size for each depot and also the mean production of glycerol per number of adipocytes or per unit surface area or volume.

The number and size of adipocytes in 0.5 ml of cell suspension varies with age and type of deposit, but the differences tend to diminish after 75 days. In fact at 180 days they become very small, except for neck fat, which still contains cells smaller than the other tissues.

The glycerol produced by 10^6 adipocytes changes considerably with the type of deposit between 50 and 75 days. This is very noticeable when comparing an internal fat such as perirenal and a subcutaneous fat such as that of the neck. At 180 days the differences between deposits diminish and the latter become similar in adipocyte number and lipolytic activity.

As expected, the evolution of the mean diameter of adipocytes for the different deposits is inversely related to that of their number.

The lipolytic activity data with respect to unit surface membrane area show that for each age and deposit the values are similar but the neck subcutaneous fat again distinguishes itself from the others by a greater activity at 50 days.

The lipolytic activity at the three ages studied, expressed in terms of glycerol released per unit volume, is different for internal fats (perirenal and lesser omentum) and subcutaneous fat (interscapular and neck). This trend, already observed by relating lipolytic activity to the weight

TABLE 3   Effect of PGH addition (10 mIU per 2 ml incubation medium) on the lipolytic activity per number of cells, unit surface area and volume of adipocytes isolated from the four adipose deposits removed from rabbits of various ages

Age (days)	Adipose deposit	Glycerol released (μmoles/ 3 hours)*	Sample cell number*	Glycerol released/ 10^6 adipocytes (μmoles)	Mean adipocyte diameter (μm)*	Glycerol released/ dm^2 (μmoles)	Glycerol released/ ml (μmoles)
	Perirenal	2.163 ± 0.25	173400 ± 13946	12.47	83.54 ± 0.46	5.7	40.8
	Lesser omentum	2.465 ± 0.29	276630 ± 9316	8.91	71.36 ± 1.30	5.6	46.8
50	Inter-scapular	2.252 ± 0.28	316620 ± 69096	7.11	64.50 ± 1.60	5.4	50.6
	Neck	2.421 ± 0.13	576870 ±104567	4.19	43.36 ± 2.54	7.1	99.0
	Perirenal	1.488 ± 0.31	92010 ± 6531	16.17	107.60 ± 6.48	4.4	24.80
	Lesser omentum	1.560 ± 0.33	143340 ± 10094	10.88	96.90 ± 2.27	3.7	22.95
75	Inter-scapular	1.842 ± 0.21	184350 ± 29252	9.9	83.34 ± 4.41	4.6	32.97
	Neck	1.719 ± 0.16	476720 ± 114888	3.60	60.00 ± 2.63	3.2	31.90
	Perirenal	1.364 ± 0.16	80308 ± 33411	16.98	123.71 ± 5.44	3.4	17.13
	Lesser omentum	1.722 ± 0.20	101580 ± 42201	16.95	115.68 ± 4.68	4.0	20.92
180	Inter-scapular	1.620 ± 0.24	78925 ± 43796	20.52	115.88 ± 5.34	4.8	25.20
	Neck	2.116 ± 0.06	163500 ± 78605	12.94	95.44 ± 5.26	4.5	28.43

* Mean ± S.E.

of lipids in grams (c.f. Table 1), is clearly demonstrated in this case. One may conclude that the subcutaneous deposits are mainly geared toward lipolysis whilst internal fats are more active lipogenically, a relationship prevailing in ruminants (Ingle et al., 1972).

ACKNOWLEDGEMENTS

We wish to thank Mrs. Thérèse Chery, Mrs. Odette Moulierac and Mr. Bouthier for skilled technical assistance. The English manuscript has been kindly revised by Dr. O. Z. Sellinger.

REFERENCES

Björntorp, P. and Karlsson, M. (1970): Triglyceride synthesis in human subcutaneous adipose tissue cells of different sizes. Europ. J. clin. Invest., 1, 112.

Derry, D. M., Morrow, E., Sadre, N. and Flattery, K. V. (1972): Brown and white fat during the life of the rabbit. Develop. Biol., 27, 204.

Dole, V. P. and Meinertz, H. (1960): Microdetermination of long chain fatty acids in plasma and tissues. J. biol. Chem., 235, 2595.

Fain, J. S., Kovacev, V. P. and Scow, R. O. (1965): Effect of growth hormone and dexamethasone on lipolysis and metabolism in isolated fat cells of the rat. J. biol. Chem., 240, 3522.

Folch, J., Lees, M. and Sloane-Stanley, G. H. (1957): A simple method for the isolation and purification of total lipids from animal tissues. J. biol. Chem., 226, 497.

Ingle, D. L., Bauman, D. E. and Garrigus, U. S. (1972): Lipogenesis in the ruminant; in vitro study of tissue sites, carbon sources and reducing equivalent generation for fatty acid synthesis. J. Nutr., 102, 609.

Knittle, J. L. and Ginsberg-Fellner, F. (1972): Effect of weight reduction on in vitro adipose tissue lipolysis and cellularity in obese adolescents and adults. Diabetes, 21, 754.

Raben, M. S. (1965): Regulation of fatty acid release with particular reference to pituitary factors. In: Handbook of Physiology, Section 5, p. 331. Editors: A.E. Renold and G.F. Cahill. American Physiological Society, Washington, D. C.

Ramachandran, J., Lee, V. and Li, C. H. (1972): Stimulation of lipolysis and cyclic AMP accumulation in rabbit fat cells by human growth hormone. Biochem. biophys. Res. Commun., 48, 274.

Rodbell, M. (1964): Metabolism of isolated fat cells. Effects of hormones on glucose metabolism and lipolysis. J. biol. Chem., 239, 375.

Romsos, D. R. and Leveille, G. A. (1972): Effect of cell size on in vitro fatty acid and glyceride-glycerol biosynthesis in rat adipose tissue. Proc. Soc. exp. Biol. (N.Y.), 141, 649.

Rudman, D. (1963): The adipokinetic action of polypeptide and amine hormones upon the adipose tissue of various animal species. J. Lipid Res., 4, 119.

Shafrir, E. and Wetheimer, E. (1965): Comparative physiology of adipose tissue in different sites and in different species. In: Handbook of Physiology, Section 5, p. 417. Editors: Renold and Cahill. American Physiological Society, Washington, D.C.

Sjöström, L., Smith, V., Krotkiewski, M. and Björntorp, P. (1972): Cellularity in different regions of adipose tissue in young men and women. Metabolism, 21, 1143.

Smith, U. (1971): Effect of cell size on lipid synthesis by human adipose tissue in vitro. J. Lipid Res., 12, 65.

Steinberg, D. and Vaughan, M. (1965): Release of free fatty acids from adipose tissue in vitro in relation to rates of triglyceride synthesis and degradation. In: Handbook of Physiology, Section 5, p. 335. Editors: Renold and Cahill. American Physiological Society, Washington, D.C.

Vezinhet, A. (1973): Effet lipolytique de l'hormone somatotrope porcine sur adipocytes de lapin. C. R. Acad. Sci. (Paris), 276, 1721.

Vezinhet, A., Charrier, J. and Dauzier, L. (1972): Evolution des taux plasmatiques d'acides gras libres et de sucres réducteurs lors d'un traitement aux hormones somatotropes porcine et bovine chez le lapin hypophysectomisé. Ann. Biol. anim., 12, 431.

Vezinhet, A. and Nouguès, J. (1974): Effet de la localisation anatomique, de la répartition en nombre et en taille des adipocytes sur le rôle lipolytique de la GH chez le lapin. Ann. Biol. anim., in press.

Wieland, O. (1957): Eine enzymatische Methode zur Bestimmung von Glycerin. Biochem. Z., 329, 313.

Stimulation of lipolysis in adipose tissue by inositol deficiency

T. Tomita, T. Maeda and **E. Hayashi**

Shizuoka College of Pharmaceutical Sciences, Oshika, Shizuoka, Japan

In rats fed an inositol-deficient diet, certain alterations in the metabolism of lipids were noted within one week at normal growth rate. The triglyceride (TG) and cholesterol levels in the liver, and the nonesterified fatty acid (NEFA) level in the serum were significantly increased in the deficient rats. The increased cholesterol level was found to be due to stimulation of cholesterol biosynthesis by inositol deficiency. The accumulation of TG in the deficient liver was caused by an increased rate of fatty acid mobilization from adipose tissue to the liver. It is suggested that a decrease of polyunsaturated fatty acids in the lipid of the deficient rats is involved in the stimulation of lipolysis.

METHODS

Rats were maintained on the basal diet with or without inositol (0.5%) for 2 weeks (control and deficient groups). After 1 week on the inositol-deficient diet, a third group of rats (Group A) were fed a 0.5% supplement of inositol and a fourth group (Group B) continued on the deficient diet for a further week but with natural instead of hydrogenated cottonseed oil.

Epididymal fat pads were incubated in situ with 4.54 μCi [1-^{14}C] palmitic acid (0.454 μCi/ μmole/2 ml) in Krebs-Ringer phosphate buffer, pH 7.4, containing glucose (5 mM) and bovine serum albumin (5%) at 37° C for 30 minutes, and then replaced on the rats. Incorporation of the activity was measured 24 hours after the operation.

RESULTS AND DISCUSSION

Table 1 shows the alterations in lipid metabolism that developed in rats fed the inositol-deficient diet for 1 week. In the inositol-deficient rats at 1 and 2 weeks of treatment, liver TG levels were

TABLE 1   The effect of inositol deficiency on serum and liver lipid levels (mean ± S.E.)

	Control (No. = 10)	Deficient (No. = 10)
Serum		
Triglycerides (mg/100 ml)	69 ±1.9	67 ± 2.8
Cholesterol (mg/100 ml)	57.7 ± 3.87	58.7 ± 2.05
NEFA (μmole/100 ml)	78 ± 2.4	99 ± 3.8*
Phospholipids (mg/100 ml)	—	—
Liver		
Triglycerides (mg/g)	13.1 ± 1.48	33.8 ± 3.11*
Cholesterol (mg/g)	4.90 ± 0.156	5.64 ± 0.194*
NEFA (μmole/g)	46.5 ± 1.54	55.5 ± 2.20*
Phospholipids (mg/g)	23.3 ± 0.89	22.3 ± 0.44

* Significant (P < 0.01)

respectively 2.6 and 5.3 times the levels in control rats. There was also a significant increase of cholesterol in the liver and of NEFA in the serum in the deficient rats.

An increased level of liver TG could be due to (1) an increase in the influx of NEFA from adipose tissue into the liver, (2) an increase in the biosynthesis of liver fatty acids and TG, (3) an inhibition of release of liver lipid into the serum, and (4) a decrease in the rate of catabolism of liver lipid. Increased mobilization of NEFA from adipose tissue to the liver was suspected to be the major cause of liver TG accumulation, as a significant increase of the serum and liver NEFA levels accompanied the increased rate of deposition of TG in the liver.

Gas chromatographic analysis of liver lipid showed that in the total lipid of the inositol-deficient rats the $C_{16:0}$ $C_{16:1}$ and $C_{18:1}$ fatty acids increased 2-3 fold, whereas a significant decrease was noted in the levels of fatty acids such as $C_{18:0}$ $C_{18:2}$ and $C_{20:4}$ (Table 2). These changes were seen in the TG fraction but not in the phospholipid fraction.

TABLE 2   Comparison of the fatty acid composition of liver and epididymal fat pad lipid

Lipid	Group	No. of rats	Fatty acids (%)						
			14:0	16:0	16:1	18:0	18:1	18:2	20:4
Liver									
Total lipid	Control	9	0.6	33.8	4.8	15.4	27.3	10.1	8.0
	Deficient	8	1.6*	45.6*	9.1*	5.1*	35.2*	2.5*	1.7*
Triglycerides	Control	10	1.6	43.9	7.8	1.5	33.9	11.3	—
	Deficient	9	2.2	51.0*	11.9*	1.3	31.2	2.3*	—
Phospholipids	Control	10	0.1	23.2	2.1	32.4	3.0	8.5	32.0
	Deficient	9	0.2	21.4	2.1	33.6	3.4	7.1	32.3
Epididymal fat pads									
Triglycerides	Control	10	4.4	42.4	13.0	1.2	28.8	10.2	—
	Deficient	9	4.3	42.3	13.0	1.2	28.6	10.6	—

* Significant (P < 0.01)

Furthermore, the fatty acid composition of the increased liver lipid of the deficient rats was similar to that of the lipid in adipose tissue and to that of the dietary fat (Fig. 1). This evidence suggests that the increased liver TG level in the deficient rats is due to an increase in the accumulation of fatty acids from adipose tissue by the liver.

TABLE 3   Mobilization of fatty acids from epididymal fat pads specifically labeled with [1-^{14}C] palmitic acid (mean ± S.E.)

Group	No. of rats	Distribution of radioactivity as percentage of total activity taken up		
		Serum	Liver	Epididymal fat pads
Control	7	0.060 ± 0.0081 (0.96 ± 0.13)	0.80 ± 0.097 (0.77 ± 0.058)	31.8 ± 2.36
Deficient	7	0.072 ± 0.0071 (1.19 ± 0.14)	2.18 ± 0.330* (1.01 ± 0.054)*	32.6 ± 1.86

The numbers in parentheses indicate percentage distribution of radioactivity/mmole fatty acid.
* Significant (P < 0.01)

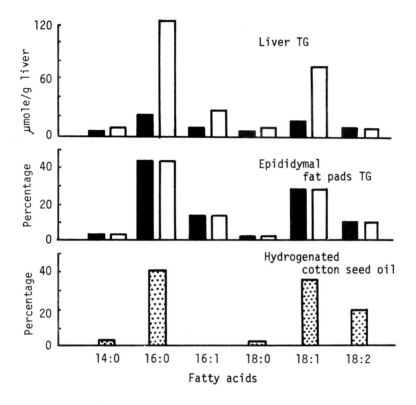

Fig. 1. Fatty acid composition of liver and epididymal fat pad triglyceride (TG) and hydrogenated cottonseed oil. Black bars = control, white bars = inositol deficient, stippled bars = hydrogenated cottonseed oil.

Epididymal fat pads were specifically labeled in situ with [1-^{14}C] palmitic acid by the method of Stein and Stein (1961) and mobilization of radioactive fatty acid from adipose tissue was measured. The activity in the liver of the deficient rats was 2.7 times that in control rats (Table 3), indicating that fatty acid mobilization from adipose tissue is stimulated by inositol deficiency.

Replacement of the hydrogenated cottonseed oil used in the diet by natural cottonseed oil, which is abundant in $C_{18:2}$ fatty acid, reversed the alterations in lipid metabolism in the inositol-deficient rats, i.e. TG and cholesterol accumulation in the liver and increase in serum NEFA. Administration of natural cottonseed oil for 1 week to rats that had been maintained on the inositol-deficient diet for 1 week inhibited further deposition of TG in the liver and decreased the high level of serum NEFA. Inositol administration to rats previously maintained on the inositol-deficient diet not only inhibited further deposition of lipid but also caused removal of the accumulated TG in the liver and a return to normal TG levels (Table 4). These results indicate that inositol deficiency affects liver TG metabolism both by stimulating lipolysis in adipose tissue and by decreasing release of deposited liver lipid. As dietary polyunsaturated fatty acids prevented the increase in liver TG deposition, the decreased content of polyunsaturated fatty acids observed in the deficient rats might be involved in the stimulation of lipolysis in adipose tissue.

Hegsted et al. (1973) have recently reported that the development of an intestinal lipodystrophy in the inositol-deficient gerbil was prevented by feeding highly unsaturated oils. Carreau et al. (1972) also observed that the sensitivity to noradrenaline of adipose tissue from rats varied

TABLE 4   Reversal by inositol and polyunsaturated fatty acids of the alteration in lipid levels caused by inositol deficiency (mean ± S.E.)

	Control (No. = 10)	Deficient (No. = 10)	Group A (No. = 10)	Group B (No. = 8)
Serum				
Triglycerides (mg/100 ml)	49.0 ± 4.55	51.3 ± 4.37	66.0 ± 3.72	—
Cholesterol (mg/100 ml)	71.0 ± 2.7	76.0 ± 2.7	75.0 ± 2.3	—
NEFA (μmole/100 ml)	73.0 ± 6.3	94.0 ± 7.8**	80.0 ± 2.8	66.0 ± 4.3
Liver				
Triglycerides (mg/g)	12.5 ± 0.98	66.3 ± 7.08**	9.9 ± 0.84	49.4 ± 2.74**
Cholesterol (mg/g)	5.36 ± 0.186	7.49 ± 0.344**	5.37 ± 0.149	6.35 ± 0.314**
NEFA (μmole/g)	54.8 ± 1.67	50.0 ± 1.62	55.2 ± 0.94	—
Phospholipids (mg/g)	24.8 ± 0.49	21.7 ± 0.87*	—	—

* Significant ($P < 0.05$)
** Significant ($P < 0.01$)

according to the type of dietary fat, and Levy (1971) found that phosphatidylinositol restored noradrenaline responsiveness of solubilized myocardial adenylate cyclase from cat. In view of these results, the effect of inositol deficiency on the responsiveness of adipose tissue to various lipolytic hormones is now being investigated in vitro.

REFERENCES

Carreau, J. P., Lariette, C., Counis, R. and Ketevi, P. (1972): Biochim. biophys. Acta (Amst.), 280, 440.
Hegsted, D. M., Hayes, K. C., Gallagher, A. and Hanford, H. (1973). J. Nutr., 103, 302.
Levy, G. S. (1971): J. biol. Chem., 246, 7405.
Stein, Y. and Stein, O. (1961): Biochim. biophys. Acta (Amst.), 54, 555.

Cellular alterations underlying insulin and glucagon insensitivity in large adipocytes

J.N. Livingston and **D.H. Lockwood**

Johns Hopkins University School of Medicine, Baltimore, Md., U.S.A.

Adipose tissue from obese individuals is less sensitive to insulin stimulation (Salans et al., 1968) and possibly to stimulation by lipolytic hormones (Gries et al., 1972) than fat from normal weight subjects. It has been postulated that the hormone resistance is related to fat cell enlargement which characteristically accompanies obesity (Salans et al., 1968). A similar situation exists in large adipocytes isolated from large adult rats. The responses of these cells to stimulation by both insulin (Salans and Dougherty, 1971) and glucagon (Manganiello and Vaughan, 1972) are much less in magnitude than responses found in small cells from young animals. We have used this animal model to gain insight into possible alterations that may contribute to hormone-resistant states in human obesity.

Insulin stimulation of glucose oxidation in large rat adipocytes produced only half the response elicited in small adipocytes. Insulin resistance in large cells was found at all concentrations of the hormone used, including amounts as high as 1500 $\mu U/ml$. The sensitivity of large and small cells to spermidine stimulation of glucose oxidation was also determined. This polyamine has been shown to produce insulin-like effects in fat tissue by apparently acting at membrane sites which are separate from insulin receptors but which share a common pathway with insulin-mediated responses (Lockwood et al., 1971). Therefore, by virtue of its site of action, spermidine allows investigation of processes that follow insulin-receptor interaction. Studies of the effect of spermidine on glucose oxidation indicated that the sensitivity of both types of cells to stimulation by this agent paralleled closely their relative sensitivity to insulin. The resistance of large cells to polyamine stimulation was as great as that found with stimulation by insulin, suggesting that the insulin receptor is not involved in the alteration(s) underlying insulin resistance. This was directly demonstrated by measuring the capability of large and small cells to 'specifically' bind ^{125}I-insulin. Results of these studies indicate that insulin binding, expressed on a per cell basis, was similar in both cell types at all concentrations of labeled hormone used (10^{-11} to 10^{-9} M). Therefore, the total binding capacity as well as the affinity between insulin and the insulin receptor are not diminished in large insulin-resistant fat cells.

An assessment of glucose metabolism was made in isolated fat cells to determine if insulin insensitivity was related to defective glucose transport and/or oxidation. The effect of various glucose concentrations on glucose oxidation indicated that large cells did not have as great a capacity as small cells to metabolize glucose to CO_2. This alteration in large cells was apparent only at relatively high glucose concentrations (> 10 mM). It was further shown that, under the conditions used to determine insulin and spermidine sensitivity, the diminished capacity did not contribute to the resistant state. Insulin and polyamine sensitivity was measured in an assay system that contained 0.3 mM glucose. Even with maximum stimulation by either agent, the enhanced responses in CO_2 production did not verge upon the maximum capacities of large or small cells to metabolize glucose.

Further evidence that the modification in glucose metabolism is not responsible for insulin resistance is provided by studies of the antilipolytic effect of insulin in these cells. Large adipocytes required greater concentrations of insulin to inhibit theophylline-induced lipolysis than was required by small cells. These experiments were performed in the absence of glucose in the incubation medium, which indicates that another cellular system under insulin influence is affected by the hormone-resistant state. Since this effect of insulin is not dependent on glucose transport or oxidation, it appears that an alteration exists elsewhere in large cells in addition to the modification found in glucose metabolism.

A possible cause of the resistance may involve partial uncoupling of the insulin receptors from intracellular effector systems (glucose transport and lipolysis) within large cells. The possibility of hampered transmission of the 'signal' generated by interaction of insulin with the insulin receptor may be related to dilution of the receptors over the surface of large cells. Although the number of receptors is equal regardless of cell type, the greater surface area (two-fold increase) of large cells results in a diminished number of receptors per unit of surface area. This type of alteration, i.e. a modification within the plasma membrane near the insulin receptor, would provide an explanation for insensitivity of two separate systems in large cells.

Large rat adipocytes are markedly resistant to the lipolytic effect of glucagon (Manganiello and Vaughan, 1972). The degree of resistance in large cells to stimulation by this hormone is much greater than the insensitivity observed during insulin stimulation. Studies of possible alterations responsible for glucagon resistance involved determination of the lipolytic capacity and glucagon-binding capability of large cells as well as their level of phosphodiesterase activity and its influence on glucagon-mediated lipolysis.

Maximum stimulation with glucagon, even in small sensitive cells, produces a relatively mild lipolytic response when compared to maximum stimulation by epinephrine, dibutyryl cyclic AMP or theophylline. The use of these agents to stimulate maximum lipolysis permitted comparison of the lipolytic capacity of large and small cells. The lipolytic response to dibutyryl cyclic AMP (3 mM) and theophylline (2 mM) was as great in large cells as that found in small cells. Although epinephrine (1 μg/ml) stimulation was slightly less in large cells, the amount of glycerol released was many times greater than the maximum glucagon response. These studies suggest that enzymes directly involved in hormone-mediated lipolysis do not constitute sites of alterations which contribute to glucagon resistance.

Phosphodiesterase is indirectly involved in glucagon-mediated lipolysis by virtue of its role in cyclic AMP degradation. Studies of both high- and low-affinity components of phosphodiesterase activity were performed in homogenates of large and small cells. These studies demonstrated that both types of activity exist in large cells and that the K_m's for each component were similar to those found in small cells. However, the maximum amount of activity for each component in large cell homogenates was twice that found in homogenates of small cells. The possible effect of elevated phosphodiesterase activity on glucagon-mediated lipolysis was examined in intact cells by the use of aminophylline, an inhibitor of phosphodiesterase activity. As expected, approximately twice the concentration of aminophylline was required by large cells to produce measurable stimulation of lipolysis than was required by small cells. When aminophylline was used to inhibit 'excess' phosphodiesterase activity in large cells, their response to maximum glucagon stimulation increased from 20% (no inhibitor) to 65% of the maximum glucagon response found in small cells in the presence of aminophylline. This increase strongly suggests that elevated phosphodiesterase activity is one alteration which contributes to the glucagon-resistant state. The finding, however, that correction of the elevated activity does not restore completely the sensitivity of resistant cells to glucagon stimulation suggests the presence of an additional alteration.

Glucagon regulation of metabolism in adipose and liver tissue is mediated by interaction of the hormone with specific glucagon receptors located on the plasma membrane (Rodbell et al., 1971). The capability of large and small fat cells to associate with this hormone was determined by ^{125}I-glucagon binding studies. 'Specific' binding was assayed both in intact cells and in particles prepared from homogenates of isolated cells. The amount of binding was expressed per cell numbers (DNA), per amount of ^{125}I-insulin binding, or per mg of protein in the particulate fraction. Regardless of how expressed, binding of ^{125}I-glucagon was less in large cells or in particles isolated from large cells. Saturation of binding was approached in large and small cells at labeled hormone concentrations in the range of 4×10^{-8} M, suggesting that the affinity of the glucagon receptor for glucagon is similar. Exact affinity constants for binding were difficult to obtain because of rapid ^{125}I-glucagon degradation by the adipocytes. However, the amount degraded was similar in both cell types, indicating that glucagon resistance in large cells is not caused by an increase in glucagon inactivation.

^{125}I-glucagon binding by large cells was approximately 65% that of small cells. This amount of binding agrees closely with the amount of glucagon-stimulated lipolysis in large cells when excess phosphodiesterase activity is inhibited. From these studies it appears that resistance of large cells to glucagon stimulation is adequately explained by an elevation in the levels of phosphodiesterase and by a diminished capacity of these cells to interact with glucagon.

REFERENCES

Gries, F.A., Berger, M., Neumann, M., Priess, H., Liebermeister, H., Hesse-Wortmann, C. and Jahnke, K. (1972): Effects of norepinephrine, theophylline and dibutyryl cyclic AMP on in vitro lipolysis of human adipose tissue in obesity. Diabetologia, 8, 75.

Lockwood, D.H., Lipsky, J.J., Meronk, F. and East, L.E. (1971): Actions of polyamines on lipid and glucose metabolism of fat cells. Biochem. biophys. Res. Commun., 44, 600.

Manganiello, V. and Vaughan, M. (1972): Selective loss of adipose cell responsiveness to glucagon with growth in the rat. J. Lipid Res., 13, 12.

Rodbell, M., Krans, H.M., Pohl, S.L. and Birnbaumer, L. (1971): The glucagon-sensitive adenyl cyclase system in plasma membranes of rat liver. J. biol. Chem., 246, 1861.

Salans, L.B. and Dougherty, J.S. (1971): The effect of insulin upon glucose metabolism by adipose cells of different sizes. J. clin. Invest., 50, 1399.

Salans, L.B., Knittle, J.L. and Hirsch, J. (1968): The role of adipose cell size and adipose tissue insulin sensitivity in the carbohydrate intolerance of human obesity. J. clin. Invest., 47, 153.

Increased glucose metabolism and insulin sensitivity in large adipocytes

F.A. Gries, A. Koschinsky and **L. Herberg**

Second Medical Clinic and Diabetes Research Institute, University of Düsseldorf, Düsseldorf, Federal Republic of Germany

It is generally accepted that the insulin effect on glucose metabolism of fat cells is reduced in obese subjects. As obesity is characterized by an increase in fat cell volume, it has been postulated that the decreased insulin sensitivity is a consequence of increased cell size. It may be questioned, however, whether this is a causal relationship or whether increased cell volume and decreased insulin sensitivity are both independently caused by a still unknown genetic or metabolic factor. The effect of cell size on insulin sensitivity can only be studied in cells of an individual donor, obtained from a single location and differentiated according to their size.

MATERIALS AND METHODS

In the following experiments fat cells of male C57BL/6J ob/ob mice, with free access to food or fasted for 48 hours, and weighing 60-80 g, were studied. By using a differential flotation technique, two populations of epididymal fat cells of significantly different diameters were obtained from each animal. Fat cells were prepared according to Rodbell (1964), washed and suspended in Krebs-Ringer bicarbonate buffer containing 4 g/100 ml bovine albumin and 1.1

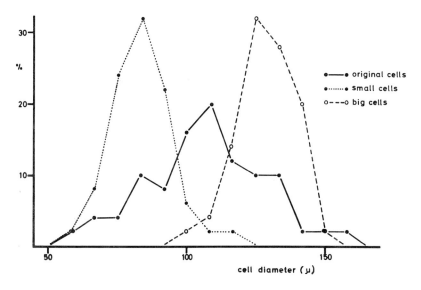

Fig. 1. Frequency distribution of the diameters of isolated epididymal fat cells from one C57BL/6J ob/ob mouse weighing 70-80 g. The difference between small and large cell size is statistically significant (P< 0.001).

mM glucose. The cells were repeatedly floated in this medium, with successive changes in albumin concentration to obtain large cells in the upper phase and small cells in the lower phase.

The whole procedure lasted less than 90 minutes and yielded metabolically active cells. The distribution of cell diameters in the original pool and the fractions of small and large cells is shown in Figure 1. The technique was reproducible and resulted in significantly different fractions in each of 14 experiments. The frequency distribution of all small and large cells studied is shown in Figure 2.

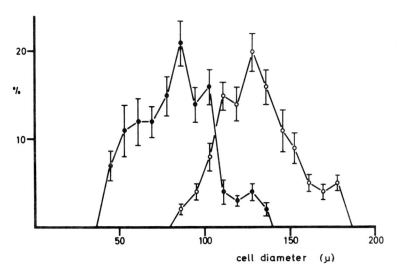

Fig. 2. Frequency distribution of the diameters (mean ± S.E.M.) of isolated epididymal fat cells from fed C57BL/6J ob/ob mice weighing 60-70 g. Black circles represent the small cell fraction, open circles the large one. Number of animals = 9.

Insulin effects were studied during a 1 hour incubation in Krebs-Ringer bicarbonate buffer with the addition of 4 g/100 ml of bovine albumin low in fatty acids and 1.1 mM glucose containing [1-^{14}C] glucose (0.4 μCi/ml). Insulin was added as indicated. Incorporation of label into CO_2 and total lipid was measured according to standard techniques (Gries and Steinke, 1967). Data are expressed as mg glucose equivalent metabolized per 10^5 cells. Statistics for cells of different animals were calculated by Student's *t* test, for small and large cells of one animal, and for insulin effects, by Wilcoxon's test for paired comparisons.

RESULTS

The results of experiments on cells of fed animals are shown in Figure 3. Basal and stimulated incorporation of glucose into CO_2 and total lipid was greater in large than in small cells, the difference being significant for both CO_2 and total lipid production under all conditions except CO_2 production in the presence of 1 μU/ml of insulin. An increased insulin effect on large cells was also observed with the minimal effective dose. As little as 1 μU of insulin/ml increased glucose metabolism in large cells. A significant effect was seen in small cells only after the addition of 1000 μU/ml.

In fasted animals, the cell diameter was a little smaller than in fed ones (Fig. 4). Again glucose metabolism of large cells was greater than that of small cells; however, the difference was statistically significant only for total lipid synthesis in the presence of 10 and 100 μU/ml of insulin.

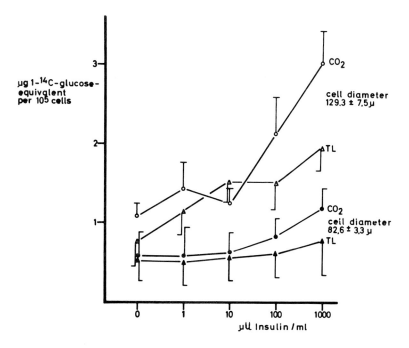

Fig. 3. In vitro insulin effects on incorporation of [1-^{14}C] glucose (mean ± S.E.M.) into CO_2 and total lipid (TL) by small (black circles and triangles) and large (open circles and triangles) epididymal fat cells from C57BL/6J ob/ob mice weighing 70-80 g fed ad lib. Number of animals = 9.

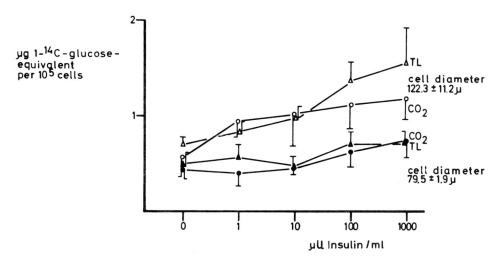

Fig. 4. In vitro insulin effects on incorporation of [1-^{14}C] glucose (mean ± S.E.M.) into CO_2 and total lipid (TL) by small (black circles and triangles) and large (open circles and triangles) epididymal fat cells from fasted C57BL/6J ob/ob mice weighing 60-70 g. Number of animals = 9.

Insulin effects were markedly decreased in the fat cells of fasted animals. The increment of glucose incorporation was significant only for total lipid synthesis at 100 and 1000 $\mu U/ml$ in either cell fraction.

DISCUSSION

Studies by Salans et al. (1968), DiGirolamo (1972) and others suggested that the decrease in insulin sensitivity of adipose tissue in obesity is a consequence of increased cell size. However, it is known that insulin sensitivity depends on various factors, such as age or nutritional state. It is impossible, therefore, to relate the changes in glucose metabolism, as observed in the study of different animals, directly to cell size.

The technique used in this study permits investigation in vitro of cells of different sizes which have been exposed to identical humoral influences in vivo. The results clearly demonstrate that basal glucose metabolism in large cells is greater than in small cells. In addition, the stimulation by insulin and the hormone sensitivity are not decreased but rather increased in large cells when compared with small cells. Similar conclusions were drawn from studies of human adipocytes by Björntorp and Karlsson (1970).

Basal glucose metabolism depends on the rate of passive diffusion of glucose through the cell membrane and the capacity of the metabolic pathways. Englhardt et al. (1971) showed that increased fat cell volume is not only a consequence of greater lipid storage but also of increased plasma space, protein content and increased enzyme concentration. These results offer a possible explanation of the enhanced basal metabolism of big cells. However, they do not explain the increased insulin sensitivity. Whether this is due to an increased number of insulin receptors or their greater affinity for the hormone, or to improved transformation of the initial hormone-receptor reaction into facilitated glucose transport, is at present speculative.

The experiments with fasted animals confirm the well known decrease in glucose metabolism and insulin sensitivity after withdrawal of food, but, even in this situation, the difference between small and large cells is still present, indicating that changes in substrate supply for a short space of time can modify but not abolish the greater insulin sensitivity of large cells.

In conclusion, decreased insulin sensitivity of adipose tissue in obesity is not the consequence of increased cell volume. Among cells of a single tissue specimen, large cells exhibit greater basal glucose metabolism and greater insulin sensitivity, even under different conditions.

REFERENCES

Björntorp, P. and Karlsson, M. (1970): Triglyceride synthesis in human subcutaneous adipose tissue cells of different sizes. Europ. J. clin. Invest., 1, 112.

Englhardt, A., Gries, F.A., Liebermeister, H. and Jahnke, K. (1971): Size, lipid and enzyme content of isolated human adipocytes in relation to nutritional state. Diabetologia, 7, 51.

DiGirolamo, M. (1972): Fat cell size, metabolic capacities and hormonal responsiveness of adipose tissue in spontaneous obesity in the rat. Israel J. med. Sci., 8, 807.

Gries, F.A. and Steinke, J. (1967): Comparative effects of insulin on adipose tissue segments and isolated fat cells of rat and man. J. clin. Invest., 46, 1413.

Rodbell, M. (1964): Metabolism of isolated fat cells. Effects of hormones on glucose metabolism and lipolysis. J. biol. Chem., 239, 375.

Salans, L.B., Knittle, J.L. and Hirsch, J. (1968): The role of adipose cell size and adipose tissue insulin sensitivity in the carbohydrate intolerance of human obesity. J. clin. Invest., 47, 153.

Studies on the influence of cell size and rat weight on the lipogenic action of insulin in isolated adipocytes

J.H. Nielsen and **F.M. Hansen**

Research Laboratory, Steno Memorial Hospital, Gentofte, Denmark

It has been recognized for several years that adipose tissue from old fat rats is less responsive to the stimulatory effect of insulin on glucose metabolism than adipose tissue from young lean rats (see review by Rudman and DiGirolamo, 1967). However, the reason for this is still poorly understood. The work of Hirsch, Salans and coworkers has drawn attention to the importance of adipose cell size, which was found to increase continuously with increasing rat weight, while the number of fat cells seemed to remain constant from the age of about 15 weeks (250-300 g body weight) (Knittle and Hirsch, 1968; Hirsch and Han, 1969). Salans and Dougherty (1971) concluded that large fat cells were less responsive to insulin than small cells when comparing cell samples taken from animals of different weights. However, recent studies by Björntorp and Sjöstrom (1972) and Gliemann and Vinten (1972) on cells of different sizes from the same animal have demonstrated that both basal and insulin-stimulated glucose metabolism increase with increasing cell size, indicating that the insulin responsiveness is the same for all cells within an individual animal.

We have extended the studies of Gliemann and Vinten (1972) to a wide range of rat weights and in addition studied some conditions which are known to influence glucose and lipid metabolism in adipose tissue of young rats in vitro. By the use of a filtration technique of osmium tetroxide-fixed adipocytes, lipogenesis was studied in cells of different sizes obtained from the same animal. The details will be described elsewhere (Hansen et al., 1974).

Adipocytes prepared by collagenase digestion of tissue were incubated with 0.55 mM [U-^{14}C] glucose for 2 hours with and without addition of a maximally stimulating insulin concentration (10^4 μU per ml). After incubation, osmium tetroxide was added and the fixed cells were separated according to size by filtration through polyamide filters of decreasing mesh. The number of cells and the radioactivity incorporated were determined on each filter and basal and insulin-stimulated lipogenesis were expressed as c.p.m. per cell.

In Figure 1 typical results with cells from one rat weighing 145 g and another weighing 510 g are shown. In both instances basal and insulin-stimulated lipogenesis increased with increasing cell size. The insulin responsiveness, expressed as percent increase above basal lipogenesis, was approximately the same for all cells within the same animal. However, the cells obtained from the fat rat showed a lower response to insulin than those from the lean rat. By comparing a large number of animals the insulin responsiveness was found to decrease abruptly when rat weight exceeded about 300 g, and in Figure 2 the results from all experiments are summarized.

In order to see whether the experimental conditions were responsible for the difference in responsiveness, epididymal fat tissue from 2 rats weighing 165 g and 400 g was mixed and handled together during the whole procedure of collagenase digestion, incubation with [^{14}C] glucose, osmium fixation and fractionation according to size. Lipogenesis and size distribution are shown in Figure 3. The cells from each of the two animals retained their difference in responsiveness to insulin. The difference cannot therefore be explained by factors transferable in vitro from one cell population to the other. From these studies it can be concluded that the change in insulin responsiveness is not a result of cell size per se, but depends on the 'internal milieu' of the rat.

The term 'lipogenesis' expresses the sum of glucose-carbon incorporated into both glyceride-glycerol and glyceride-fatty acids. The proportion between these has been shown to vary

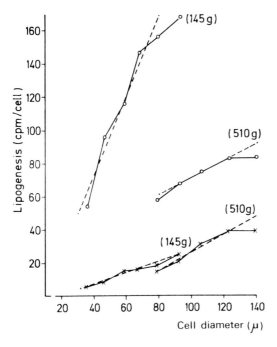

Fig. 1.   Lipogenesis in fat cells of different sizes from two rats weighing 145 and 510 g.

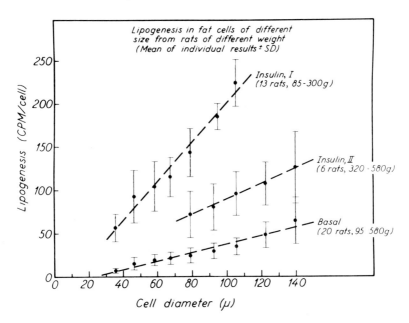

Fig. 2.   Lipogenesis in fat cells of different sizes from rats of varying weights (mean of individual results ± S.D.).

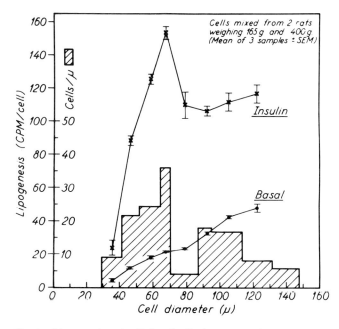

Fig. 3. Lipogenesis and cell size distribution (mean of three samples ± S.E.M.).

considerably with rat weight (DiGirolamo and Rudman, 1968; Romsos and Leveille, 1972; Goldrick et al., 1972; DiGirolamo and Mendlinger, 1972). This was confirmed in our study, as shown in Table 1, where the insulin-stimulated incorporation of ^{14}C from $[1-^{14}C]$ glucose into glyceride-fatty acids is expressed as a percentage of the incorporation into total glycerides for two weight groups of rats fed ad lib and one group of the same age as those weighing 300-450 g

TABLE 1

Feeding condition	Weight range	No. of rats	% glyceride-fatty acid synthesis (mean and range)
Ad lib	100-150 g	10	85 (67-96)
Ad lib	300 450 g	15	4 (0-10)
Restricted diet	220-240 g	5	75 (57-91)

but kept on a restricted diet. It can be seen that the decrease in insulin responsiveness of cells from fat rats was accompanied by a decrease in fatty acid synthesis. Restriction of diet to 8 g chow per day followed by 3 days feeding ad lib before sacrifice was found to maintain fatty acid synthesis, insulin responsiveness and the low fat cell diameter, as also found by Goldrick et al. (1972).

The effect on fat cells of agents which are known to increase do novo fatty acid synthesis from glucose in lean rats was tested. The results are showh in Table 2, where total lipid synthesis is expressed as a percentage of the basal rate and fatty acid synthesis as a percentage of total lipid synthesis from $[1-^{14}C]$ glucose.

TABLE 2

Experiment	Addition (mM)	Glucose (mM)	Lipid (L) and fatty acid (F) synthesis							
			Lean rats				Fat rats			
			Basal		Insulin		Basal		Insulin	
			L	F	L	F	L	F	L	F
I	—	0.5	*100*	20	1258	76	*100*	2	401	3
	—	10.0	421	48	1546	87	431	2	489	9
II	—	5.5	*100*	45	322	87	*100*	0	112	1
	acetate (7.4)	5.5	105	41	389	98	80	0	99	1
	DNP (0.15)	5.5	98	29	359	96	62	0	79	0
III	—	0.55	*100*	21	1543	83	*100*	0	192	1
	ACTH (10^{-6})	0.55	102	5	2543	54	93	2	93	0

High glucose concentration is known to be similar to insulin in its effect on the pattern of glucose metabolism (Ho and Jeanrenaud, 1967). Acetate and DNP are known to stimulate the incorporation of glucose-carbon into glyceride-fatty acids (Saggerson, 1972a) and lipolytic agents to stimulate total lipogenesis from glucose, though predominantly the glycerol moiety (Saggerson, 1972b). None of these agents increased basal or insulin-stimulated fatty acid synthesis in cells from fat rats and only the high glucose concentration increased basal lipogenesis similarly in the two weight groups. Flatt (1970) has suggested that the rate of lipogenesis in adipose tissue may be limited by the ability of the tissue to use excess ATP produced when glucose is converted to fatty acids. This does not seem to explain the limited rate of fatty acid synthesis in fat cells of fat rats, as acetate, DNP and ACTH induce energy-consuming metabolic processes but do not stimulate lipogenesis.

Enzymatic studies on pig adipose tissue by Anderson and Kauffman (1973) showed a decline in the activities of malic enzyme and glucose-6-phosphate dehydrogenase, both leading to decreased fatty acid synthesis. Similar changes may occur in the rat. Changes in diet composition and feeding schedules have recently been shown to induce an increase in the rate of fatty acid synthesis (Goldrick et al., 1972). We did not succeed in increasing fatty acid synthesis by a change to meal feeding of two fat rats for two weeks. However, further studies are needed.

Recent studies on insulin receptors have shown that the binding of [125I] insulin did not differ significantly between fat cells obtained from lean and fat rats, which indicates that the resistance to the metabolic effects of insulin is located at a step beyond the receptor (Livingston et al., 1972). This is in accordance with the present study, from which it is concluded that, not the cell size per se, but changes in the enzymatic pattern are responsible for the decreased responsiveness to insulin in the normal adult rat fed ad lib.

REFERENCES

Anderson, D.B. and Kauffman, F.G. (1973):   Cellular and enzymatic changes in porcine adipose tissue during growth. J. Lipid Res., 1 , 160.
Björntorp, P. and Sjöström, L.(1972):   The composition and metabolism in vitro of adipose tissue fat cells of different sizes. Europ. J. clin. Invest., 2, 78.
Flatt, J.P. (1970):   Conversion of carbohydrate to fat in adipose tissue: an energy-yielding and, therefore, self-limiting process. J. Lipid Res., 11, 131.
DiGirolamo, M. and Mendlinger, S. (1972):   Glucose metabolism and responsiveness to bovine insulin by adipose tissue from three mammalian species. Diabetes, 21, 1151.

DiGirolamo, M. and Rudman, D. (1968): Variations in glucose metabolism and sensitivity to insulin of the rat's adipose tissue in relation to age and body weight. Endocrinology, 82, 1133.

Gliemann, J. and Vinten, J. (1972): Glucose metabolism and insulin sensitivity of single fat cells. Israel J. med. Sci., 8, 34.

Goldrick, R.B., Hoffmann, C.C. and Reardon, M. (1972): Studies on lipid and carbohydrate metabolism in the rat. Effects of diet on the metabolism and the responses in vitro to insulin of epididymal adipose tissue and hemidiaphragms. Aust. J. exp. Biol. med. Sci., 50, 289.

Hansen, F.M., Nielsen, J.H. and Gliemann, J. (1974): The influence of body weight and cell size on lipogenesis and lipolysis of isolated rat fat cells. Submitted for publication.

Hirsch, J. and Han, P.W. (1969): Cellularity of rat adipose tissue: effects of growth, starvation and obesity. J. Lipid Res., 10, 77.

Ho, R.J. and Jeanrenaud, B. (1967): Insulin-like action of ouabain. I. Effect on carbohydrate metabolism. Biochim. biophys. Acta (Amst.), 144, 61.

Knittle, J.L. and Hirsch, J. (1968): Effect of early nutrition on the development of rat epididymal fat pads: cellularity and metabolism. J. clin. Invest., 47, 2091.

Livingston, J.N., Cuatrecasas, P. and Lockwood, D.H. (1972): Insulin insensitivity of large fat cells. Science, 177, 626.

Romsos, D.R. and Leveille, G.A. (1972): Effect of cell size on in vitro fatty acid and glyceride-glycerol biosynthesis in rat adipose tissue. Proc. Soc. exp. Biol. (N.Y.), 141, 649.

Rudman, D. and DiGirolamo, M. (1967): Comparative studies on the physiology of adipose tissue. Advanc. Lipid Res., 5, 35.

Saggerson, E.D. (1972a): The regulation of glyceride synthesis in isolated white-fat cells. The effects of acetate, pyruvate, lactate, palmitate, electron acceptors, uncoupling agents and oligomycin. Biochem. J., 128, 1069.

Saggerson, E.D. (1972b): The regulation of glyceride synthesis in isolated white-fat cells. The effects of palmitate and lipolytic agents. Biochem. J., 128, 1057.

Salans, L.B. and Dougherty, J.W. (1971): The effect of insulin upon glucose metabolism by adipose cells of different size. Influence of cell lipid and protein content, age and nutritional state. J. clin. Invest., 50, 1399.

Effects of denervation on the metabolism and the response to glucagon of white adipose tissue of rats

P. Lefebvre, A. Luyckx* and Z. M. Bacq

Division of Diabetes, Institute of Medicine and Laboratory of Physiopathology, University of Liège, Liège, Belgium

During the past few years considerable attention has been devoted to the interrelationships between glucagon and catecholamines. It has been shown that glucagon stimulates the release of catecholamines from the adrenal medulla (Dresse and Lefebvre, 1961; Farah and Tuttle, 1960; Lawrence, 1965, 1967; Lefebvre et al., 1968; Lefebvre and Dresse, 1961), and that catecholamines stimulate glucagon secretion (Iversen, 1971; Leclercq-Meyer and Brisson, 1970), a similar effect being observed after pancreatic nerve stimulation in both atropinized and nonatropinized anesthetized dogs (Marliss et al., 1972). In addition, under many physiologic conditions in which glucagon secretion is stimulated, a concomitant stimulation of the adrenergic system is also observed. Hypoglycemia (Buchanan et al., 1969; Luyckx and Lefebvre, 1972a; Ohneda et al., 1969) and muscular exercise (Böttger et al., 1971; Lefebvre et al., 1972), for example, both of which are potent stimulators of glucagon secretion, are also accompanied by an increase in adrenal catecholamine release (Duner, 1954; Von Euler and Luft, 1952) or a stimulation of the sympathetic nervous system (Vendsalu, 1960). Moreover, exercise-induced glucagon secretion has recently been demonstrated to be secondary to adrenergic stimulation (Luyckx and Lefebvre, 1972b). Under these circumstances stimulation of adipose tissue lipolysis is observed, a phenomenon to which both catecholamines (see review in Jeanrenaud, 1968) and glucagon (see review in Lefebvre, 1972 and Lefebvre and Luyckx, 1967) may contribute. The purpose of the present study was to investigate the effect of denervation on the metabolic changes induced in white adipose tissue by glucagon. The following experiments were designed to assess the extent to which lipid mobilization is affected by the suppression of sympathetic innervation, whose direct role (Jeanrenaud, 1968) in controlling the mobilization of fat has been brought into question since the investigations of Cantu and Goodman (1967).

MATERIALS AND METHODS

Male rats of the Wistar strain were used in all studies. The animals weighed 180-200 g at the time of study. They were fed standardized pellets (Hesby, Liège) and given water ad lib.

Unilateral denervation of the lumbar fat body was performed according to Cantu and Goodman (1967) under Nembutal anesthesia (60 mg/kg body weight). Denervation was performed at random on the left or right side. The animals were housed 7-10 days after denervation and prior to adipose tissue studies.

After a 24- or 48-hour fast (see below), the rats were killed by decapitation and the lumbar fat deposits carefully dissected out and weighed. Pieces of lumbar fat weighing 100-120 mg were then incubated in 4 ml Krebs-Ringer bicarbonate buffer containing 1 mg glucose and 30 mg bovine serum albumin (fraction V. Armour Co.) per ml. Incubations were performed in a Labline Dubnoff apparatus for 4 hours at 37.6°C. Determinations on the medium were made according to published methods for free fatty acids (FFA) (Dole and Meinertz, 1960), glycerol (Wieland, 1963) and glucose (Hoffmann, 1937).

---

* Chargé de Recherches du Fonds National de la Recherche Scientifique, Belgium

RESULTS

Effect of unilateral denervation on lumbar fat weight

Unilateral denervation after 7-10 days resulted in a significant increase in weight of the lumbar fat body. This increase was already significant after an overnight fast (+ 18.7 ± 3.4% in favor of the denervated tissue) and became much more pronounced after a 48-hour fast (+ 30 ± 3.6%).

Effect of unilateral denervation of lumbar fat on adipose tissue metabolism studied in vitro

Basal glycerol and FFA release were not affected by denervation, regardless of whether the adipose tissue was taken from rats fasted overnight (Table 1) or from rats fasted for 48 hours (Table 2). Prolonged fasting, however, resulted in significantly increased glycerol and FFA

TABLE 1 Effects of denervation of lumbar fat on glucagon-induced lipolysis in overnight-fasted rats*

	Glycerol release (μmoles/g/4 hours)	FFA release (μmoles/g/4 hours)
I. Control adipose tissue	8.0 ± 1.0 (6)	4.7 ± 1.0 (6)
II. Control adipose tissue + glucagon (0.5 μg/ml)	22.7 ± 3.0 (6)	20.6 ± 2.3 (6)
III. Denervated adipose tissue	9.6 ± 0.5 (6)	4.2 ± 1.0 (5)
IV. Denervated adipose tissue + glucagon (0.1 μg/ml)	19.9 ± 3.1 (6)	13.8 ± 1.6 (6)
Statistical comparison	III versus I: N.S.** II versus I: $P < 0.01$ IV versus III: $P < 0.02$ IV versus II: N.S.**	I versus III: N.S.** II versus I: $P < 0.01$ IV versus III: $P < 0.01$ IV versus II: $P < 0.05$

* Results are expressed as mean ± S.E.M. The number of rats is indicated in parentheses.
** N.S. = not statistically significant.

TABLE 2 Effect of denervation of lumbar fat on glucagon-induced lipolysis in 48 hour-fasted rats*

	Glycerol release (μmoles/g/4 hours)	FFA release (μmoles/g/4 hours)
I. Control adipose tissue	17.1 ± 1.4 (10)	10.3 ± 2.2 (10)
II. Control adipose tissue + glucagon (0.1 μg/ml)	30.9 ± 1.9 (10)	30.5 ± 3.5 (10)
III. Denervated adipose tissue	14.3 ± 1.5 (11)	11.6 ± 1.5 (11)
IV. Denervated adipose tissue + glucagon (0.1 μg/ml)	25.7 ± 1.9 (11)	22.2 ± 1.7 (11)
Statistical comparison	III versus I: N.S.** II versus I: $P < 0.01$ IV versus III: $P < 0.01$ IV versus II: N.S.**	III versus I: N.S.** II versus I: $P < 0.01$ IV versus III: $P < 0.01$ IV versus II: $P < 0.05$

* Results are expressed as mean ± S.E.M. The number of rats is indicated in parentheses.
** N.S. = not statistically significant.

release in both normal and denervated adipose tissue. As shown in Table 3, glucose uptake by denervated fat tissue was significantly greater than uptake by normally innervated tissue (+52%; $P < 0.01$).

Response of denervated adipose tissue to glucagon

Both normal and denervated adipose tissue responded to glucagon with an increase in glycerol and FFA release as well as glucose uptake (Tables 1 and 2). The magnitude of the stimulation of glycerol release was similar in normally innervated and denervated tissue. In contrast, denervation resulted in a significantly reduced FFA mobilization in response to glucagon. Glucose uptake by denervated adipose tissue in the presence of glucagon was significantly increased (Table 3).

TABLE 3   Effect of denervation of lumbar fat on adipose tissue glucose uptake in 48 hour-fasted rats*

	Glucose uptake (μmoles/g/4 hours)	Statistical comparisons
I. Control adipose tissue	12.6 ± 0.83 (10)	—
II. Control adipose tissue + glucagon (0.1 μg/ml)	16.8 ± 1.27 (10)	II versus I: $P < 0.02$
III. Denervated adipose tissue	19.1 ± 1.72 (11)	III versus I: $P < 0.01$
IV. Denervated adipose tissue + glucagon 0.1 (μg/ml)	24.8 ± 2.33 (11)	IV versus II: $P < 0.01$ IV versus III: $0.05 < P < 0.1$

* Results are expressed as mean ± S.E.M. The number of rats is indicated in parentheses.

DISCUSSION

The present investigation demonstrates that denervation causes some hitherto unreported changes in the metabolism of rat lumbar white adipose tissue. As compared with normally innervated contralateral tissue, denervated adipose tissue consumed more glucose and, in response to glucagon, released less FFA. As other investigators have found with brown (Clement, 1950: Confalonieri et al., 1961; Forn et al., 1970; Hausberger, 1934) and white (Cantu and Goodman, 1967) adipose tissue, we observed that denervation resulted in a significant increase in adipose tissue mass. The difference in weight, which was already present after an overnight fast, was more marked after a 48-hour fast. This is attributable to the decreased fat mobilization, in response to fasting, of the denervated adipose tissue (Cantu and Goodman, 1967). When incubated in vitro, both normally innervated and denervated adipose tissue exhibited similar basal glycerol and FFA release. In contrast, glucose uptake was significantly greater in denervated adipose tissue. Even more striking differences between the two types of tissue were found when their response to glucagon was studied: while both tissues responded to glucagon with an increase in glycerol and FFA release, the FFA release was significantly less in denervated adipose tissue. As in the case of glucose uptake under basal conditions, denervated adipose tissue exhibited a greater glucose uptake when incubated with glucagon. These metabolic responses of denervated adipose tissue exposed to glucagon – identical glycerol release, increased glucose uptake, and reduced FFA release – strongly suggest a relative increase in reesterification. Indirect confirmation of this can be found in the data of Cantu and Goodman (1967), who showed that denervation did not affect incorporation of ^{14}C from [U-^{14}C] glucose into the fatty acid moiety of triglycerides. It is therefore likely that in denervated tissue the extra glucose uptake produced α-glycerophosphate, permitting (re)esterification of FFA, rather than providing acetyl-COA for fatty acid synthesis. An attempt to evaluate reesterification on the basis of the relative amounts of glycerol and FFA released into the incubation medium (Table 2) yielded the following figures: under basal conditions, 77.9% of the FFA mobilized by the lipolytic process (3 times the glycerol release) was reesterified by normal adipose tissue and 75.2% was reesterified by denervated adipose tissue. In the presence of glucagon, the percentage dropped to 67.1% for normal adipose tissue ($-$ 10.8%) but decreased only to 71.2% ($-$ 4%) for denervated adipose tissue*. Adipose tissue nerves may affect fat mobilization by regulating the rate of blood flow to, and hence the rate of removal from, adipose tissue (Cantu and Goodman, 1967) rather than by promoting direct stimulation of lipolysis as suggested by others (Correll, 1963; Fredholm, 1970; Nash et al., 1961; Weiss and Maickel, 1964). We should like to propose an alternative explanation. Adrenergic innervation may exert some degree of control over FFA (re)esterification by producing changes in glucose uptake. Denervation, by increasing glucose uptake by adipose tissue, would therefore favor (re)esterification. This mechanism must be considered in any attempt to explain the classic observation of an increase in adipose tissue weight after denervation. In addition, the present study suggests that innervation may exert some control over the lipolysis-reesterification balance of adipose tissue submitted to the influence of circulating lipolytic hormones such as glucagon.

REFERENCES

Böttger, I., Faloona, G.R. and Unger, R.H. (1971): The effect of intensive physical exercise on pancreatic glucagon secretion (Abstract). Diabetes, 20 (Suppl. 1), 339.

Buchanan, K.D., Vance, J.E., Dinstl, K. and Williams, R.H. (1969): Effect of blood glucose on glucagon secretion in anesthetized dogs. Diabetes, 18, 11.

Cantu, R.C. and Goodman, H.M. (1967): Effect of denervation and fasting on white adipose tissue. Amer. J. Physiol., 212, 207.

Clement, G. (1950): Quelques aspects de la physiologie du tissu adipeux; mobilisation des graisses de réserve. Ann. Nutr. (Paris), 4, 295.

Confalonieri, C., Mazzuchelli, M.V. and Slechter, P. (1961): The nervous system and lipid metabolism of adipose tissue. I. Influence of denervation on lipid mobilization processes, lipid synthesis and lipopexia in the interscapular adipose body of the rat. Metabolism, 10, 324.

Correll, J.W. (1963): Adipose tissue ability to respond to nerve stimulation in vitro. Science, 140, 387.

Dole, V.P. and Meinertz, H. (1960): Microdetermination of long-chain fatty acids in plasma and tissues. J. biol. Chem., 235, 2595.

Dresse, A. and Lefebvre, P. (1961): Nouvelle mise en évidence de la libération par le glucagon de l'adrénaline surrénalienne. C.R.Soc. Biol. (Paris), 155, 1168.

Duner, H. (1954): The effect of insulin hypoglycemia on the secretion of adrenalin and noradrenalin from the suprarenal of the cat. Acta physiol. scand., 32, 63.

Farah, A. and Tuttle, R. (1960): Studies on the pharmacology of glucagon. J. Pharmacol. exp. Ther., 129, 49.

Forn, J., Gessa, G.L., Krisna, G. and Brodie, B.B. (1970): Increased lipolytic response to norepinephrine in isolated brown fat cells after sympathetic denervation. Life Sci., 9, 429.

Fredholm, B.B. (1970): Studies on the sympathetic regulation of circulation and metabolism in isolated canine subcutaneous adipose tissue. Acta physiol. scand., Suppl. 354, 1.

Hausberger, F.X. (1934): Über die Innervation der Fettorgane. Z. mikr. anat. Forsch., 36, 231.

Hoffmann, W.S. (1937): A rapid photoelectric method for the determination of glucose in blood and urine. J. biol. Chem., 120, 51.

Iversen, J. (1971): Adrenergic receptor for the secretion of immunoreactive glucagon and insulin from the isolated perfused rat pancreas (Abstract). Diabetologia, 7, 485.

Jeanrenaud, B. (1968): Adipose tissue dynamics and regulation revisited. Ergebn. Physiol., 60, 57.

Lawrence, A.M. (1965): A new provocative test for pheochromocytoma. Ann. intern. Med., 63, 905.

Lawrence, A.M. (1967): Glucagon provocative test for pheochromocytoma. Ann. intern. Med., 66, 1091.

Leclercq-Meyer, V. and Brisson, G.R. (1970): In vitro release of glucagon (IRG) and insulin (IRI) from pancreatic tissue of duct-ligated rats (Abstract). Diabetologia, 6, 636.

Lefebvre, P. (1972): Glucagon and lipid metabolism. In: Glucagon. Molecular Physiology, Clinical and Therapeutic Implications, p. 109. Editors: P. Lefebvre and R.H. Unger. Pergamon Press, Oxford.

---

* These calculations were made on individual experimental values and therefore differ slightly from those which could be calculated on the basis of the mean values indicated in Table 2.

Lefebvre, P., Cession-Fossion, A., Luyckx, A., Lecomte, J. and Van Cauwenberge, H. (1968): Interrelationships glucagon-adrenergic system in experimental and clinical conditions. Arch. int. Pharmacodyn., 172, 394.

Lefebvre, P. and Dresse, A. (1961): Influence du glucagon sur le taux des catécholamines surrénaliennes chez le rat. C.R. Soc. Biol. (Paris), 155, 412.

Lefebvre. P. and Luyckx, A. (1967): Glucagon et métabolisme du tissu adipeux. In: Journées Annuelles de Diabétologie de l'Hôtel-Dieu, p. 67. Flammarion, Paris.

Lefebvre, P., Luyckx, A. and Federspil, G. (1972): Muscular exercise and pancreatic function in rats. Israel J. med. Sci., 8, 390.

Luyckx, A. and Lefebvre, P. (1972a): Changes in insulin and glucagon secretion related to concentrations of metabolic substrates in the isolated perfused rat pancreas. In: Hormones pancréatiques. Hormones de l'Eau et des Electrolytes. Colloque INSERM Paris, 17-19 May 1972, p. 99. INSERM.

Luyckx, A. and Lefebvre, P. (1972b): Role of catecholamines in exercise-induced glucagon secretion in rats (Abstract). Diabetes, (Suppl. 1) 21, 334.

Marliss, E.B., Girardier, L., Seydoux, J., Kanazawa, Y., Wollheim, C. and Porte Jr., D. (1972): The autonomic nervous system and pancreatic hormone secretion: effect of nerve stimulation and catecholamines. In: Hormones pancréatiques. Hormones de l'Eau et des Electrolytes. Colloque INSERM Paris, 17-19 May 1972, p. 129. INSERM.

Nash, C.W., Smith, R.P. and Paoletti, R. (1961): In situ study of fatty acid release from adipose tissue. Pharmacologist, 3, 55.

Ohneda, A., Aguilar-Parada, E., Eisentraut, A.M. and Unger, R.H. (1969): Control of pancreatic glucagon secretion by glucose. Diabetes, 18, 1.

Vendsalu, A. (1960): Studies on adrenaline and noradrenaline in human plasma. Acta physiol. scand., Suppl. 173, 1.

Von Euler, U.S. and Luft, R. (1952): Effect of insulin on urinary excretion of adrenalin and noradrenalin. Metabolism, 1, 528.

Weiss, B. and Maickel, R.P. (1964): Pharmacological demonstration of sympathetic innervation of adipose tissue. Pharmacologist, 6, 172.

Wertheimer, E. (1972): Stoffwechselregulation. V. Mitteilung. Regulationen im Hungerstoffwechsel. Die Abhängigkeit der Muskelglykogendepots vom Nervensystem. Pflügers Arch. ges. Physiol., 215, 779.

Wieland, O. (1963): Glycerol. In: Methods of Enzymatic Analysis, p. 211. Editor: H.U. Bergmeyer. Academic Press, New York and London.

DISCUSSION

Foa: Does denervation modify the catecholamine content of adipose tissue and, if so, to what extent may these changes explain your very interesting results?

Lefebvre: Preliminary studies by Dr. Cession-Fossion (Department of Physiology, University of Liège) on our material have indicated a 60-80% reduction in the catecholamine content of denervated adipose tissue. I have, however, to state that these results are relatively inconstant, probably because of the low basal catecholamine content of the lumbar fat. We have to extend our studies and look at the catecholamine content of normal and denervated adipose tissue after adrenergic stimulation, such as in muscular exercise. We did not attempt to correlate the changes we have reported with the adipose tissue catecholamine content data.

Lipotropic Peptides A and B from pig pituitary glands*

P. Schwandt, P. Weisweiler, H. Ruschewski and **R. Hagen**

First University Medical Clinic, Munich, Federal Republic of Germany

Since the elucidation of the primary structure of ß- and *A*-lipotropins (LPH) from sheep and pig pituitary glands (Chretien et al., 1972; Chretien and Li, 1967; Graf et al., 1971; Li et al., 1966) and the description of a radioimmunoassay for ß-LPH (Desranleau et al., 1972), the presence of special polypeptides in the pituitary gland with potent lipolytic action is no longer in doubt. Li calls them the 'eighth anterior lobe hormone'.

We wish to report on the biological action of two peptides isolated from pig pituitary glands, referred to as Peptide A and Peptide B. In our laboratory, these peptides were purified from the acid acetone extract of fresh pituitaries by a combination of column chromatography and preparative acrylamide gel electrophoresis. Both are homogeneous by the criteria of chromatography, disc electrophoresis and ultracentrifugation. The amino acid composition and the molecular weight suggest that Peptide A could correspond to γ-LPH while Peptide B seems to be related to ß-LPH. End group analyses and cleavages of the peptide chains are in progress for a complete comparison with published data on lipotropins.

The investigations carried out so far with our two peptides (bioassay, radioimmunoassay) do not reveal any significant activity of the other lipolytically active pituitary hormones, except melanocyte-stimulating hormone (MSH) (Table 1). This is in good agreement with data from the literature. The high MSH activity could be due to the common amino acid sequence of

TABLE 1   Pituitary hormone activities of Peptide A and Peptide B

	Peptide A	Peptide B	Tested by:
MSH	1×10^6 U/g	1.3×10^4 U/g	Dr. Sandow (Frankfurt)
ACTH	12 mU/mg	1.3 mU/mg	Dr. Mueller (Munich)
GH	1 : 1.000.000**		Dr. Quabbe (Berlin)
TSH	< 25 mU/mg	< 25 mU/mg	Dr. Louwrens (Oss)
	$< 1 \times 10^{-5}$ ng/mg***	< 1 : 5000***	Dr. Schams (Munich)
LH	6 μg/mg	< 1 : 1000***	Dr. Schams (Munich)
	1 : 5.000*		Dr. Rayford (Bethesda)

Radioimmunologic determinations: *pig, **human, ***bovine.

MSH and a part of LPH. The highly purified state of our lipotropic peptides and the lack of other hormone activities provide evidence that the lipolytic activity is not due to contamination. Immunological and immunohistological investigations. (Dr. Hachmeister, Giessen) with a specific antiserum to Peptide A did not show cross-reactivity with other pituitary hormones so far investigated. However, commercial thyroid-stimulating hormone (TSH) preparations seem to be contaminated with lipotropins (Weisweiler and Schwandt, 1973).

* Supported by Deutsche Forschungsgemeinschaft (SFB 51).

The detailed investigations of Rudman and Girolamo (1967) have shown the usefulness of species spectrum analyses of lipolytic hormones. Peptide A has been lipolytically active in nearly all species tested so far (Table 2). Peptide B was nearly as potent as Peptide A on rabbit adipose tissue in vitro with a similar log dose-response curve (Fig. 1). With rat, mouse and duck adipose tissue Peptide A gave log dose-response curves with the same slope for free fatty acid (FFA) and glycerol production.

TABLE 2   Lipid-mobilizing activity of Peptide A in different species (minimal effective dose in μg/ml or μg/kg)*

Species	Site of adipose tissue	In vitro	In vivo
Rabbit	Perirenal	0.008	10.0
Rat	Epididymal	0.01	
Guinea pig	Epididymal	1.0	
Man	Subcutaneous	10.0	
Pig	Subcutaneous	not active	23.0
Duck		1.0**	10.0**

  * Lipolysis was measured in vitro as previously described (Schwandt et al., 1968).
  ** For these results we are grateful to Dr. P. Desbals, Limoges.

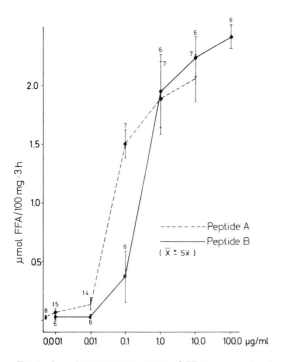

Fig. 1. Log dose-response curves of FFA release after incubation of 100 mg rabbit perirenal adipose tissue with different concentrations of Peptides A and B. The figures indicate the number of animals. For methodological details see Schwandt et al. (1968).

After intravenous injection of 25, 50, 75, 100 and 150 μg/kg into the minipig, FFA and glycerol rose within 10 minutes, reaching maximal concentrations after 60 minutes and normalizing 4 hours post injection. The log dose-response curves for both metabolites 30, 60, 90 and 120 minutes after injection were identical. In contrast to the in vitro stimulation of lipolysis in other species, the lipolytic effect of Peptide A in minipigs increased with age. In 156-week-old minipigs, FFA and glycerol concentrations 60 minutes after the injection of 100 μg/kg were significantly higher than in animals aged 14 weeks. The plasma glucose concentration in minipigs after 100 μg/kg of Peptide A paralleled the FFA increase in the first 20 minutes only. It then fell significantly below preinjection levels, with a minimum 60 minutes after injection. Whereas FFA returned to normal concentrations within the next 150 minutes, glucose levels again showed an increase.

The decrease of plasma glucose concentration seems to be in good agreement with the results of Bielmann et al. (1972), who found that ß-LPH from sheep pituitaries enhanced the conversion of both [1-^{14}C] and [6-^{14}C] glucose to CO_2 and fatty acids in rat epididymal fat pads in vitro. We investigated this further by enzyme activity measurements in liver and adipose tissue of minipigs and rabbits 60 minutes after intravenous injection of Peptide A, when plasma glucose concentration was minimal. Activities of hexokinase, glucokinase, phosphofructokinase, glucose-6-phosphate dehydrogenase, 6-phosphogluconate dehydrogenase and phosphoglucomutase did not differ from those in control animals. These experiments, which are scarcely comparable to the investigations of Bielmann et al., suggest that tissue metabolism is not responsible for the decrease of plasma glucose concentration. It is more likely that there is a ß-cell cytotropic effect either of ketone bodies or of LPH itself. Fussgänger et al. (1973) did find a direct insulinotropic action of lipotropic pituitary peptides in the isolated perfused pancreas of the rat.

According to Lis et al. (1972) sheep ß-LPH stimulated adenyl cyclase of rat and rabbit isolated fat cells. We have reported previously (Schwandt et al., 1971) that phosphodiesterase activity of intact human and rat adipose tissue is partly blocked by the Peptide A precursor fraction J (Fig. 2). Post-heparin lipoprotein lipase, on the contrary, is not altered in the minipig after injection of Peptide A.

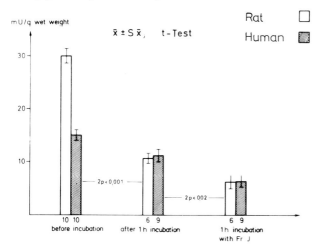

Fig. 2. Phosphodiesterase activity in human (subcutaneous) and rat (epididymal) adipose tissue before and after incubation of 100 μg/ml fraction J (Schwandt et al., 1971).

In summary, the pig pituitary lipotropins Peptide A and Peptide B are comparable to γ- and ß-LPH and essentially free from other pituitary hormone activities, except MSH activity. By stimulation of the adenylate cyclase system, both peptides have a potent lipolytic effect in the 6 species tested so far. The minimal effective dose of Peptide A in the rabbit was 8 ng/ml in vitro and 10 μg/kg in vivo, and Peptide B was nearly as active in this species. The hypoglycemia following an initial increase in serum glucose concentration in the minipig after injection of Peptide A was not reflected in changes in the activity of enzymes of glucose metabolism in adipose tissue and liver.

REFERENCES

Bielmann, P., Chretien, M. and Gattereau, A. (1972): Lipogenic activity of a potent lipolytic hormone: sheep β-lipotropin (β-LPH). II. Further effects of sheep β-LPH on specifically labeled glucose and the localization of lipogenic active center of the molecule. Horm. metab. Res., 4, 22.

Chretien, M., Gilardeau, C. and Li, C.H. (1972): Revised structure of sheep beta-lipotropic hormones. Int. J. Peptide Prot. Res., 4, 263.

Chretien, M. and Li, C.H. (1967): Isolation, purification and characterization of γ-lipotropic hormone from sheep pituitary glands. Canad. J. Biochem., 45, 1163.

Desranleau, R., Gilardeau, C. and Chretien, M. (1972): Radioimmunoassay of bovine beta-lipotropic hormone. Endocrinology, 91, 1004.

Fussgänger, R.D., Schleyer, M. and Pfeiffer, E.F. (1973): Correlative insulin and glucagon secretion of the perfused rat pancreas following pituitary peptides. Acta endocr., (Kbh.) Suppl. 173, 105.

Graf, L., Barat, E., Cseh, G. and Sajgo, M. (1971): Amino acid sequence of porcine β- lipotropic hormone. Biochim. biophys. Acta (Amst.), 229, 276.

Li, C.H., Barnafi, L., Chretien, M. and Chung, D. (1966): Isolation and structure of β-LPH from sheep pituitary glands. In: Proceedings, VI Pan-American Congress of Endocrinology, Mexico City, 1965, p. 349. Editor: C. Gual. ICS 112. Excerpta Medica, Amsterdam.

Lis, M., Gilardeau, C. and Chretien, M. (1972): Fat cell adenylate cyclase by sheep β-lipotropic hormone. Proc. Soc. exp. Biol. (N.Y.), 139, 680.

Rudman, D. and DiGirolamo, M. (1967): Comparative studies on the physiology of adipose tissue. In: Advances of Lipid Research, p. 35. Editors: R. Paolotti and D. Kritchevsky. Academic Press, New York and London.

Schwandt, P., Knedel, M. and Lindlbauer, R. (1968): Experimentelle Untersuchungen zum Verfahren der in vitro Lipolyse. Z. klin. Chem., 6, 81.

Schwandt, P., Weisweiler, P., Krone, H. and Doerr. H.W. (1971): Zur Wirkung eines lipidmobilisierenden Peptids aus Schweinehypophysen bei verschiedenen Spezies. Klin. Wschr., 49, 437.

Weisweiler, P. and Schwandt, P. (1973): Lipolytic action of porcine TSH caused by lipotropin contamination. Acta endocr., (Kbh.) Suppl. 173, 106.

GENERAL DISCUSSION

Goldrick: Some of the discrepancies in the data reported by Drs. Livingston, Herberg and Nielsen may simply reflect differences in methodology. The rates of incorporation of glucose into CO_2, glycerol and fatty acids by adipose tissue do not parallel each other either in the basal state or in the presence of insulin during growth in the rat. Dr. Livingston measured the rate of glucose oxidation, whereas Drs. Herberg and Nielsen measured the rate of incorporation of glucose into triglycerides. The latter thus represents the sum of two independent parameters of metabolism and these are not necessarily comparable to the use of glucose oxidation as an index of insulin responsiveness.

I would like to raise the possibility that comparisons of fat cells of different sizes from the same fat depot do not eliminate the effects of age on metabolism. There is no way of determining the age of a fat cell, so this question cannot be answered at present. However, it is possible that the largest adipocytes in a given fat depot represent the first cells to differentiate into storage cells during the development of the adipose organ. If this is so, then the large cells may be older than their smaller counterparts.

Chlouverakis: I must congratulate Drs. Herberg and Nielsen for the most elegant work they reported. I myself, working with a system similar to that of Dr. Nielsen, found that when taken from the same specimen of adipose tissue large cells responded to insulin at least as well as small cells. I would like also to suggest that Dr. Nielsen's experiments support Dr. Herberg's conclusion that the insulin sensitivity of large cells is greater than that of small cells. In Figure 2 (p. 94), in which rate of lipogenesis was plotted against cell diameter, the regression line relating these two variables was 'flat' for basal conditions but had a positive, steep slope under insulin stimulation. Thus, when comparing rats of similar age, the larger the cell, the greater its insulin sensitivity. This would become obvious if results were expressed in histogram form.

Legros: It has been demonstrated that neutral insulin (Novoactrapid), at concentrations ranging from 4 to 40 mU/ml, enhances oxygen uptake of a marine unicellular alga, *Acetabularia mediterranea,* grown in darkness. This stimulatory effect, varying from 30 to 90% of the basal value, depends on the age of the cells and on the duration of incubation in darkness. As demonstrated by the negative results obtained with performic acid-oxidized insulin and by inactivation of the hormone by anti-insulin serum, this effect can be considered a specific one. It is observed with the nucleated as well as anucleated fragments.

Ouabain, at concentrations ranging from 10^{-3} to 10^{-3} M, transiently depresses oxygen uptake of the cells. There is a competitive effect between insulin (40 mU/ml) and ouabain (10^{-5} M). Preincubation of cells in a medium containing insulin (40 mU/ml for 1 hour) nullified the inhibitory effect of the glycoside drug.

Fixation of ^{125}I-insulin (Sorin) at a concentration of 2 nM has been studied in the absence and presence of various concentrations of unlabelled hormone. After a 20 to 30 minutes' delay, binding kinetics reach a plateau. The level of this plateau, corresponding to 25% to *f*0% of the total initial c.p.m., is reduced to 10% in the presence of cold insulin (400 mU/ml). There is a competition between ^{125}I-insulin and ouabain for the binding at the cell membrane level.

It may be concluded from these experiments that the metabolic effect of insulin on *Acetabularia mediterranea* is related to its previous binding to the cell membrane.

Adipose tissue cellularity and metabolism

Regulation of adipose tissue mass in genetically obese rodents*

G. A. Bray, D. Luong and D. A. York

Harbor General Hospital, Torrance and UCLA School of Medicine, Los Angeles, Calif., U.S.A.

Genetically transmitted obesity has been described in several strains of rodents (Table 1) (Bray and York, 1971a). The modes of inheritance include expression through a single gene in both dominant and recessive modes, polygenic expression, and inbred lines. In spite of this variety of genetic types, all of these obese animals have a number of features in common. Food intake is uniformly increased compared to lean littermates. Qualitative differences in the response to food are seen in the desert rodents and in rats which are susceptible to a high-fat diet. The *Psammomys obesus* (Hackel et al., 1966) usually subsists on vegetation available in the desert clime. When they are captured and fed a calorically dense laboratory chow, they overeat and become obese. Abnormalities such as hyperglycemia and hyperinsulinemia can be prevented by limiting their food intake to the same caloric amount as eaten by animals with access to green vegetables. The effects of qualitative changes in diet are also observed when rats are fed a high-fat diet. Certain strains are unable to regulate food intake appropriately and will become grossly obese when fed a high-fat diet (Schemmel et al., 1970). In other strains, however, a high-fat diet does not lead to obesity.

In addition to hyperphagia, most obese rodents show hyperglycemia and hyperinsulinemia. The exception is the fatty rat, which shows increased insulin but little or no increase in blood glucose (York et al., 1972b; Stern et al., 1972; Zucker and Antoniades, 1972). All animals, however, show an elevation in the circulating concentrations of insulin and an increase in the size of the islets of Langerhans with the exception of the mouse called 'diabetes' (db/db). Development of ketosis is unusual and is observed only in the db/db mouse and in the two desert rodents, Acomys and Psammomys (Bray and York, 1971a). The final abnormality common to all genetic obesities is the enlarged depots of fat. All of the obese rodents have an increase in the size of fat cells (Bray and York, 1971a; Johnson and Hirsch, 1972), and some of them, specifically the fatty rat (Lemonnier, 1971b; Johnson et al., 1971) and the obese mouse (Johnson and Hirsch, 1972), have an increase in the total number of adipocytes.

The animals to be considered in detail below are members of a strain which appeared as a spontaneous mutant in the laboratories of Zucker and Zucker (1961). These animals are the only known strain of rats with genetically transmitted obesity (Table 1). The abnormality is inherited as an autosomal recessive: that is, when both parents are carriers for the gene, neither one is obese, but obesity will be present in approximately one-quarter of the offspring of these matings. The phenotypic expression of this trait is not manifest at birth. Indeed, it cannot be recognized for the first 2 to 3 weeks of life. By that time, the animals are hyperphagic and show increased body fat and elevated levels of plasma insulin (Zucker and Antoniades, 1972; Zucker, 1972). The temporal sequence for the onset of identifiable abnormalities in the fatty rat and obese mouse is shown in Table 2.

To provide a more detailed perspective on the features of these animals, they have been compared with the widely studied strain of mouse called 'obese' (ob/ob; synonym – obese-hyperglycemic), which was described in 1950 by Ingalls et al. (Table 3). The body weight in both animals carrying the two genes for corpulence can be more than 100% greater than corre-

* Supported in part by Grants AM 42-15165 and RR 425 from the National Institutes of Health.

TABLE 1   Genetically transmitted obesity

Mode of inheritance	Designation	Gene symbol	Hyper-phagia	Hyper-glycemia	Hyper-insulinemia	Ketosis	Hyper-trophic	Hyper-plastic hyper-trophic
Dominant	Yellow mouse	A^y, A^{vy}, A^{iy}	+	+	+	—	+	
Recessive	Obese mouse	ob	+	++	++	—		+
	Diabetes mouse	db	++	+ ➤ ++	++ ➤ +	+	+	
	Adipose mouse	db^{ad}*						
	Fatty rat	fa	+	—	+	—		+
Polygenic	New Zealand obese	NZO	+	±	+	—		+
	Japanese KK	KK**	+	+	+	—		
Hybrid	Wellesley mouse	C₃Hfl	+	+	+	—		
Desert rodents	*Acomys cahirinus*		+	+	+	+		
	Psammomys obesus		+	+	+	+	+	

* Diabetes and adipose are alleles.
** Polygenic or dominant with reduced penetrance.

TABLE 2   Temporal sequence for appearance of abnormalities in fatty rats (fa/fa) and obese mice (ob/ob)

Age (days)	Abnormalities	
	Fatty	Obese
Birth	None	None
7	None	None
14	↑ Body fat	↑ Size fat cells; Glycosuria ↓ Oxygen consumption
21	↑ Insulin	↑ Body fat
28	↑ Fat cell size	
35		↑ Insulin

TABLE 3   Comparison of some features of the fatty rat and the obese mouse

	Fatty	Obese
Gene symbol	fa	ob
Synonym	Zucker	Obese-hyperglycemic
Body weight	↑↑	↑↑
Food intake	↑	↑
Plasma: Insulin	↑	↑↑
Glucose	Normal or SL ↑	↑
Cholesterol	SL ↑	↑
Triglycerides	↑↑	↑
Urine volume	↑	—
Thyroid function	Impaired	Normal (?)
Reproductive function	Somewhat impaired	Severely impaired
Pancreas	↑ islets size and number	↑ islets size and number
Adipose tissue: Cell size	↑	↑
Fatt cell number	↑	↑

SL ↑ = slightly increased, ↑ = increased, ↑↑ = greatly increased.

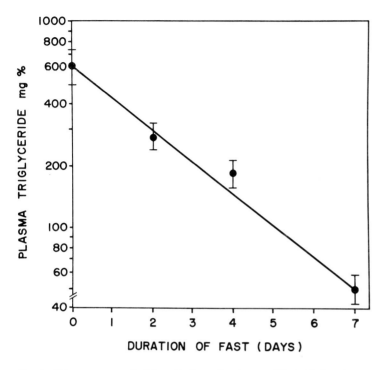

Fig. 1. Disappearance of triglyceride from the plasma of fasted fatty rats.

sponding lean littermates. Both sexes are equally affected and over 90% of the increased weight is stored as triglyceride. Hyperphagia is present in both strains and food intake may be increased as much as 50% above that of lean controls (Zucker and Zucker, 1962; Barry and Bray, 1969; York and Bray, 1971; Bray and York, 1972). This difference in food intake is more obvious in the fatty rat but is also observed in the obese mouse (Fuller and Jacoby, 1955). The fatty rat does not show the finickiness of appetite which characterizes the obesity associated with bilateral injury to the ventromedial nucleus in the hypothalamus. In contrast, a finicky appetite has been observed in the obese mouse (Fuller and Jacoby, 1955). Plasma insulin is markedly elevated in the obese mouse (Bray and York, 1971a; Stauffacher et al., 1967). It is also elevated during the dynamic phase of obesity in the fatty rat but falls off with advancing age (Lemonnier, 1971a; Stern et al., 1972; Zucker and Antoniades, 1972; York and Bray 1973a). The magnitude of hyperinsulinemia is greater in the obese mouse than in the fatty rat. In both animals, there is enlargement of the islets of Langerhans (York et al., 1972b; Bleisch et al., 1952). As might be expected, studies in vitro have shown the release of insulin to be augmented in the fatty rat (Stern et al., 1972).

The lactescence in the serum of the fatty rat has been a prominent feature of these animals. Elevated triglycerides are also found in the obese mouse, but to a lesser degree than in the fatty rat. On the usual laboratory diet, plasma triglycerides of fatty rats are often above 1000 mg/100 ml of serum. Table 4 shows the concentration of triglycerides in three groups of animals. In rats of normal weight, triglycerides of less than 100 mg/100 ml are usually observed. Both the genetically obese rats and rats with obesity following hypothalamic injury (lesioned-obese) have markedly elevated levels of triglyceride. With fasting, the concentration of triglyceride falls to normal with a half-time of about 2 days (Fig. 1). If the genetically obese and lesioned-obese rats are fasted to bring triglyceride levels down to normal and then pair-fed to normal animals (i.e. given the same amount of daily food as eaten by the lean animal on the previous day), the trigly-

TABLE 4 Plasma triglycerides of normal, genetically obese and hypothalamic obese rats*

	No. of animals	Body weight (g)**	Plasma triglycerides (mg/ 100 ml)**
Normal	6	262 ± 6	76 ± 13
Genetically obese	3	538 ± 40	616 ± 118
Lesioned-obese	3	496 ± 30	598 ± 119

* Tail-vein blood was collected in tubes containing EDTA after a 4-hour fast. The rats were 4 months old at the time of this experiment.
** Mean ± S.E.

TABLE 5 Effects of pair-feeding on plasma triglycerides of lean and obese rats

Treatment	Plasma triglycerides (mg/ 100 ml)*		
	Lean	Genetically obese	Lesioned obese
Fasting	45.8 ± 7.3	157.7 ± 44	60.2 ± 6.6
Pair-fed 7 days	28.5 ± 3.0	115.9 ± 16	91.8 ± 16.0
Pair-fed 14 days with 10% fructose	58.9 ± 13.0	187.3 ± 33	167.5 ± 15.2

* Mean ± S.E.M. for 3 animals.

TABLE 6 Effect of diet on plasma triglycerides

Diet	Plasma triglycerides (mg/ 100 ml)*			
	No.	Normal rats	No.	Fatty rats
Fat-free	12	99.2 ± 19.8	12	98.7 ± 13.1
25% corn oil	12	65.5 ± 6.9	12	274.2 ± 26.8

* Mean ± S.E.M.

cerides remain near normal (Table 5). Addition of 10% fructose to the drinking water produced an increase in the level of triglyceride in all three groups. The effects of exogenous dietary fat on the level of triglyceride has also been examined. When genetically obese rats are fasted to bring their triglycerides down to normal levels and then fed a fat-free diet, there is no significant increase in the concentration of plasma triglyceride (Table 6). Substitution of a 25% corn oil diet for the fat-free diet produces a sharp increase in the concentration of triglycerides in fatty rats but a drop in the concentration in normal rats. These observations suggest that a major portion of the circulating triglyceride is of dietary origin. The work of Zucker and Zucker (1962) and Schonfeld and Pfleger (1971) suggest that hepatic triglycerides also play an important additional role in the high levels of circulating lipoproteins in the fatty rat.

Urine output is higher in fatty rats than in lean controls (York and Bray, 1971). Thyroid function is normal in the obese mouse (Goldberg and Mayer, 1952) but impaired in the fatty rat

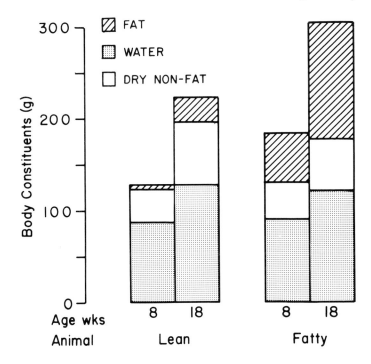

Fig. 2. Body composition of lean and fatty rats at 8 and 18 weeks of age. Water, fat, and lean body weight are plotted as weight attributable to each component.

(Bray and York, 1971b; York et al., 1972a). Reproductive function is severely impaired in the obese mouse; females are uniformly sterile and have never been known to ovulate or develop corpora lutea without prior hormonal treatment (Bray and York, 1971a; Runner, 1952). The fatty rat has estrous cycles, but the frequency is less than that observed in lean animals (Saiduddin et al., 1973). To recapitulate: the fatty rat is characterized by marked elevations in triglycerides, and impairment of thyroid function, but only a modest loss in reproductive function; the obese mouse, on the other hand, has marked hyperinsulinemia and on occasion marked hyperglycemia, whereas thyroid function is normal and reproductive function severely impaired. Both the fatty rat and the obese mouse show hypertrophy and hyperplasia of the fat organ, and it is this relationship that we will now examine in detail for the fatty rat.

The increase in body weight of fatty rats probably begins at birth, but they cannot be distinguished from lean animals with certainty until 2 to 3 weeks of age (Table 2). After this time, there is a rapid increase in body fat and total body weight in the fatty rat (Zucker and Zucker, 1963; Johnson et al., 1971; Bray et al., 1973). Increasing quantities of body fat make up the preponderance of the weight difference between lean and fatty rats. This is shown in studies by Zucker (1967) and Johnson et al., (1971) and in recent data from our laboratories (Bray et al., 1973) (Fig. 2). Between 8 and 18 weeks of age, the weight of the lean rats increased from 128 to 228 g. Fat, which represented less than 5% of the body weight of lean animals at 8 weeks of age, increased to 15% by 18 weeks. Corresponding increases were observed in body water and lean body tissue. In contrast, the fatty rat at 8 weeks of age weighed nearly 50 g more than the lean rat, and almost all of this increase was due to excess triglyceride. Between 8 and 18 weeks of age there was a striking increase in fat tissue. By 18 weeks of age, nearly 40% of total body weight was accounted for by triglyceride. An analysis of fatty acids in these depots has shown no differences between fatty rats and obese animals with hypothalamic lesions (Bray, 1969). The fatty rat, however, does show hypercellularity and hypertrophy of the fat cells (Bray, 1969; Lemonnier, 1971b; Johnson et al., 1971). The increased numbers of fat cells are found in subcutaneous and retroperitoneal fat pads. In contrast, animals that became obese after hypothalamic lesions showed hypertrophy of their fat cells but no increase in the total number of such cells (Hirsch and Han, 1969; Johnson et al., 1971). In these latter animals, essentially all of the extra triglyceride was stored by hypertrophy of previously existing adipocytes. Thus, the fatty rat represents an example of hyperplastic-hypertrophic obesity.

The functional characteristics of this adipose tissue have been examined in a number of studies (Bray, 1969; Bray et al., 1970; Zucker, 1972; York and Bray, 1973a, b). The attractiveness of this tissue for study resides in its grossness and its accessibility. It is tempting to think that the recessively inherited defect which makes these animals fat might manifest itself in the adipose tissue. Indeed, this premise has been at the foundation of much work on the fatty rat as well as on the obese mouse. Some data on lipolysis in lean and fatty rats are summarized in Table 7. This compares lipolysis in lean rats aged 6 and 18 weeks with data from fatty rats of similar ages. Body weight in the fatty rat at 6 weeks of age was only slightly above that of the 6-

TABLE 7   Body weight, fat cell size, lipolysis and lipogenesis at 6 and 18 weeks of age in lean and fatty rats

	Lean		Fatty	
Age (weeks)	6	18	6	18
Body weight (g)	114	223	140	303
Fat cell volume (nl)	0.023	0.36	0.36	1.58
Glycerol release (μmoles/10⁶ cells/2 hours)				
Basal	0.19	1.29	0.57	4.36
+ epinephrine (1 μg/ml)	0.28	4.76	0.88	5.00
+ glucose (5.6 mM)	0.24	1.47	0.69	5.40
+ epinephrine and glucose	0.52	8.45	1.19	4.94
Conversion of radioactivity from [U ¹⁴C] glucose into fatty acids				
Basal	7.7	900	459	948
+ insulin (1 mU/ml)	51.9	2190	790	1520

week-old lean rats, but by 18 weeks there was a difference of nearly 80 g in body weight (York and Bray, 1973a). The fat cells of the 6-week-old fatty rats were as large as those observed in the 18-week-old lean animals. Lipolysis, as measured by the release of glycerol, was approximately 3 times higher in 6-week-old fatty rats and bore approximately the same relationship in 18-week-old fatty rats. Although there had been an increase in the rate of lipolysis between 6 and 18 weeks of age, the relative rate remained comparable in lean and fatty rats. Lipolysis was stimulated by epinephrine in fat cells from both groups of animals. The degree of stimulation at 6 weeks of age was approximately 3-fold in tissue from fat or lean rats, but at 18 weeks of age epinephrine had a somewhat greater effect in adipose tissue from lean rats. Basal lipolysis in the presence of glucose was not significantly altered. In the presence of glucose and epinephrine, however, hydrolysis of triglyceride was enhanced to a significantly greater degree in adipose tissue of lean animals, whereas no significant enhancement occurred in adipose tissue from fatty rats at either age.

One interpretation of these data is that basal rates of lipolysis might reflect differences in food intake or fat cell size and that the alterations may result from obesity rather than playing any important role in its development. Two experiments have been performed to test this hypothesis. In the first, fatty rats were pair-fed the same quantity of food as eaten by lean rats for a period of 10 weeks. At the end of that time, the size of the adipocytes and the release of glycerol in the basal and epinephrine-stimulated states were measured. In the second experiment, the body weight of the fatty rat was reduced by starvation for 11 weeks from an initial body weight

TABLE 8   Effect of dietary control on lipolysis and lipogenesis in lean and fatty rats

	Lean	Fatty		
		Fed ad lib	Pair-fed	Reduced-refed
Body weight (g)	223	303	246	290
Fat cell size (nl)	0.36	1.58	1.14	0.18
Glycerol release (μmoles/10⁶ cells)				
Basal	1.29	4.36	2.37	0.69
+ epinephrine (1 μg/ml)	6.00	9.36	6.24	1.08
Conversion of radioactivity from [U ¹⁴C]				
glucose into fatty acids (μmoles/10⁶ cells/2 hours)				
Basal	0.90	0.95	0.75	0.052
+ insulin (1 mU/ml)	2.19	1.52	1.17	0.044

of 582.5 g (reduced-refed) (York and Bray, 1973a). Prior to each biopsy, they were refed ad lib
for 72 hours. These two studies are summarized in Table 8. The size of adipocytes was enlarged
in the fatty rats fed ad lib and was significantly reduced in pair-fed fatty rats, although the cells
of this latter group were still 3 times larger than those of lean rats. With prolonged weight
reduction (reduced-refed rats), the adipocytes of the fatty rat were reduced to half the size of the
adipocytes from lean rats (Fig. 3). In the reduced-refed fatty rats, fat cells were half as large as
those of lean animals, and the rate of glycerol release was also half as great. In the pair-fed and
ad lib-fed fatty rats, the fat cells were larger and so were the rates of basal glycerol release.

Stimulation of lipolysis by epinephrine was greater in adipocytes from lean rats than in any of
the groups of fatty rats. Even prolonged pair-feeding or reduction in body weight did not restore
epinephrine-stimulated lipolysis to normal. In summary, the basal lipolysis in the fatty rat is

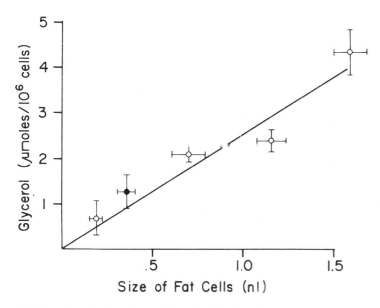

Fig. 3.   Glycerol release from adipocytes of various sizes.

related to the volume of the fat cells. Stimulated lipolysis occurs in the presence of epinephrine, of dibutyryl cyclic 3',5'-AMP, of aminophylline, and of thyrotropin. However, the magnitude of the response in both young and old fatty animals appears to be impaired. Such observations suggest the possibility of impaired formation of cyclic AMP by the adipocytes of these rats.

To examine this possibility, the concentration of cyclic AMP was measured in the adipocytes of 12-month-old female fatty rats and of lean rats of the same age, sex and strain. For these experiments, adipocytes were prepared by collagenase digestion and the cells washed and aliquoted into Krebs-Ringer phosphate buffer containing albumin buffered to pH 7.4. After a 20-minute preincubation, 0.1 μg of isoproterenol was added to each ml of medium and the incubation continued for varying intervals up to 40 minutes. The reaction was stopped by boiling the tissue, and the concentration of the cyclic nucleotide was measured by a radioimmunoassay system. The results are shown in Figure 4. The size of adipocytes was measured using an ocular

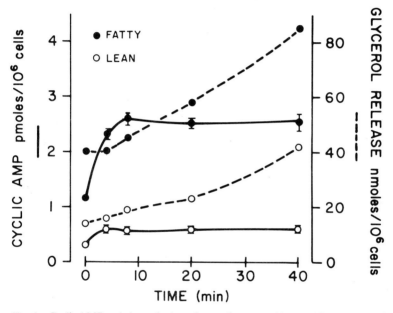

Fig. 4.   Cyclic AMP and glycerol release from adipocytes of lean and fatty rats. Cyclic AMP in pmoles/ 10⁶ cells is shown with solid lines and the corresponding glycerol concentrations in dashed lines.

micrometer (Bray, 1970). There was a significantly higher concentration of cyclic AMP in the untreated fat cells from fatty rats than in those from lean rats, which might account for the increased rate of basal lipolysis observed in this Figure and in our earlier experiments. Within 4 minutes of adding isoproterenol there was a significant increase in the concentration of cyclic AMP in cells from both groups of animals. The concentration remained elevated throughout the period of incubation and was significantly higher in the tissue from the fatty rats at all times. Thus, there seems to be no reduction in the concentration of cyclic AMP in the basal state nor during its rise after exposure to isoproterenol.

Reesterification of fatty acids by the fat cells from fatty rats is also consistently enhanced. Reesterification requires a source of glycerol-3-phosphate. Current evidence suggests that the majority of the glycerol-3-phosphate required for reesterification comes from glycolysis, since little or no free glycerol is incorporated into triglyceride by adipose tissue from the fatty rat (Bray et al., 1970). The formation of glycerides when glucose is absent from the incubation medium implies intracellular stores of glycogen which can be used for this purpose. When rats were fasted to deplete them of glycogen, essentially no reesterification occurred in fat cells from fatty rats whether incubated in the presence or absence of glucose. This implies that intracellular stores are used preferentially. Attempts to measure intracellular concentrations of glycerol-3-phosphate in fat have been unsuccessful, although it is readily measured in liver. Thus, fat cells from fatty rats do not accumulate measurable amounts of glycerol-3-phosphate which might stimulate glyceride formation. Measurements of glycerol-3-phosphate hydrolysis at alkaline and acidic pH showed no differences from the activity in tissues of normal rats. Thus, enhanced esterification of fatty acids would appear to result from the increased quantities of intracellular glycogen which are available to form glycerol-3-phosphate.

Lipogenesis has also been examined in the fatty rat at 6 and 18 weeks of age. In contrast with the data discussed previously, lipogenesis shows some striking differences with age (Table 7). At 6 weeks of age, fatty acid formation from [U-^{14}C] glucose by adipocytes from fatty rats was more than 50-fold greater than in adipocytes from lean animals. Thus, even at this early age, lipogenesis is markedly enhanced. With growth, the basal rate of lipogenesis rises sharply in lean animals, and by 18 weeks of age, adipocytes from fat and lean animals converted ^{14}C from glucose into fatty acids at comparable rates (York and Bray, 1973a). Stimulation of lipogenesis by insulin also differed sharply in young lean and fatty rats. The addition of insulin to adipocytes from lean animals stimulated the conversion of radioactivity from glucose into fatty acids more than 6-fold, whereas the effect was less than 2-fold in adipocytes from fatty rats. With increasing age, the magnitude of the insulin effect declined in the lean animals, but at both ages it remained greater in the lean than in the fatty rats. These data suggest that the young fatty rat converts more glucose to fatty acids in the absence of exogenous insulin.

Enhanced lipogenesis requires increased quantities of NADPH, which are formed by the pentose cycle and via 'malic enzyme'. Changes in the activity of these enzymes should be particularly prominent in the young animals, but such measurements have not been made yet. In older animals, the activity of hepatic malic enzyme is increased (Table 9). Whether changes in any of

TABLE 9   Activity of some hepatic enzymes in lean and fatty rats*

	Activity (nmoles/min/mg protein) (mean ± S.E.M.)	
	Lean (8)	Fatty (8)
Fructosediphosphate aldolase	14.02 ± 1.79	17.9 ± 1.98
Glucose-6-phosphate isomerase	103.8 ± 11.49	126 ± 12.8
Phosphofructokinase	1.46 ± 0.34	1.56 ± 0.38
Glycerol-3-phosphate dehydrogenase	4.84 ± 1.71	2.65 ± 0.54
Glyceraldehydephosphate dehydrogenase	3.54 ± 0.42	4.56 ± 0.59
Triosephosphate isomerase	365 ± 36	419 ± 50
Phosphoglycerate kinase	2.08 ± 0.33	1.77 ± 0.21
Pyruvate kinase	1.24 ± 0.23	2.64 ± 1.21
Malate dehydrogenase	810 ± 168	778 ± 119
Lactate dehydrogenase	2.79 ± 0.48	1.90 ± 0.25
Malic enzyme	1.21 ± 0.17	2.37 ± 0.30**

* All animals were pair-fed 7 days prior to sacrifice.
** $P < 0.005$.

the enzymes which have been measured related directly to the underlying biochemical defect is uncertain. The adaptive nature of most of these enzymes makes it likely that they are responding to the enhanced supply of nutrients which accompanies the hyperphagia rather than causing the increased food intake of these animals.

CONCLUSION

By 6 weeks of age, lipogenesis is enhanced in adipocytes of fatty rats. As the animals become older, the differences are reduced. These data suggest, however, that overall lipogenesis is enhanced throughout life. Initially this is produced by greater lipogenesis per cell, whereas later in life the increased lipogenesis is due to the increased numbers of fat cells as well. Basal lipolysis is enhanced in fatty rats mainly because they have larger fat cells. Stimulated lipolysis was greater in the obese animals because basal lipolysis was greater. On a percentage basis lipolysis was greater in the lean animals. Thus, the major difference between lean and fatty rats appears to be the augmented lipogenesis early in life.

REFERENCES

Barry, W.S. and Bray, G.A. (1969): Plasma triglycerides in genetically obese rats. Metabolism, 18, 833.
Bleisch, V.R., Mayer, J. and Dickie, M.M. (1952): Familial diabetes mellitus in mice, associated with insulin resistance, obesity and hyperplasia of the islands of Langerhans. Amer. J. Path., 28, 369.
Bray, G.A. (1969): Studies on the composition of adipose tissue from the genetically obese rats. Proc. Soc. exp. Biol. (N.Y.), 131, 1111.
Bray, G.A. (1970): Measurement of subcutaneous fat cells from obese patients. Ann. intern. Med., 73, 565.
Bray, G.A., Mothon, S. and Cohen, A.S. (1970): Mobilization of fatty acids in genetically obese rats. J. Lipid Res., 11, 517.
Bray, G.A. and York, D.A. (1971a): Genetically transmitted obesity in rodents. Physiol. Rev., 51, 598.
Bray, G.A. and York, D.A. (1971b): Thyroid function of genetically obese rats. Endocrinology, 88, 1095.
Bray, G.A. and York, D.A. (1972): Studies on food intake of genetically obese rats. Amer. J. Physiol., 223, 176.
Bray, G.A., York, D.A. and Swerdloff, R.S. (1973): Genetic obesity in rats. I. The effects of food restriction on body composition and hypothalamic function. Metabolism, 22, 435.
Fuller, J.L. and Jacoby Jr., G.A., (1955): Central and sensory control of food intake in genetically obese mice. Amer. J. Physiol., 183, 279.
Goldberg, R.C. and Mayer, J. (1952): Normal iodine uptake and anoxic resistance accompanying apparent hypometabolism in the hereditary obese hyperglycemic syndrome. Proc. Soc. exp. Biol. (N.Y.), 81, 323.
Hackel, D.V., Fröhman, L., Mikat, E., Lebovits, H.E., Schmidt-Nielsen, K. and Kinney, T.D. (1966): Effect of diet on the glucose tolerance and plasma insulin levels of the sand rat. Diabetes, 15, 105.
Hirsch, J. and Han, P.W. (1969): Cellularity of rat adipose tissue: effects of growth, starvation and obesity. J. Lipid Res., 10, 77.
Ingalls, A.M., Dickie, M.M. and Snell, G.D. (1950): Obese, new mutation in the mouse. J. Hered., 41, 317.
Johnson, P.R. and Hirsch, J. (1972): Cellularity of adipose depots in six strains of genetically obese mice. J. Lipid Res., 13, 2.
Johnson, P.R., Zucker, L.M., Cruce, J.A.F. and Hirsch, J. (1971): Cellularity of adipose depots in the genetically obese Zucker rat. J. Lipid Res., 12, 706.
Lemonnier, D. (1971a): Hyperinsulinism in genetically obese rats. Hormone metab. Res., 3, 287.
Lemonnier, D. (1971b): Sex difference in the number of adipose cells from genetically obese rats. Nature (Lond.), 231, 50.
Runner, M.N. (1952): Study of the ovarian and pituitary function of the obese mouse. In: Conference on the Obese Mouse, Bar Harbor, Maine, 1952, p. 6.
Saiduddin, S., Bray, G.A., York, D.A. and Swerdloff, R.S. (1973): Reproductive function in the genetically obese fatty rat. Endocrinology, 93, 1251.
Schemmel, R., Mickelsen, O. and Gill, J.L. (1970): Dietary obesity in rats: body weight and fat accretion in seven strains of rats. J. Nutr., 100, 1041.
Schonfeld, G. and Pfleger, B. (1971): Overproduction of very low-density lipoproteins by livers of genetically obese rats. Amer. J. Physiol., 220, 1178.
Stauffacher, W.W., Lambert, E., Vecchio, D. and Renold, A.E. (1967): Measurements of insulin activities in pancreas and serum of mice with spontaneous ('obese' and 'New Zealand obese') and induced (gold thioglucose obesity) and hyperglycemia with considerations on the pathogenesis of the spontaneous syndrome. Diabetologia, 3, 230.
Stern, J., Johnson, P.R., Greenwood, M.R.C., Zucker, L.M. and Hirsch, J. (1972): Insulin resistance and pancreatic insulin release in the genetically obese Zucker rat. Proc. Soc. exp. Biol. (N.Y.), 139, 66.
York, D.A. and Bray, G.A. (1971): Regulation of water balance in genetically obese rats. Proc. Soc. exp. Biol. (N.Y.), 136, 798.
York, D.A. and Bray, G.A. (1973a): Genetic obesity in rats. II. The effect of food restriction on the metabolism of adipose tissue. Metabolism, 22, 443.
York, D.A. and Bray, G.A. (1973b): Adipose tissue metabolism in 6-week-old fatty rats. Hormone metab. Res., 5, 355.

York, D.A., Hershman, J.M., Utiger, R.D. and Bray, G.A. (1972a): Thyrotropin secretion in genetically obese rats. Endocrinology, 90, 67.

York, D.A., Steinke, J. and Bray, G.A. (1972b): Hyperinsulinemia and insulin resistance in genetically obese rats. Metabolism, 21, 277.

Zucker, L.M. (1972): Fat mobilization in vitro and in vivo in the genetically obese Zucker rat 'fatty'. J. Lipid Res., 13, 234.

Zucker, L.M. (1967): Some effects of caloric restriction and deprivation on the obese hyperlipemic rat. J. Nutr., 91, 247.

Zucker, L.M. and Antoniades, H.N. (1972): Insulin and obesity in the Zucker genetically obese rat 'fatty'. Endocrinology, 90, 1320.

Zucker, L.M. and Zucker, T.F. (1961): Fatty, a new mutation in the rat. J. Hered., 52, 275.

Zucker, T.F. and Zucker, L.M. (1963): Fat accretion and growth in the rat. J. Nutr., 80, 6.

Zucker, T.F. and Zucker, L.M. (1962): Hereditary obesity in the rat associated with high serum fat and cholesterol. Proc. Soc. exp. Biol. (N.Y.), 110, 165.

DISCUSSION

Foa: Was hyperinsulinism present in your 6-week-old animals in whom lipogenesis was very active, or, in other words, what came first, hyperinsulinism or obesity?

Bray: Insulin was elevated by 6 weeks in fatty rats. Food intake is also enhanced by this age. Although we do not have an unequivocal answer, in all likelihood the food intake is enhanced very early, with obesity and hyperinsulinemia following.

Alterations of hexokinase isoenzymes by insulin, glucose and dexamethasone in adipose tissue incubated in vitro *

R.S. Bernstein

Department of Medicine and Institute of Human Nutrition, Columbia University College of Physicians and Surgeons and Medical Service, St. Luke's Hospital Center, New York, N.Y., U.S.A.

Although much information is available about the actions of insulin in adipose tissue, there is little data to explain the variations in tissue responsiveness to this hormone in disorders such as obesity, starvation, infection, and adrenal and pituitary diseases. Livingston et al. (1972) have recently demonstrated that the obese adipocyte has the same number of insulin receptors as the normal fat cell. Thus it is likely that the well-known decrease in insulin-stimulated glucose utilization in enlarged fat cells is due either to alterations in a postulated second messenger system for insulin or to abnormalities in the mechanism for glucose metabolism. Since the adipocyte insulin resistance in rats with either enlarged fat cells (Salans and Dougherty, 1971; Di Girolamo and Rudman, 1968; Bernstein and Kipnis, 1973a,b) or glucocorticoid excess (Fain et al., 1963; Yorke, 1967; Leboeuf et al., 1962) involves all pathways of glucose metabolism, the diminished glucose utilization must be mediated by a step prior to the branching of the pathways. The only prior steps are transport of glucose into the cell and phosphorylation by the hexokinase (HK) isoenzyme system.

Prompted by the studies of Katzen (1967), which have demonstrated that insulin-sensitive tissues are characterized by a high ratio of HK type II (HK-II) to type I (HK-I), we investigated the role of these isoenzymes in altered adipocyte insulin sensitivity (Bernstein and Kipnis, 1973a,b). Using a fluorimetric micromethod for HK isoenzymes, we have demonstrated that HK-II is diminished during aging, starvation and glucocorticoid treatment in vivo, whereas HK-I usually remains constant. The HK-II levels are very closely correlated with the rates of glucose oxidation and lipogenesis from glucose in the presence of insulin, whereas the HK-I levels show a weak correlation only with lipogenesis in the absence of insulin. Thus it seemed pertinent to study the interrelationships of insulin and this enzyme system in greater depth. Since in vivo manipulations all entail multiple hormonal and nutritional alterations, we turned to an in vitro system in which individual variables could be separately altered.

In previous in vitro studies of HK, prolonged fasting of the animals was necessary before incubation in order to demonstrate an insulin effect (Hansen et al., 1967, 1970; Borrebaek, 1967). In addition, total HK activity diminished during the course of incubation under most circumstances. Thus, although double-labeling experiments demonstrated that insulin increased amino acid incorporation into HK, the cells were in a net catabolic state with regard to the enzyme. Smith (1971) has shown that adipose tissue incubated in TC 199 medium with the addition of HEPES buffer can be kept viable for up to a month. Therefore we used this medium and were able to study adipose tissue under more normal metabolic conditions.

Epididymal fat pads were removed under sterile conditions from 6 to 9 male Wistar rats weighing 150-175 g. The fat pads were placed in isotonic saline in a Petri dish and cut into approximately 20 mg pieces. The pieces were pooled and 4-6 pieces were incubated overnight in 10 ml of the culture medium with 0.5% bovine serum albumin in a 25 ml polypropylene Ehrlenmeyer flask. The tissue was removed the next morning, weighed, homogenized and centrifuged at $800 \times g$. The supernatant was kept on ice and HK was assayed the same day by

* Supported by NIH Grant No. AM08107.

our fluorimetric method (Bernstein and Kipnis, 1973a). At least 5 replicates of each treatment were performed in each experiment.

In Figure 1, it can be seen that adipose tissue removed from fed rats and incubated overnight in the presence of a physiologic concentration of glucose without insulin was able to maintain preincubation levels of both HK isoenzymes. In the presence of higher concentrations of glucose and massive amounts of insulin, HK-I was increased by 55% and HK-II by 125% above the levels in the control incubation.

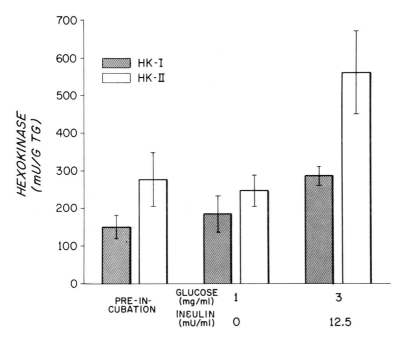

Fig. 1. Effect of 18-hour incubation on the activities of hexokinase isoenzymes in adipose tissue pieces. Results are mean ± S.E.M.

Insulin dose-response curves for overnight incubation of adipose tissue from both fed animals and animals sacrificed after a 24-hour fast are shown in Figure 2. In both of these studies, 3 mg glucose/ml was in the incubation medium. In the tissue from the fasted animals, a maximal increase in HK-II was seen with 100 μU insulin/ml. This is identical to the concentration of insulin necessary for maximal short-term insulin stimulation of glucose utilization (Rodbell, 1964). HK-I in the fasted adipose tissue rose gradually with increasing insulin up to the highest dose of 1000 μU/ml. Higher doses (not shown here) did not further elevate HK-I. The maximal HK-II value in the fat from fed animals was seen after incubation with 100-300 μU insulin/ml, but the rise in isoenzyme activity was much smaller than in fasted animals and not statistically significant. This is probably due to a higher basal level of HK-II in the cells, since fasting for 1 day has been shown to reduce adipose tissue HK-II to 50% of control activity (Bernstein and Kipnis, 1973a). HK-I did not rise in the presence of insulin in the fat from the fed animals. In most subsequent experiments we have used fat from 1-day-fasted animals in order to magnify the response. However, the results are qualitatively similar in fed and fasted fat pads.

Because of the similarity of the amounts of insulin necessary to increase HK-II and to stimulate glucose utilization, it was important to determine whether the insulin effect on HK-II was dependent on a direct action of the hormone or due solely to the increase in glucose uptake by the cells. Therefore incubations were performed in a glucose-free medium in the presence and

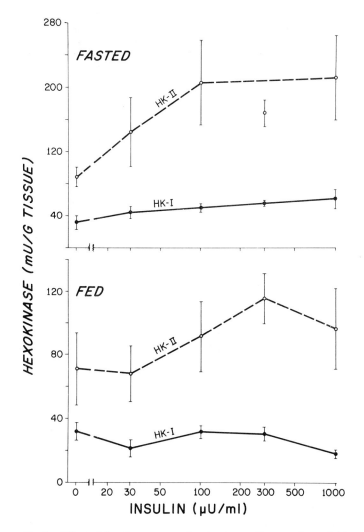

Fig. 2. Effect of insulin concentration in the medium on the activities of hexokinase isoenzymes in adipose tissue pieces from fed and fasted rats incubated 18 hours with 3 mg glucose/ml. Results are mean ± S.E.M.

absence of 300 μU insulin/ml. As can be seen in Table 1, HK-II was not increased by insulin in the tissue from fasted animals. This is not due to the depletion of glycogen or other endogenous energy sources by fasting, since insulin was also ineffective in the absence of glucose in the fed fat pads. In contrast, both fasted and fed adipose tissue had a statistically significant 60-70% increase in HK-I activity in the presence of insulin.

To further delineate the effects of glucose and insulin, an incubation was performed at 3 concentrations of glucose (0, 0.3 and 3.0 mg/ml) and 2 levels of insulin (0 and 300 μU/ml). Table 2 shows that increasing glucose concentration, either with or without insulin, causes a dramatic increase in activity of both isoenzymes. There was a 2-fold stimulation of HK-II by insulin in both 0.3 and 3 mg glucose/ml, but once again there was no stimulation in the absence of glucose. Roughly similar percent increases in HK-I in the presence of insulin were seen at all three glucose levels. There was a 3-fold increase in glucose uptake due to insulin during this incubation at both glucose concentrations. The glucose uptake in 3 mg glucose/ml without

TABLE 1   Effects of insulin (300 μU/ml) on HK isoenzymes in adipose tissue incubated 18 hours in a glucose-free medium

	HK-I (mU/g tissue)	HK-II (mU/g tissue)
Fasted animals		
– Insulin	26.3 ± 4.9	51.9 ± 11.4
+ Insulin	42.0 ± 5.8	61.3 ± 16.2
Fed animals		
– Insulin	15.7 ± 2.0	56.7 ± 7.4
+ Insulin	26.4 ± 3.5	66.1 ± 6.3

TABLE 2   Effects of insulin (300 μU/ml) on HK isoenzymes and glucose utilization of fat pad pieces incubated in varying glucose concentrations

	18-hour incubation			2-hour incubation	
	HK-I (mU/g tissue)	HK-II (mU/g tissue)	Glucose uptake (mg/g tissue)	Glucose→CO_2 (mg/g tissue)	Glucose→TG (mg/g tissue)
0 Glucose					
– Insulin	8.6 ± 3.8	16.3 ± 5.7	—	—	—
+ Insulin	11.8 ± 3.6	7.8 ± 6.1	—	—	—
0.3 mg/ml Glucose					
– Insulin	12.5 ± 5.5	37.5 ± 5.4	3.2 ± 0.5	0.25 ± 0.03	0.33 ± 0.05
+ Insulin	27.5 ± 5.2	73.5 ± 8.6	9.9 ± 0.9	1.04 ± 0.10	0.70 ± 0.04
3.0 mg/ml Glucose					
– Insulin	23.6 ± 7.4	55.9 ± 9.9	10.6 ± 2.9	0.58 ± 0.08	0.66 ± 0.04
+ Insulin	38.5 ± 3.9	117.1 ± 11.9	29.5 ± 1.6	2.03 ± 0.22	2.02 ± 0.25

insulin was roughly the same as that in 0.3 mg glucose/ml with insulin, as were the HK-I levels. The HK-II activity was higher in the low glucose incubation with insulin than in the high glucose incubation without insulin, but the difference was not quite significant.

The similarity of the glucose uptake pattern to the HK isoenzymes is consistent with the possibility that the insulin effect on HK is secondary to increased glucose uptake. However, the increased uptake over this period might be due in part to the altered HK activity. Thus we performed a 2-hour incubation of adipose tissue pieces in the same concentrations of glucose and insulin as in the previous experiment, and measured glucose utilization. In this short-term experiment, both insulin and high glucose concentrations stimulated glucose conversion to CO_2 and triglycerides, and the effects were synergistic. Insulin was the more potent stimulus for glucose oxidation, as it was for increasing HK-II activities in the longer incubation, and both insulin and increased glucose were equally effective in stimulating lipogenesis.

In preliminary experiments, we have investigated the substances which might be able to replace glucose in mediating the insulin effect. During incubations with insulin in a glucose-free medium, fructose, which is a substrate of HK and is utilized by adipose tissue, permitted a stimulation of HK-II activity. In contrast, galactose, glucose-6-phosphate, glycerol, acetate and pyruvate do not mediate an increase in HK-II above baseline.

In contrast to insulin, the effects of the synthetic glucocorticoid, dexamethasone, are clear-cut (Fig. 3). This experiment was performed in fat from fasted animals with 3 mg glucose/ml in the incubation medium. Large amounts of dexamethasone cause a diminution of both HK-isoenzymes, both in the presence and absence of insulin. Previous authors have shown (Yorke,

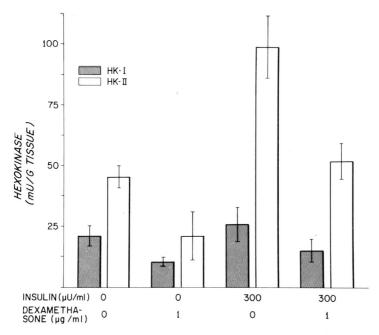

Fig. 3. Effects of high concentrations of dexamethasone and insulin in the medium on the activities of hexokinase isoenzymes in adipose tissue pieces incubated 18 hours with 3 mg glucose/ml. Results are mean ± S.E.M.

1967; Leboeuf et al., 1962; Czech and Fain, 1972), and we have confirmed, that in the presence of insulin and high glucose concentrations there is no inhibition of glucose uptake during incubation with dexamethasone. Thus the effect on HK is not secondary to decreased glucose utilization. In Figure 4, a dose-response curve of dexamethasone effect on HK isoenzymes is shown. The incubation was performed with 3 mg glucose and 300 μU insulin/ml. A maximal inhibition of HK-II was achieved with 0.03 μg dexamethasone/ml, whereas no diminution of HK-I was seen with less than 0.1 μg/ml. In other experiments the concentration of dexamethasone necessary for maximal inhibition of HK-I was 1 μg/ml. The inhibition of glucose uptake by dexamethasone during the last 2 hours of a 4-hour incubation without insulin is shown in Figure 5. The maximal effect was achieved with 0.01–0.03 μg/ml. This is comparable to the value of 0.016 μg/ml found by Fain et al. (1963), and almost identical to the amount of steroid necessary for reduction of HK-II. The reduction of glucose uptake is not mediated by HK, since no consistent change in either isoenzyme can be demonstrated during a 4-hour incubation.

Thus a differential effect on the individual HK isoenzymes has been demonstrated. Insulin was able to stimulate HK-I activity in the absence of an added energy source. In contrast, its effect on HK-II has not been separable in these studies from the stimulation of utilization of glucose or fructose, both of which are substrates of the enzyme. Dexamethasone causes a diminution of HK-II activity under conditions in which glucose utilization does not change. A similar effect on HK-I is seen with much larger amounts of dexamethasone. Since the conditions under which dexamethasone inhibits adipose tissue glucose utilization acutely in vitro are unphysiological (i.e. insulin lack or very low glucose concentrations) (Czech and Fain, 1972), it is unlikely that the insulin-resistance in fat induced by glucocorticoids is mediated through this mechanism. In contrast, dexamethasone inhibits HK-II activity in the presence of physiologic concentrations of glucose and insulin, both in vivo (Bernstein and Kipnis, 1973b) and in vitro, and the decreased HK-II in vivo is associated with decreased adipocyte insulin sensitivity. Thus it is possible that HK-II mediates the insulin resistance of glucocorticoid excess.

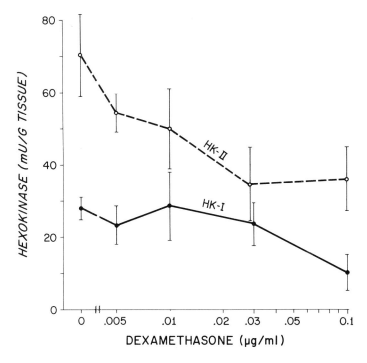

Fig. 4. Effect of dexamethasone concentration in the medium on the activities of hexokinase isoenzymes in adipose tissue pieces incubated 18 hours with 3 mg glucose and 300 μU insulin/ml. Results are mean ± S.E.M.

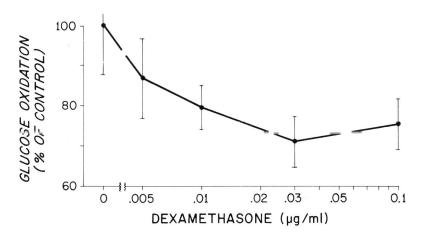

Fig. 5. Effect of dexamethasone concentration on rates of glucose oxidation in adipose tissue pieces during the final 2 hours of a 4-hour incubation in Krebs-Ringer bicarbonate buffer with 0.5 mg glucose/ml and no insulin. Results are mean ± S.E.M. pooled from 3 experiments, and expressed as the percentage of the quantity of [1-^{14}C] glucose converted to $^{14}CO_2$ in the absence of dexamethasone.

REFERENCES

Bernstein, R.S. and Kipnis, D.M. (1973a): Regulation of rat hexokinase isoenzymes. I. Assay and effect of age, fasting and refeeding. Diabetes, 22, 913.

Bernstein, R.S. and Kipnis, D.M. (1973b): Regulation of rat hexokinase isoenzymes. II. Effects of growth hormone and dexamethasone. Diabetes, 22, 923.

Borrebaek, B. (1967): Adaptable hexokinase activity in epididymal adipose tissue studied in vivo and in vitro. Biochim. biophys. Acta (Amst.), 141, 221.

Czech, M.P. and Fain, J.N. (1972): Antagonism of insulin action on glucose metabolism in white fat cells by dexamethasone. Endocrinology, 91, 518.

Di Girolamo, M. and Rudman, D. (1968): Variations in glucose metabolism and sensitivity to insulin of the rat's adipose tissue in relation to age and body weight. Endocrinology, 82, 1133.

Fain, J.N., Scow, R.O. and Chernick, S.S. (1963): Effects of glucocorticoids on metabolism of adipose tissue in vitro. J. biol. Chem., 238, 54.

Hansen, R., Pilkis, S.J. and Krahl, M.E. (1967): Properties of adaptive hexokinase isoenzymes of the rat. Endocrinology, 81, 1397.

Hansen, R., Pilkis, S.J. and Krahl, M.E. (1970): Effect of insulin on the synthesis in vitro of hexokinase in rat epididymal adipose tissue. Endocrinology, 86, 57.

Katzen, H.M. (1967): The multiple forms of mammalian hexokinase and their significance to the action of insulin. Advanc. Enzym. Regulation, 5, 335.

Leboeuf, B., Renold, A.E. and Cahill, G.F. (1962): Studies on rat adipose tissue in vitro. IX. Further effects of cortisol on glucose metabolism. J. biol. Chem., 237, 988.

Livingston, J.N., Cuatrecasas, P. and Lockwood, D.H. (1972): Insulin insensitivity of large fat cells. Science, 177, 626.

Rodbell, M. (1964): Metabolism of isolated fat cells. I. Effect of hormones on glucose metabolism and lipolysis. J. biol. Chem., 239, 375.

Salans, L.B. and Dougherty, J.W. (1971): The effect of insulin upon glucose metabolism by adipose cells of different size. J. clin. Invest., 50, 1399.

Smith, U. (1971): Studies of human adipose tissue in culture. I. Incorporation of glucose and release of glycerol. Anat. Rec., 169, 97.

Yorke, R.E. (1967): The influence of dexamethasone on adipose tissue metabolism in vitro. J. Endocr., 39, 329.

DISCUSSION

Berger: If the changes of hexokinase activity you have reported have any significance as far as the glucose flux is concerned, one would expect an accumulation of glucose within the tissue. Did you make any attempts to determine the glucose spaces in this respect?

Bernstein: You are correct in saying that there must be intracellular accumulation of glucose if the alterations in hexokinase have any physiological significance. Unfortunately, the glucose space is very difficult to measure in adipose tissue because intercellular water is a very small fraction of tissue water. However, we are currently setting up methods to make this determination.

Foa: I need clarification because I believe I detected in your very interesting communication the implied statement that the activity of hexokinase I correlates with glucose oxidation, while the activity of hexokinase II correlates with lipogenesis. If my understanding is correct, I would like to know whether the tissue distinguishes between glucose-6-phosphate derived from the action of isoenzyme I and glucose-6-phosphate derived from the action of isoenzyme II.

Bernstein: It is not true that hexokinase I correlates with glucose oxidation and hexokinase II with lipogenesis. The quantities of glucose utilised in these pathways have been compared with the stimulation of hexokinase activity in an 18-hour incubation. Indeed, hexokinase II correlates with both lipogenesis and glucose oxidation, whereas hexokinase I is not associated with either activity in the presence of insulin.

It is known that there are two metabolically distinct pools of glucose-6-phosphate in liver and muscle, but these have not been identified in adipose tissue and there is no evidence that hexokinase is associated with them.

The effect of 3-oxymethylglucose on glucose metabolism of the isolated fat cell*

H.M.J. Krans and **C.L.M. Arkesteijn**

Department of Endocrinology and Metabolic Diseases and Department of Chemical Pathology, University Hospital, Leyden, The Netherlands

For the study of glucose transport mechanisms 2-deoxyglucose and 3-oxymethylglucose (3-OMG) are used as non-metabolizable sugar analogues. 2-Deoxyglucose has an inhibitory effect on glycolytic enzymes. 3-OMG is a non-metabolizable sugar which does not interfere with glycolysis but can be taken up by muscle and adipose tissue (Kohn and Clausen, 1971; Bihler, 1968; Crofford, 1967; Crofford and Renold, 1965a). Inside the cell 3-OMG has no affinity for hexokinase and remains in an unbound state. After preloading the fat pads with 3-OMG the release of 3-OMG is enhanced by insulin (Crofford, 1967; Clausen, 1969). Insulin stimulates glucose uptake in adipose tissue and isolated fat cells (Rodbell, 1964a). This can only be measured indirectly by the stimulation of the production of $^{14}CO_2$ or [^{14}C] triglycerides from [^{14}C] glucose. We hoped that we could use [^{14}C] 3-OMG as a means for obtaining more information about the effect of insulin and other substances on the transport of glucose through the membrane.

Isolated fat cells were prepared according to Rodbell (1964a) from male rats weighing 150 g. The cells or fat pads were incubated in Krebs-Ringer bicarbonate buffer (pH 7.4). [^3H] insulin was used as a marker to calculate the extracellular space. After incubation the fat cells were filtered on millipore filters, dried and counted in a liquid scintillation counter. The fat pads were dried, homogenized and the fat was extracted with methanol/heptane. The ^{14}C was counted in the water phase. The [^{14}C] counts determined the total intra-plus extracellular [^{14}C] 3-OMG space. From these data the intracellular 3-OMG space was calculated.

Incubation of fat cells with [^{14}C] 3-OMG in the presence or absence of glucose did not result in the formation of $^{14}CO_2$, [^{14}C] glycerol or [^{14}C] fatty acids, indicating that 3-OMG was not metabolized by the fat cells. We could not measure any uptake of 3-OMG by the isolated fat cells (Table 1). This may have been caused by the fact that in isolated fat cells the ratio of intracellular water space to extracellular water space is lower than in adipose tissue or in muscle (Crofford and Renold, 1965b). Incubation of fat cells in an identical way with [6-^{14}C] glucose demonstrated an insulin-dependent incorporation of label, indicating that the method used was valid for measuring uptake.

3-OMG inhibited the uptake of glucose by the cells, measured both as CO_2 production or

TABLE 1   Uptake of [^{14}C] 3-OMG and [6-^{14}C] glucose in isolated fat cells

Incubation time (min)	Insulin (250 μU/ml)	Uptake (d.p.m./g total lipid weight)	
		[^{14}C] 3-OMG (10 mM)	[6-^{14}C] glucose (10 mM)
5	—	18	119
60	—	28	552
5	+	31	172
60	+	23	1094

* This research was supported by the Organization for Health Research T.N.O. (GO-691-6-09).

TABLE 2

Concentration in incubation medium (mM)		Insulin (250 μU/ml)	CO_2	Triglycerides from fat cells	Triglyceride-fatty acids	Triglyceride-glycerol
Glucose	3-OMG					
1	—	—	1	1	1	1
1	10	—	0.4	0.23	—	0.10
10	—	—	3.6	2.9	7.1	3.3
10	10	—	2.4	2.3	2.7	2.3
1	—	+	6.5	6.3	10	4.9
1	10	+	1.8	1.6	—	1.6
10	—	+	20.4	15.9	43	7.2
10	10	+	14.9	11.3	30	6.3

Fat cells were incubated for 60 minutes in Krebs-Ringer bicarbonate buffer containing [1-^{14}C] glucose. The production of $^{14}CO_2$, [^{14}C] triglycerides, [^{14}C] triglyceride-fatty acids and [^{14}C] triglyceride-glycerol in standard incubation medium (1 mM glucose) was designated as 1. The production of these substances in other incubation media are given as relative numbers compared to the standard incubation. Results are the mean of 3 triplicate experiments.

incorporation into fat (Table 2). The degree of inhibition was dependent on the ratio between the 3-OMG concentration and the glucose concentration in the incubation medium. Insulin (250 μU) enhanced the glucose uptake, but did not change the relative inhibition of glucose uptake by 3-OMG. The substitution of 3-OMG for glucose did not change the total amount of transported substrate in the isolated fat cells. It has been reported that in the soleus muscle glucose inhibits the 3-OMG uptake and 3-OMG inhibits the uptake of glucose (Kohn and Clausen, 1971). Preincubation with 3-OMG did not change the relative inhibition of glucose transport in a positive or negative way (Table 3). This excludes the possibility that 3-OMG impedes glucose transport by blocking some important sites for glucose transport. Glucose and 3-OMG seem to share the same saturatable transport system in a competitive way. For fat cell ghosts a competition between 3-OMG and glucose in membrane transport has been reported although the studies were done at much higher glucose concentrations (Carter et al., 1972).

3-OMG cannot be used for the study of membrane transport in isolated fat cells. If this is due to the small intracellular space, it is difficult to understand why the uptake of 3-OMG found in adipose tissue cannot be demonstrated in isolated fat cells. This raises some doubts about the interpretation of the findings reported for adipose tissue.

There are three possible explanations. (1) Isolated fat cells are prepared using collagenase. This enzyme is contaminated with other proteolytic enzymes, which may cause the loss of some glucose 'receptors'. This may result in a smaller intracellular space for 3-OMG. (2) The non-fat cells in the fat pad (which account for about half of the protein (Rodbell, 1964b)) are responsible

TABLE 3   Effect of preincubation of fat cells with 3-OMG

3-OMG (10 mM)	Insulin (250 μU/ml)	CO_2*
not added	—	1
0 min	—	0.5
30 min	—	0.6
not added	+	6.0
0 min	+	1.9
30 min	+	1.9

At 30 minutes 1 mM glucose was added (containing [1-^{14}C] glucose 0.5 μCi/ μmole). Results are mean of 3 triplicate experiments.
* See note to Table 2.

for the greater 3-OMG uptake. (3) The apparently greater space in adipose tissue may be caused by the fact that part of the 3-OMG does not enter the cell, but sticks to or is bound by the outside of the membranes. It is known that insulin decreases the adhesiveness of glucose to fat cell membranes (Robinson et al., 1972). If this also holds for 3-OMG it could explain the apparent increase of 3-OMG release by insulin in the efflux experiments on adipose tissue (Crofford, 1967; Clausen, 1969). Another, less probable, explanation is that the extracellular diffusion from the extracellular volume to the incubation medium is influenced by insulin.

Note
After the preparation of this manuscript we performed short term incubation studies using the method of Gliemann et al. (1972). We determined the 3-OMG uptake every 10 seconds. Incubating in 10 mM 3-OMG, the uptake of 3-OMG could be demonstrated to reach an equilibrium in about 2 minutes maximum. The 3-OMG space in isolated adipocytes is only 1-2% and is small compared to the 3-OMG space in fat pads. The accessibility of the isolated adipocytes is much greater than in fat pads.

We conclude from these findings that in the interpretation of the data reported extracellular diffusion in fat pads has to be taken into account, especially if one tries to draw conclusions from quantitative data.

ACKNOWLEDGEMENT

The skilful technical assistance of Miss M. Jansen is acknowledged.

REFERENCES

Bihler, I. (1968): The action of cardiotonic steroids on sugar transport in muscle in vitro. Biochim. biophys. Acta (Amst.), 163, 401.
Carter Jr., J.R., Avruch, J. and Martin, D.B. (1972): Glucose transport in plasma membrane vesicles from rat adipose tissue. J. biol. Chem., 247, 2682.
Clausen, T. (1969): The relationship between the transport of glucose and cations across cell membranes in isolated tissues. V. Stimulating effect of ouabain, K^+- free medium and insulin on efflux of 3-O-methyl-glucose from epididymal adipose tissue. Biochim. biophys. Acta (Amst.), 183, 625.
Crofford, O.B. (1967): Countertransport of 3-O-methylglucose in incubated rat epididymal adipose tissue. Amer. J. Physiol., 212, 217.
Crofford, O.B. and Renold, A.E. (1965a): Glucose uptake by incubated rat epididymal adipose tissue. Characteristics of the glucose transport system and action of insulin. J. biol. Chem., 240, 3237.
Crofford, O.B. and Renold, A.E. (1965b): Glucose uptake by incubated rat epididymal adipose tissue. Rate-limiting steps and site of insulin action. J. biol. Chem., 240, 14.
Gliemann, J., Osterlind, K., Vinten, J. and Gammeltoft, S. (1972): A procedure for measurement of distribution spaces in isolated fat cells. Biochim. biophys. Acta (Amst.), 286, 1.
Kohn, P.G. and Clausen, T. (1971): The relationship between the transport of glucose and cations across cell membranes in isolated tissues. VI. The effect of insulin, ouabain and metabolic inhibitors on the transport of 3-O-methylglucose and glucose in rat soleus muscle. Biochim. biophys. Acta (Amst.), 225, 277.
Robinson Jr., C.A., Boshell, B.R. and Reddy, W.J. (1972): Insulin binding to plasma membranes. Biochim. biophys. Acta (Amst.), 290, 84.
Rodbell, M. (1964a): Metabolism of isolated fat cells. I. Effects of hormones on glucose metabolism and lipolysis. J. biol. Chem., 239, 375.
Rodbell, M. (1964b): Localization of lipoprotein lipase in fat cells of rat adipose tissue. J. biol. Chem., 239, 753.

DISCUSSION

Ailhaud: Did you measure the K_i for 3-OMG and if so, how does it compare to the apparent K_m for glucose uptake and metabolism?

Krans: No, I have not measured the K_m yet.

Ailhaud: Since your cells can be loaded quite quickly with 3-OMG, did you assay the efflux rate of 3-OMG?

Krans: We tried to do some efflux experiments but it is very difficult to design the experiment in such a way that you get reliable results. We have an indication that in 1 minute 40-50% of the 3-OMG has left the cell. The same result has been found by Gliemann (personal communication).

Permeability of adipose tissue to glucose and fructose during starvation

E. Couturier and **M.G. Herrera**

Department of Nutrition, Harvard University School of Public Health, Boston, U.S.A.

Starvation in mammals induces a rapid deterioration of glucose tolerance and a diminished sensitivity to insulin (Tütso, 1923). This is also observed in isolated adipose tissue incubated in vitro in the presence of glucose (Truehart and Herrera, 1971). [U-^{14}C] glucose incorporation into CO_2 and total lipid is indeed markedly diminished after a 24-hour fast as compared to values obtained with adipose tissue from fed animals. It has been shown (Hollifield and Parson, 1965) that glucose-6-phosphate dehydrogenase and 6-phosphoglucoronic dehydrogenase have a diminished activity during fasting. But very few data are available on the permeability of the cell membrane in this condition.

Two protocols were used to approach this problem. First, the inhibition produced by phloretin in epididymal adipose tissue from fed rats was compared to that produced in adipose tissue from fasted animals. The rats were Sprague-Dawley males weighing 100 to 130 g at the time of sacrifice and were either fed on a Purina Chow diet or fasted for 48 to 72 hours. Epididymal adipose tissue was incubated for 2 hours at 37° C in a Krebs-Ringer bicarbonate buffer containing 20 g/l albumin, 1 mM glucose and 0.1 μCi/ml [U-^{14}C] glucose (sp. act. 24 μCi/mg). The contralateral fat pad of each rat was incubated in the same medium but with 1 mM phloretin dissolved in absolute ethanol added (the ethanol concentration of the control and the phloretin flasks was 0.5% of the medium). After 2 hours incubation, CO_2 was liberated from the medium by the addition of 2N. H_2SO_4, collected in hyamin hydroxide and assayed for radioactivity in a liquid scintillation spectrometer (Nuclear, Chicago). The lipids of the tissue were isolated by a modification of the method of Folch et al. (1957) and the radioactivity was counted by the same procedure, using Bray's solution (Bray, 1960) as liquid scintillator. The fat-free remnant was dried to constant weight, allowing us to express the results as μ atoms [U-^{14}C] glucose (or fructose in fructose experiments) per 100 mg fat-free dry weight incorporated into CO_2 and total lipid. The data are expressed as mean ± S.E.M. Results are shown in Table 1. The inhibition produced by phloretin in fat tissue from fed animals is the same as that produced in fat tissue from fasting rats. Evidence has been presented by Crofford and Renold (1965) that glucose crosses the cell membrane of adipocytes by a stereospecific and mobile carrier. Phloretin acts at the cell membrane surface by blocking this system of transport. Our results indicate that the carriers available in the cell membrane for the transport of glucose are not quantitatively modified during fasting. In a second protocol, the metabolism of glucose was compared to that of fructose in epididymal adipose tissue from fed and fasted animals. In fructose experiments glucose and [U-^{14}C] glucose were replaced in the incubation medium by fructose and [U-^{14}C] fructose at the same concentrations. Results are shown in Table 2. For the same amounts of glucose or fructose (1 mM) in the incubation medium, the metabolism of glucose is 4 to 5 times more active than that of fructose. This difference is explained by a lower affinity of fructose for hexokinase and also by a lower efficiency of the transport system of fructose at low levels of this sugar in the incubation medium. In fasting, the metabolism of glucose is sharply reduced, whereas no significant effect could be elicited on fructose. Since at 1 mM transport is the rate-limiting step of sugar metabolism, the lack of effect of fasting implies that the transport system for fructose is not modified by starvation. The difference in glucose and fructose metabolism during fasting observed in our experiments is in accordance with the concept that, in the absence of insulin, glucose and fructose utilize a different system of

TABLE 1 [U-^{14}C] glucose incorporation into CO_2 and total lipid by paired epididymal adipose tissue from fed and fasted rats in the presence and absence of 1 mM of phloretin

Treatment group	μ atoms of [U-^{14}C] glucose per 100 mg fat-free dry weight incorporated into CO_2 and total lipid in 2 hours		
	without phloretin	with phloretin	% inhibition due to phloretin
Fed (n = 9)	45.6 ± 2.9	19.2 ± 4.3	58
Fasted 72 hours (n = 10)	5.0 ± 0.6	1.7 ± 0.3	66
P			n.s.

n = number of animals.
P = significance of difference from fed in the same column.

transport across the cell membrane (Froesch and Ginsberg, 1962). The fall in glucose metabolism, contrasting with the maintenance of a normal metabolism of fructose, suggests that the permeability of the adipocytes to glucose is diminished in fasting. These preliminary results confirm that the glucose transport system and the insulin-independent transport system for fructose are different, and indicate that they behave differently in starvation.

CONCLUSIONS

In the absence of insulin the metabolism of fructose by the isolated fat pad incubated in vitro is not altered by starvation. In contrast, under the same conditions, the metabolism of glucose is sharply reduced. As glucose and fructose in the adipocytes cross the cell membrane by a different carrier, and are then both phosphorylated by hexokinase before entering the glycolytic pathway, it is highly probable that one of the important alterations in glucose metabolism during fasting is at the level of transport into the cell. Phloretin produces the same inhibition of glucose metabolism in epididymal adipose tissue from fed and fasted animals, suggesting that the alteration of cell permeability is not due to a quantitative modification of the transport system of the cell membrane.

TABLE 2 Effect of fasting on glucose and fructose metabolism by epididymal adipose tissue from fed and fasted rats

Treatment group	μ atoms of [U-^{14}C] glucose or fructose per 100 mg fat-free dry weight incorporated into CO_2 and total lipid in 2 hours			
	Glucose	No. of experiments	Fructose	No. of experiments
Fed	49.3 ± 4.9	9	11.8 ± 1.1	12
Fasted 55 to 72 hours	15.5 ± 1.7	13	9.3 ± 1.3	13
P	< 0.001		n.s.	

P = significance of difference from fed in the same column.

ACKNOWLEDGEMENT

The technical assistance of Miss Ellen Marston is gratefully acknowledged.

REFERENCES

Bray, G.A. (1960): A simple efficient liquid scintillator for counting aqueous solutions in a liquid scintillation counter. Analyt. Biochem., 1, 279.

Crofford, O.B. and Renold, A.E. (1965): Glucose uptake by incubated rat epididymal adipose tissue. J. biol. Chem., 240, 3237.

Folch, J., Lees, M. and Sloane Stanley, G.H. (1957): A simple method for the isolation and purification of total lipids from animal tissues. J. biol. Chem., 226, 497.

Froesch, E.R. and Ginsberg, J.L. (1962): Fructose metabolism of adipose tissue. I. Comparison of fructose and glucose metabolism in epididymal adipose tissue of normal rats. J. biol. Chem., 237, 3317.

Hollifield, G. and Parson, W. (1965): In: Handbook of Physiology: Adipose Tissue, Section 5, Chapter 40. American Physiological Society, Washington, D.C.

Tiitso, M. (1923): Influence of nutritive condition on initial fall in blood sugar after insulin. Proc. Soc. exp. Biol. (N.Y.), 23, 40.

Truehart, P.A. and Herrera, M.G. (1971): Decreased response to insulin in adipose tissue during starvation. Diabetes, 20, 46.

Regulation of adipose tissue mass in obese hyperglycemic mice

C. Chlouverakis

Department of Medicine, State University of New York at Buffalo, E.J. Meyer Memorial Hospital, Buffalo, N.Y., U.S.A.

The obesity of the obese-hyperglycemic mice (ob/ob), like that of animals with hypothalamic obesity, shows two distinct phases: the dynamic phase in which the adipose tissue of the animal is expanding, and the static phase in which the expanded adipose tissue is relatively stable (Christophe, 1961; Mayer, 1960). The static phase can be regarded as a new steady state in which the expanded adipose tissue is regulated, though higher than normal.

The mechanism by which this regulation is achieved is still unclear. Fuller and Jacoby (1955) showed that the response of ob/ob mice to manipulation of the diet, such as dilution with an inert substance, alteration of its palatability or its fat content, is similar to that of hypothalamic obese animals, thus suggesting a hypothalamic involvement. Whether this involvement is primary or secondary to a metabolic abnormality is still debatable, but in favor of a primarily faulty hypothalamus is the observation that metabolic abnormalities of ob/ob mice subside with body weight reduction (Chlouverakis and White, 1969) and the fact that obesity precedes the development of such metabolic features as hyperinsulinemia, hyperglycemia and insulin resistance (Chlouverakis et al., 1970).

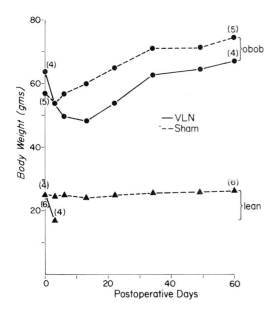

Fig. 1. Body weight (g) of ob/ob mice and their lean littermates subjected to bilateral electrolytic lesions of the VLN or to 'sham' operation (at day 0). Numbers in brackets indicate number of surviving animals at each time point.

To determine whether the two hypothalamic nuclei, ventrolateral (VLN) and ventromedial (VMN), which are primarily responsible for the regulation of food intake, are functional in ob/ob mice, an experiment was performed in which bilateral electrolytic lesions were placed in either of these two nuclei. Damage of the VLN was accompanied by a period of aphagia in both lean and ob/ob mice which led to the death of all the lean animals. The ob/ob animals, however, survived this period and finally stabilized at a body weight which was considerably lower than that of the 'sham' operated animals (Fig. 1). Blood glucose and serum insulin levels were slightly reduced (Table 1).

TABLE 1   Blood glucose and serum immunoreactive insulin before and 2 months after bilateral lesions in the ventrolateral nucleus (VLN) or 'sham' operation

	No. of animals	Blood glucose (mg/100 ml, mean ± S.E.M.)		Serum insulin (μU, mean ± S.E.M.)	
		Before operation	2 months after operation	Before operation	2 months after operation
VLN-lesioned	4	190.0 ± 27.9	156.5 ± 29.8	3605 ± 596	2935 ± 982
'Sham' operated	5	188.3 ± 12.4	221.6 ± 58.4	3910 ± 733	4238 ± 1048

Thus, the response of ob/ob mice to bilateral VLN lesions appears to be normal. Furthermore, their excess adiposity, by protecting them during the early postoperative period, facilitates their recovery. The final stabilization of the body weight of lesioned ob/ob mice at a level lower than that of control mice is compatible with the view that the VLN acts as the low set point controller in the regulation of body weight (Powley and Keesey, 1970).

Bilateral electrolytic lesions of the VMN of ob/ob and lean control mice enhanced the body weight gain of lean but not of ob/ob mice, thus suggesting that the VMN of ob/ob mice behaves as if it were already 'lesioned' and that the electrolytic damage failed to further affect body weight (Fig. 2). However, carcass analysis revealed that there were body composition changes in

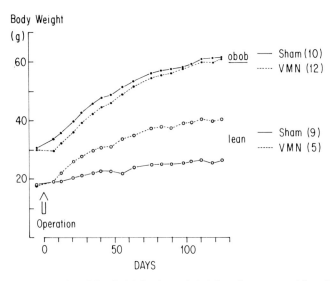

Fig. 2.   Body weight of ob/ob mice and their lean littermates subjected to bilateral electrolytic lesions of VMN or 'sham' operation. Numbers in brackets indicate number of animals.

TABLE 2 Percentages of body fat and body water, serum glucose and serum insulin (mean ± S.E.M.) of ob/ob and lean mice lesioned in the ventromedial nucleus (VMN) and 'sham' operated control mice

	No. of animals	% body fat	% body water	Serum glucose (mg/100 ml)	Serum insulin (µU/ml)
VMN ob/ob	12	54.4 ± 2.1	25.6 ± 0.9	180.0 ± 14.9	823 ± 216
'Sham' ob/ob	10	48.1 ± 1.9	28.3 ± 0.7	194.0 ± 15.6	1319 ± 345
VMN lean	5	23.2 ± 4.0	44.4 ± 4.1	167.4 ± 5.4	114 ± 51
'Sham' lean	9	8.0 ± 0.8	59.6 ± 1.2	142.7 ± 3.9	36 ± 5

the VMN-lesioned ob/ob mice consistent with an increased adiposity. Blood glucose and serum insulin were affected by the lesions only in the lean mice (Table 2). These findings are consistent with the presence of a functional VMN in the ob/ob mice and are in basic agreement with the conclusions of Coleman and Hummel (1972). The increased adiposity of the VMN-lesioned ob/ob mice in the absence of a parallel increase of their circulating insulin is consistent with the view that excess adiposity in these animals can occur without significant hyperinsulinemia (Chlouverakis and White, 1969).

The demonstration of a functional VMN and VLN in ob/ob mice suggests that regulation of body fat in these animals is normal, but that it is achieved at a higher level.

To determine whether the amount of fat per adipocyte or the total amount of body fat is the regulated variable, an experiment was performed in which ob/ob mice were subjected to partial lipectomy. Fifty-two days after the operation, lipectomized animals had failed to put on extra weight to compensate for the body fat removed (Fig. 3). In addition, the body composition, serum glucose and insulin did not differ between lipectomized and 'sham' operated ob/ob mice (Table 3). However, when the amount of fat gained by each animal between operation ('sham' or lipectomy) and sacrifice was estimated by calculating the difference between the amount of fat found at sacrifice and the theoretical amount of body fat present immediately after the

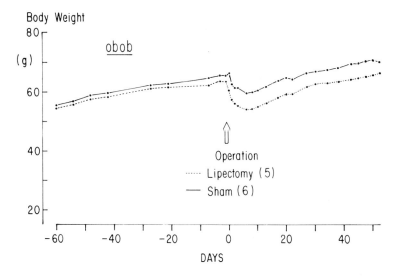

Fig. 3. Body weight (g) of ob/ob mice subjected to partial lipectomy or to 'sham' operation (at day 0). Numbers in brackets indicate number of animals.

TABLE 3   Body composition, serum glucose and insulin (mean ± S.E.M.) at sacrifice of lipectomized and 'sham' operated ob/ob mice

	No. of animals	% body fat	% body water	% body protein	Serum glucose (mg/100 ml)	Serum insulin (μU/ml)
Lipectomized	6	54.2 ± 2.9	29.3 ± 0.8	3.2 ± 0.3	212 ± 8	4426 ± 548
'Sham'	5	51.9 ± 2.0	31.4 ± 1.3	3.5 ± 0.2	197 ± 14	4298 ± 502
P		n.s.	n.s.	n.s.	n.s.	n.s.

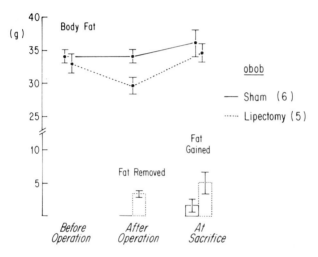

Fig. 4.   Body fat (g) of ob/ob mice subjected to partial lipectomy or to 'sham' operation just prior to operation (calculated), immediately after operation and at sacrifice. At the bottom of the figure, the amount of fat removed from the lipectomized animals and that gained by both groups is indicated. Numbers in brackets indicate number of animals.

operation, animals of the lipectomized group gained 5.0 ± 1.6 g of body fat, whereas 'sham' operated animals gained only 1.5 ± 1.1. The difference beween the amount of fat gained by the lipectomized and the 'sham' operated animals was statistically significant (P < 0.05) (Fig. 4).

Whether this compensatory increase of body fat in the lipectomized animals was achieved by hypertrophy of the remaining adipose tissue or by formation of new adipocytes cannot be determined on the basis of the present experiments. In either case, however, the hypothesis that the fat content per adipocyte is the regulated variable does not seem to be true for the following reasons: if compensation in the lipectomized ob/ob mice occurred by enlargement of the existing adipocytes alone, one must infer that the fat content per cell increased postoperatively, therefore it is not 'constant' and cannot be the regulated variable. Even accepting the possibility of a compensatory increase in the number of adipose cells, so that the fat content per adipocyte remained constant postoperatively, there is still no support for the view that the fat content per adipocyte is the controlled variable, since the lipectomized ob/ob mice, having a 'normal' fat content per adipocyte postoperatively, need not have expanded their adipose tissue. Thus, these data support the hypothesis that the total body fat is subject to homeostatic regulation in ob/ob mice.

REFERENCES

Chlouverakis, C., Dade, E.F. and Batt, R.A.L. (1970): Glucose tolerance and time sequence of adiposity, hyperinsulinemia and hyperglycemia in obese-hyperglycemic mice (ob/ob). Metabolism, 19, 687.

Chlouverakis, C. and White, P.A. (1969): Obesity and insulin resistance in the obese-hyperglycemic mouse (ob/ob). Metabolism, 18, 998.

Christophe, J. (1961): Contribution à la biochimie des obésités expérimentales. Arcia, Brussels.

Coleman, D.L. and Hummel, K.P. (1972): Comparison of the obesity syndromes of obese (ob/ob) and diabetic (db/db) mice. Diabetologia, 8, 49.

Fuller, J.L. and Jacoby Jr., G.A. (1955): Central and sensory control of food intake in genetically obese mice. Amer. J. Physiol., 183, 279.

Mayer, J. (1960): The obese hyperglycemic syndrome of mice as an example of 'metabolic' obesity. Amer. J. clin. Nutr., 8, 712.

Powley, T.L. and Keesey, R.E. (1970): Relationship of body weight to the lateral hypothalamic feeding syndrome. J. comp. physiol. Psychol., 70, 25.

Cellularity of the different cell compartments in white adipose tissue of mice in chronic starvation, refeeding and in two types of obesity

L. Rakow

II Department of Pathology, Centre for Biology and Theoretical Medicine, University of Ulm, Ulm, Federal Republic of Germany

Today it is commonly accepted that, just as there are specific cells in other tissues (liver cell, heart muscle cell), the adipocyte represents the specific cell of adipose tissue. Like all other tissues adipose tissue also contains nonspecific cells, the so-called stromal cells (cells of connective tissue and blood vessels). This fact is of great importance since under pathophysiological conditions such as starvation and obesity these cell compartments will react in different ways (Beneke and Rakow, 1973). Under these conditions the behaviour of adipocytes is of great interest since changes in the mass of adipose tissue can depend on changes in the number of adipocytes as well as in the mass of the single adipocyte.

To investigate which alterations within adipose tissue occur in obesity, two experimental models were employed: aurothioglucose-induced obesity (Brecher and Waxler, 1949) of NMRI/Han. mice and genetically transmitted obesity of the C57BL/6J strain (Ingalls et al., 1950). Aurothioglucose was given as a single dose (intraperitoneal injection) of 800 mg/kg body weight. Six weeks later the experimental animals were 40% heavier than the control animals. The body weight of obese C57BL/6J mice increased within a year up to double the value of that of their lean littermates.

The investigations were performed with epididymal fat pads. The fresh weight of the fat pads in the obese animals was more than 3 times higher than in the lean controls. The question remains as to whether this increase of adipose tissue mass is due to an increase in fat cell number or to an enlargement of the single adipocyte. To estimate the mass of the single adipocyte, the diameter of isolated fat cells was measured and the volume estimated. Since the average specific weight of lipid is 0.9 g/ml the mass of the single adipocyte may be calculated. The average fat cell diameter in the fat pads of obese mice was 1.5 times greater than in lean animals, and therefore the average fat cell mass was over 3 times higher than in the fat pads of lean mice. This fact alone could account for the increase of the fat pad mass.

Under certain conditions the DNA content of a tissue is an index of its cell count. The fat pads of the obese animals contained 3 times as much DNA as the fat pads of lean mice. This is in contrast to the fact that the increase of adipose tissue mass could be explained by an equal increase of the mass of the single adipocytes. An increase of DNA in adipose tissue could be due to four factors: (1) increase of fat cell number, (2) increase of DNA content of the fat cell nuclei, (3) increase of stromal cell number, (4) increase of DNA content of stromal cell nuclei.

The DNA content of the cell nuclei of adipocytes and stromal cells was first investigated. This was accomplished by cytophotometric measurements of cell nuclei after Feulgen staining. The results showed that adipocytes in both lean and obese mice had exclusively diploid nuclei. On the other hand, the stromal cells in obese mice, when compared to those in lean animals, exhibit a higher number of interdiploid-tetraploid nuclei. This indicates an increase in the number of stromal cells in obesity. By a radioautographic method (pulse loading with [^{3}H] TdR) it was found that the adipocytes in both lean and obese mice had exclusively unlabelled nuclei. In contrast the stromal cells of obese mice and, to a lesser extent, those of lean mice had a certain number of labelled nuclei.

In obesity the adipocyte compartment within the adipose tissue is stable with regard to proliferation. The adipocytes only show an increase of their mass; this causes the increase of adipose tissue mass. The stromal cell compartment within the adipose tissue is unstable with regard to proliferation. The stromal cells show an increase in their number, leading to the increased DNA content of adipose tissue. Under certain circumstances (see scheme below), one can estimate the cell counts of adipocytes and stromal cells by computation.

Mode of computation of cell counts of adipocytes and stromal cells in adipose tissue (epididymal fat pads)

I. $\dfrac{\text{Mass of fat pads}}{\text{Mass of single adipocyte}}$ = Number of adipocytes

II. Number of adipocytes $\times 6.10^{-12}$ g DNA = DNA content of adipocytes

III. DNA content of fat pads — DNA content of adipocytes = DNA content of stromal cells

IV. $\dfrac{\text{DNA content of stromal cells}}{6.10^{-12} \text{ g DNA}}$ = Number of stromal cells

Two facts are important: under physiological conditions the cell count of stromal cells in adipose tissue is 3 to 4 times higher than the adipocyte cell count (Table 1); in both lean and obese mice the adipocyte count in adipose tissue always remains the same, whereas the stromal cell count in the adipose tissue of obese mice will be 3 times that in lean mice (Table 1). If the stromal cell count rises in such a manner, the amount of collagen – the specific product of fibroblasts – must be elevated. In obese mice the collagen content of adipose tissue (calculated by estimation of hydroxyproline) shows a real elevation parallel to the stromal cell count (Rakow et al., 1971c).

TABLE 1   Cell counts in epididymal fat pads of lean and obese mice

	Total cell count (10^6)	No. of adipocytes (10^6)	No. of stromal cells (10^6)	% adipocytes
NMRI				
Lean	19.7	3.2	16.5	16.2
Obese	53.4	3.2	50.2	6.0
C57BL/6J				
Lean	20.5	2.5	18.0	12.2
Obese	61.3	2.9	58.4	4.7

In obesity the mass of the adipose tissue increases. The fat cell compartment remains stable and the fat cell count does not change. Only the mass of the single adipocyte shows an increase (Beste, 1961; Sims et al., 1969; Rakow et al., 1971 a, b; Salans et al., 1971). In obesity the compartment of stromal cells within the adipose tissue is unstable and the cell count increases considerably (Rakow et al., 1971 a, b; Lemonnier, 1971).

Starvation provides another important pathophysiological condition suitable for investigation of the reactions of adipose tissue. Starvation is also of interest in the therapeutic treatment of obesity. Both lean and obese mice were starved for a period of 6 weeks followed by a period of refeeding. When lean mice were starved, a 50% decrease in their body weight occurred. During the refeeding phase there was a rapid increase of body weight. After 7 days of refeeding the body weight of the experimental animals nearly reached that of the control animals. After the starvation phase the fresh weight of the epididymal fat pads was only 5% of the value found in control animals. During the refeeding phase the fresh weight showed a rapid increase but the value found in control animals could not be reached.

In starvation the fat cell diameter was reduced to one third and the mass of the single adipocyte to 5% of the control animal value. The reduction of the mass of the single adipocyte paralleled the reduction of the mass of the whole fat organ. During the refeeding phase the fat cell diameter increased, but at the end of this phase values for diameter and mass were lower than the corresponding values from control animals.

After the starvation phase the DNA content of the epididymal fat pads was half that of the controls. During the refeeding phase it regained the control value. After starvation and during the refeeding period the cell counts of adipocytes and stromal cells were determined according to the calculations described in detail for obesity.

The adipocyte cell count does not change in starvation and refeeding (Table 2). Even long term starvation does not cause loss of adipocytes (Napolitano, 1963; Simon and Williamson, 1971; Mohr et al., 1969; Rakow et al., 1970). Reefeeding does not cause an increase in adipocyte number (Table 2). The fat cell compartment again was stable with regard to proliferation (Rakow et al., 1970). The changes in the DNA content (Durand et al., 1969) were due to changes in the stromal cell count (Rakow et al., 1970; Hubbard and Matthew, 1971). Starvation caused the loss of half of the stromal cells (Table 2). In the refeeding phase the stromal cell count returns to the control level (Rakow et al., 1970). Corresponding to these

TABLE 2   Cell counts in epididymal fat pads of lean mice

	Total cell count (10^6)	No. of adipocytes (10^6)	No. of stromal cells (10^6)	% adipocytes
NMRI				
Control	16.2	4.0	12.2	24.7
Group A	11.0	4.1	6.9	37.3
Group B	11.3	3.3	8.0	29.2
Group C	19.0	4.7	14.3	24.7
C57BL/6J				
Control	15.8	3.0	12.8	19.0
Group A	8.9	3.2	5.7	36.0
Group B	12.3	2.8	9.5	22.8
Group C	13.5	3.1	10.4	23.0

Group A = starved animals; Group B = animals starved and subsequently refed for 3 days; Group C = animals starved and subsequently refed for 7 days.

findings the amounts of collagen within the fat pads reflected the number of stromal cells. The synthesis of collagen does not occur so fast as the increase in stromal cell number. This effect is known to occur at every site where connective tissue is newly synthesized (Rakow et al., 1971c).

Obese mice (NMRI/aurothioglucose and C57BL/6J-ob/ob) have also been starved and refed (Rakow et al., 1974a, b). It is important to state that in these cases also the adipocyte count did not change, whereas the adipocyte diameter did. Starvation of obese animals did not lead to a decrease of the fat cell number (Table 3) (Salans et al., 1971; Knittle and Ginsberg-Fellner, 1972; Rakow et al., 1974a, b). The decrease of the DNA content (Durand et al., 1969) was solely due to a decrease of the stromal cell count (Table 3) (Hubbard and Matthew, 1971; Rakow et al., 1974a, b). The collagen content of the adipose tissue behaved similarly to the stromal cell count (Rakow et al., 1974a, b). During the refeeding phase the changes of the cell counts in the different cell compartments were not so distinct as during the refeeding phase of formerly lean animals (Rakow et al., 1974a, b).

Which index should be used for investigations on adipose tissue? The fresh weight is not a reliable index since differences in the nutritional state lead to differences in adipose cell number within an adipose tissue specimen. The total DNA content is also not reliable since under physiological conditions only one fourth of the DNA is from the adipocytes (Rodbell, 1964; Hirsch and Han, 1969; Rakow et al., 1970, 1971a, b; Hubbard and Matthew, 1971). The fat cell number

TABLE 3 Cell counts in epididymal fat pads of obese mice

	Total cell count (10^6)	No. of adipocytes (10^6)	No. of stromal cells (10^6)	% adipocytes
NMRI				
Control	43.6	3.2	40.4	7.3
Group A	22.7	2.8	19.9	12.3
Group B	17.3	2.5	14.8	14.5
Group C	21.6	2.5	19.2	11.5
C57BL/6J				
Control	46.6	2.9	43.7	6.2
Group A	28.2	3.1	25.1	11.0
Group B	35.5	3.1	32.4	8.7
Group C	39.5	3.1	36.4	7.8

For explanation of groups see Table 2.

does not change under pathophysiological conditions, whereas the much greater number of stromal cells shows extensive changes. Thus, the DNA content will be influenced largely by the stromal cells (Rakow et al., 1970, 1971a, b). Under pathophysiological conditions the percentage of organ-specific adipocytes within the adipose tissue varies widely in relation to the wide variations of the stromal cell count. In starvation the percentage amounts to over 35%; the value under physiological conditions varies between 20 and 25%; in severe obesity the value decreases to only 5% (Rakow et al., 1970, 1971a, b; 1974a, b).

A reliable index is the DNA content of the fat cell population (Hubbard and Matthew, 1971). To estimate this DNA content, the adipocytes must be separated from the surrounding connective tissue (Lorch and Rentsch, 1969). However, the most reliable index is the adipocyte number within the adipose tissue specimen (Faulhaber et al., 1969; Lorch and Rentsch, 1969; Hubbard and Matthew, 1971), since the metabolic activity of adipose tissue is determined by the number of organ-specific adipocytes (Rodbell, 1964).

It has been shown that adipose tissue represents an organ containing organ-specific cells (adipocytes). In addition to this cell compartment there are other nonspecific cells of the so-called stromal cell compartment. Under pathophysiological conditions such as starvation or obesity the adipocyte compartment is stable with regard to proliferation. The number of adipocytes therefore does not change; only the mass of the single adipocyte changes. In contrast, the stromal cell compartment is unstable with regard to its proliferation. Under the pathophysiological conditions mentioned above the number of stromal cells within the adipose tissue varies in a rapid and significant manner. This behaviour of the stromal cell compartment is understandable in relation to the structure and function of the adipocytes. Fibroblasts produce collagen fibers which form a network around every adipocyte. If the fat cell diameter changes, the surrounding extracellular tissue must also change. Alterations of the blood capillaries are also understandable since there is a profound structural and functional relationship between the adipocytes and the capillaries.

REFERENCES

Beneke, G. and Rakow, L. (1973): The fat organ. Acta endocr. (Kbh.), Suppl. 173, 175.

Beste, W. (1961): The size of fat cells and their dependence upon the nutritional state. Virchows Arch. path. Anat., 334, 243.

Brecher, G. and Waxler, S.H. (1949): Obesity in albino mice due to single injections of goldthioglucose. Proc. Soc. exp. Biol. (N.Y.), 70, 498.

Durand, G., Penot, E. and Bourgeaux, N. (1969): Evolution du nombre et de la taille des cellules dans les tissus de la rate adulte amaigrie à la suite d'une carence énergétique. Croissance compensatrice. Ann. Biol. anim., 9, 575.

Faulhaber, J.D., Petruzzi, E.N., Eble, H. and Ditschuneit, H. (1969): In-vitro-Untersuchungen über den Fettstoffwechsel isolierter menschlicher Fettzellen in Abhängigkeit von der Zellgrösse: Die durch Adrenalin induzierte Lipolyse. Horm. metab. Res., 1, 80.

Hirsch, J. and Han, P.W. (1969): Cellularity of rat adipose tissue: effects of growth, starvation and obesity. J. Lipid Res., 10, 77.

Hubbard, R.W. and Matthew, W.T. (1971): Growth and lipolysis of rat adipose tissue: effect of age, body weight and food intake. J. Lipid Res., 12, 286.

Ingalls, A.M., Dickie, M.M. and Snell, G.D. (1950): Obese, a new mutation in the house mouse. J. Hered., 41, 317.

Knittle, J.D. and Ginsberg-Fellner, F. (1972): Effect of weight reduction on in vitro adipose tissue lipolysis and cellularity in obese adolescents and adults. Diabetes, 21, 754.

Lemonnier, D. (1971): Sex difference in the number of adipose cells from genetically obese rats. Nature (Lond.), 213, 50.

Lorch, E. and Rentsch, G. (1969): A simple method for staining and counting isolated adipose tissue fat cells. Diabetologia, 5, 356.

Mohr, W., Beneke, G. and Balser, W. (1969): Untersuchungen am Fettgewebe von Ratten nach Mangeler-nährung und kurzzeitiger Wiederauffütterung. Virchows Arch. Abt. B, 3, 77.

Napolitano, L. (1963): The differentiation of white adipose cells. J. Cell Biol., 18, 663.

Rakow, L., Beneke, G., Mohr, W. and Brauchle, I. (1970): Veränderungen von Fett- und Bindegewebszell-zahlen des weissen Fettgewebes der Maus im chronischen Hunger und nach Wiederauffütterung. Beitr. path. Anat., 142, 38.

Rakow, L., Beneke, G., Mohr, W. and Brauchle, I. (1971a): Untersuchungen über die Zellvermehrung im weissen Fettgewebe der genetisch adipösen Maus (C57BL/6J ob/ob). Beitr. path. Anat., 143, 300.

Rakow, L., Beneke, G., Mohr, W. and Brauchle, I. (1971b): Untersuchungen über die Zellvermehrung im weissen Fettgewebe der Maus bei durch Aurothioglukose experimentell erzeugter Adipositas. Path. et Microbiol. (Basel), 37, 261.

Rakow, L., Beneke, G. and Vogt, C. (1971c): Veränderungen im Kollagengehalt des weissen Fettgewebes von Mäusen im chronischen Hunger und bei nachfolgender Wiederauffütterung sowie bei zwei Fett-suchtformen. Beitr. path. Anat., 144, 377.

Rakow, L., Beneke, G. and Vogt, C. (1974a): Vergleichende chemische Untersuchungen am weissen Fettge-webe von normalgewichtigen und fettsüchtigen C57BL/6J Mäusen im chronischen Hunger und nach Wiederauffütterung. In preparation.

Rakow, L., Beneke, G. and Vogt, C. (1974b): Änderungen der Fett- und Bindegewebszellzahlen des weissen Fettgewebes von normalgewichtigen und fettsüchtigen C57BL/6J Mäusen im chronischen Hunger und nach Wiederauffütterung. In preparation.

Rodbell, M. (1964): Localization of lipoprotein lipase in fat cells of rat adipose tissue. J. biol. Chem., 239, 753.

Salans, L.B., Horton, E.S. and Sims, E.A.H. (1971): Experimental obesity in man: Cellular character of the adipose tissue. J. clin. Invest., 50, 1005.

Simon, T.L. and Williamson, J.R. (1971): Veränderung der lipidentspeicherten Fettzellen. Arch. Path., 83, 162.

Sims, E.A.H., Kelleher, P.E., Horton, E.S., Gluck, C.M., Goldman, R.F. and Rowe, D.W. (1969): Experi-mental obesity in man. In: Physiopathology of Adipose Tissue, p. 78. Editor: J. Vague. Excerpta Medica, Amsterdam.

DISCUSSION

Björntorp: I would like to emphasize that there is a methodological difficulty here when you analyze fat cells and stromal cells. These fractions are separated by the difference in density due to different amounts of lipid in the cells. Fat cells without fat will thus be recovered in the stromal cell fraction. This fact limits the conclusions drawn from cells with or without fat as far as I can see. Histologically it is also difficult to separate these cell types.

Rakow: I have shown that the cell count of the adipocyte compartment does not change. Therefore one cannot assume that a certain number of nondetectable fat cell precursors within the stromal cell compart-ment change into lipid-laden adipocytes. It can be assumed that the count of fat cell precursors changed but the cell count of the mature adipocyte compartment did not change.

The origin and metamorphosis of the human fat cell*

W. J. Poznanski, J. Rushton and **I. Waheed**

Department of Metabolism, University of Ottawa Medical School, Ottawa Civic Hospital, Ottawa, Canada

It has been shown in experimental animals and man that the increase in fat cell size and number contributes to the formation of adipose tissue (Reh, 1953; Björntorp and Sjöström, 1971; Hirsch et al., 1966). Further studies in adult animals and man suggest, however, that the number of fat cells is constant, while changes in weight induced by starvation or overfeeding correlate well with changes in adipose cell size (Hollenberg and Vost, 1968; Sims et al., 1968). In rats exogenous factors such as nutrition determine fat cell number only during early life (Knittle and Hirsch, 1968).

In this laboratory we have recently shown that stromal cells obtained from adult fat tissue have, in vitro, the capacity to generate new cells, which eventually evolve into a mature fat cell type. Contrary to previously published studies, this finding suggests that some adults are able to generate new fat cells which could contribute to the increase of adipose tissue (Poznanski and Ng, 1971; Ng et al., 1971).

To test the feasibility of such a mechanism it is necessary to confirm that adult adipose tissue stromal cells are in fact true preadipocytes, well differentiated from other connective tissue cells and containing the apparatus necessary for transformation into mature fat cells. It is also necessary to study the behaviour of the mature fat cell since existing evidence suggests that their number remains constant (Hollenberg and Vost, 1968). This in vitro study compares the growth characteristics, morphology and biology of such 'preadipocytes' with mature fat cells obtained from the same adipose tissue and with nondifferentiated connective tissue cells obtained from adult human dermis.

MATERIALS AND METHODS

Subcutaneous fat tissue samples were obtained from patients undergoing abdominal surgery. Obese and normal patients were between 25 and 45 years of age. Adipose tissue was dispersed with collagenase according to the technique of Rodbell (1964) and the stromal cells obtained by differential centrifugation were prepared for culture according to methods described previously (Poznanski and Ng, 1971; Ng et al., 1971). Mature fat cells were placed on a coverslip, 35 × 10 mm, and covered by another glass coverslip to hold the cells in place. This 'sandwiched' specimen was then transferred into a Leighton tube where it was allowed to dry for 15-30 minutes before the addition of culture medium. Medium 199 with Hepes buffer in Hank's salts and containing 20% fetal calf serum was used at pH 7.3, and changed every second day.

Aliquots were removed from the culture flasks from time to time and the cells were counted to determine the growth curve. Staining was carried out on coverslip cultures using Oil Red O for neutral lipid and counter staining with hematoxylin. DNA content was determined by the method of Switzer and Summer (1971) and glycerol content was determined according to Garland and Randle (1962). Lipids were extracted according to the technique of Dole (1956).

* This study was partially financed by a Medical Research Council of Canada Grant (MA-4517).

RESULTS AND DISCUSSION

The single mature fat cells in culture exhibit a gradual change from the typical spherical form to a fibroblast-like form through a process of lipolysis with eventual loss of most of their lipids. In their final form the cells appear similar to those obtained from adipose tissue stroma (so-called 'preadipocytes'). On day 3 of culture most of the cells appear to have maintained their original mature form and spherical shape (Fig. 1). Several cells exhibit two nuclei and a thin **membrane encloses a central single lipid drop. On day 6 of culture cells exhibit various forms,**

Fig. 1.   Isolated mature fat cells in medium 199 + 20% fetal calf serum on days 3 *(left)* and 6 *(right)* of culture. Oil Red O stain for cellular lipid and hematoxylin for nucleus and cytoplasm. See text. (× 40.) (Reduced for reproduction 15%.)

some of them retaining their spherical shape while others become elongated. A few cells still show the large central lipid droplet but most of the remaining fat is already divided into many small droplets. In those cells where the central lipid droplet appears to have been released a large vacuole remains for some time (Fig. 1). Cells containing two nuclei are either undergoing cell division or are the result of the fusion of two cells. On day 9 many cells are already fibroblast-like in appearance but the cytoplasm is still densely populated with small lipid droplets (Fig. 2).

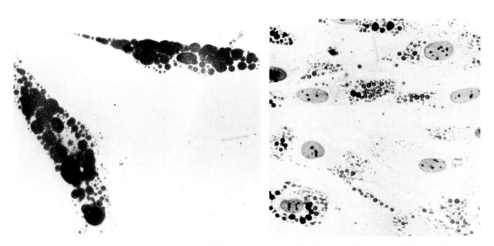

Fig. 2.   Isolated mature fat cells on days 9 *(left)* and 12 *(right)* of culture. Staining as in Figure 1. See text. (× 40.) (Reduced for reproduction 15%.)

The central large vacuole due to the release of the lipid droplet has disappeared. On day 12 the fibroblast-like cells appear to have lost most of the lipid droplets and a monolayer of cells is established (Fig. 2).

The cells can now be trypsinized and recultured on single coverslips. The growth of these cells can be compared with that of cells obtained from adipose tissue stroma as well as with cells obtained from the connective tissue of human dermis.

Fibroblast cells (skin) maintain their spindle-like morphology and fibrous growth pattern **with increased cell to cell contact (Fig. 3), but fat cell precursors tend to grow in a dispersed**

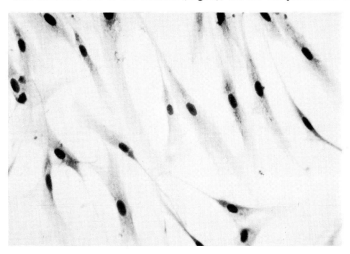

Fig. 3. Adult human skin fibroblasts in medium 199 + 20% fetal calf serum. 12 days after culture. Staining as in Figure 1. (× 25.) (**Reduced for reproduction 20%.**)

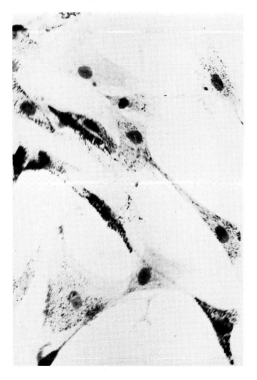

Fig. 4. Primordial fat cells 6 days after culture in medium 199 + 20% fetal calf serum. Staining as in Figure 1. (× 25.) (Reduced for reproduction 20%.)

fashion. Throughout culture, fibroblasts contain only a few fine lipid droplets spread within the cytoplasm, whereas the primordial fat cells (Fig. 4), as well as the fibroblast-like cells originating from mature fat cells (Fig. 5), show a greater number of large lipid droplets. In most of the fat cell precursors, lipids show a definite tendency to coalesce and later during growth displace the nucleus to the periphery of the cell (Fig. 5). No such phenomena were observed in the fibroblasts obtained from the dermis.

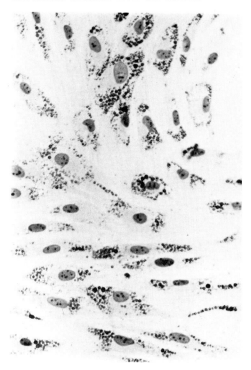

Fig. 5. Primordial fat cells 12 days after culture in medium 199 + 20% fetal calf serum. Staining as in Figure 1. (× 25.) (Reduced for reproduction 20%.)

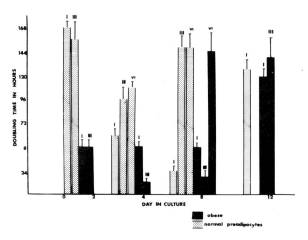

Fig. 6. Time in hours required to double cell number. Roman numerals indicate sequential passages of cells (subcultures).

The rate of cell proliferation during the experimental part of the growth curve appears similar in the two cell types. However, the doubling time for both fibroblasts and preadipocytes varies slightly from one passage of cells to another. The mean doubling time for both fibroblasts and preadipocytes is about 48 hours from the first to the third passage, but becomes longer in the later passages, possibly due to aging of the cells (Fig. 6). Doubling time of the preadipocytes obtained from the obese adipose fat tissue at the third passage is much shorter: 23 hours (Fig. 7).

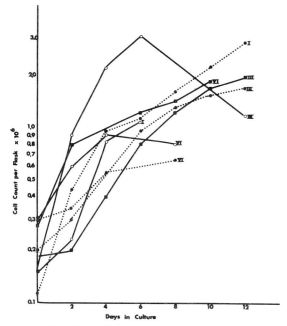

Fig. 7. Growth in different passages of obese and normal preadipocytes and adult skin fibroblasts. Semi-log scale chosen for convenience of presentation. Roman numerals indicate sequential passages of cells (subcultures). —○— = obese preadipocytes, —■— = normal preadipocytes, ···●··· = fibroblasts.

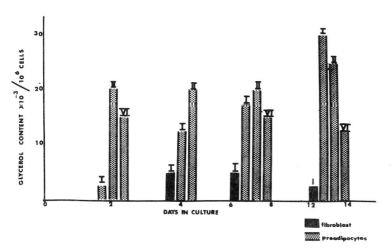

Fig. 8. Glycerol content of cells (μmoles $\times 10^{-3}$ per 10^6 cells). Roman numerals indicate sequential passages of cells (subcultures).

The glycerol content is markedly different in fibroblasts and primordial fat cells (Fig. 8). The primordial fat cells contain 6.5×10^{-3} μmoles/10^6 cells at day 4 with a continuing increase in glycerol up to day 12. The fibroblasts however contain only 2.5×10^{-3} μmoles/10^6 cells at days 8 and 10.

ACKNOWLEDGEMENTS

The authors wish to thank Mrs. N. Poznanski and Mrs. J. Moxley for editing and illustrations and Mrs. P. Cullen for secretarial work.

REFERENCES

Björntorp, P. and Sjöström, L. (1971): Number and size of adipose tissue fat cells in relation to metabolism in obesity. Metabolism, 20, 703.

Dole, P.V. (1956): A relation between non-esterified fatty acids in plasma and the metabolism of glucose. J. clin. Invest., 35, 150.

Garland, P.B. and Randle, P.J. (1962): A rapid enzymatic assay for glycerol. Nature (Lond.), 196, 987.

Hirsch, J., Knittle, L.J. and Salans, L.B. (1966): Cell lipid content and cell number in obese and non-obese human adipose tissue. J. clin. Invest., 45, 1023.

Hollenberg, C.H. and Vost, A. (1968): Regulation of DNA synthesis in fat cells and stromal elements from rat adipose tissue. J. clin. Invest., 47, 2485.

Knittle, J.L. and Hirsch, J. (1968): Effect of early nutrition on the development of rat epididymal fat pads: cellularity and metabolism. J. clin. Invest., 47, 2091.

Ng, C.W., Poznanski, W.J., Borowiecki, M. and Reimer, G. (1971): Studies of growth of adipose cell from normal and obese patients in vitro. Nature (Lond.), 231, 5303.

Poznanski, W.J. and Ng, C.W. (1971): Studies of growth of adipose cells from normal and obese patients in vitro. Ann. roy. Coll. Phys. Surg. Canad., 4, 32.

Reh, H. (1953): Die Fettzellgrobe beim Menschen und ihre Abhangigkeit vom Ernahrungszustand. Virchows Arch. path. Anat., 324, 234.

Rodbell, M. (1964): Metabolism of isolated fat cells: effects of hormones on glucose metabolism and lipolysis. J. biol. Chem., 239, 375.

Sims, E.A.H., Goldman, R.F., Gluck, C.M., Horton, E.S., Kelleher, P.C. and Rowe, D.W. (1968): Experimental obesity in man. Trans. Ass. Amer. Phycns, 81, 153.

Switzer, R.B. and Summer, G.K. (1971): A modified fluorometric method for DNA. Clin. chim. Acta, 32, 203.

Adipose tissue hyperplasia induced by small doses of insulin

L. Kazdová, P. Fábry and **A. Vrána**

Metabolism and Nutrition Research Centre, Institute for Clinical and Experimental Medicine, Prague, Czechoslovakia

Evidence from the literature on the mitogenic action of insulin in many tissues, including cultured brown fat cells (Masters, 1970), led us to examine the influence of this hormone on proliferation and cellularity of adipose tissue in vivo.

Insulin was injected intraperitoneally twice daily to growing male rats (Wistar strain) aged 8 weeks and weighing 180-200 g. Small doses of the hormone (500 μU/rat) were used which acutely stimulate lipid and glycogen synthesis in adipose tissue (Rafaelsen et al., 1965) but do not affect glycemia or food intake and do not evoke metabolic counteraction in the organism (Malaisse and Malaisse-Lagae, 1969; Takebe et al., 1969). Control animals received an equal volume of saline. It was found that repeated administration of insulin for 48-72 hours increased the incorporation of [2-^{14}C] thymidine into DNA and enhanced total DNA and RNA content in the epididymal fat pads (Kazdová and Vrána, 1970a, b). Insulin-stimulated DNA synthesis occurred in the adipocyte fraction as well as in the connective tissue and vascular cells, but was lacking in isolated blood vessels. After isolation of some subcellular fractions by preparatory centrifugation increased DNA synthesis was found in nuclei of adipose tissue only; no increase was detected in mitochondrial DNA (Kazdová et al., 1972).

Histological examination revealed that the above changes were accompanied by mitotic division (Fig. 1) and by increasing ratio of smaller fat cells. Calculation of the number of cells

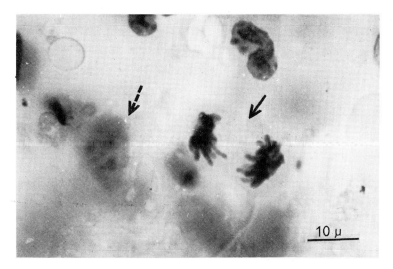

Fig. 1. Light micrograph of a squash preparation of epididymal adipose tissue of rat injected with insulin for 48 hours showing mitotic figure (solid arrow) and nuclei of fat cell (broken arrow). Adipose tissue was fixed in a mixture of ethanol and acetic acid (3:1) and stained with aceto-orcein.

in the epididymal fat pads revealed that after insulin administration the number of adipocytes as well as of stromovascular cells increased.

The results indicate that small doses of insulin injected intraperitoneally may serve as an effective stimulus for increasing the proliferation and cellularity of epididymal adipose tissue.

REFERENCES

Kazdová, L. and Vrána, A. (1970a): Insulin and adipose tissue cellularity. Horm. metab. Res., 2, 117.

Kazdová, L. and Vrána, A. (1970b): Changes in adipose tissue composition induced by insulin in rats. Physiol. bohemoslov., 19, 326.

Kazdová, L., Vrána, A. and Fábry, P. (1972): Localization of insulin-stimulated DNA synthesis in rat adipose tissue. Physiol. bohemoslov., 21, 404.

Malaisse, W. and Malaisse-Lagae, F. (1969): Chronic effect of insulin and glucagon upon islet function. Diabetologia, 5, 349.

Masters, E.M. (1970): Mitotic activity in cultured brown fat cells. Exp. Cell Res., 59, 334.

Rafaelsen, O.J., Lauris, V. and Renold, A.E. (1965): Localized intraperitoneal action of insulin on rat diaphragm and epididymal adipose tissue in vivo. Diabetes, 14, 19.

Takebe, K., Kunita, H., Sawano, S., Horuiuchi, Y. and Mashimo, K. (1969): Circadian rhythms of plasma growth hormone and cortisol after insulin. J. clin. Endocr., 12, 1630.

Body weight increase, food intake, glucose tolerance and insulin sensitivity in lipectomized rats*

G. Enzi[1], C. Tremolada[2] and A. Baritussio[1]

[1]*Clinica Medica Generale and* [2]*Semeiotica Chirurgica, University of Padua, Padua, Italy*

Obesity is frequently associated with reduced glucose tolerance and high insulin levels both in basal conditions and during glucose loading and with high free fatty acid (FFA) levels in plasma (Dole, 1956; Gordon, 1960; Opie and Walfish, 1963; Issekutz et al., 1967; Nestel and Whyte, 1968; Björntorp et al., 1969a, b). In particular, fasting insulin values correlate with body weight and better still with total body fat and fat cell size, while plasma glycerol and FFA turnover correlate with the number of adipocytes (Björntorp and Sjöström, 1971; Felig et al., 1969; Salans et al., 1970).

If a marked reduction of body fat mass is obtained by means of diet or fasting, insulin values decrease proportionally almost to normal (Salans et al., 1968; Kalkhoff et al., 1971; Stern et al., 1972). The relation between the increased adipose mass, the reduced glucose tolerance and the high insulin levels is not yet known. This could be explained by three mechanisms.

1. The increased adipose mass could be secondary to a reduced capacity to utilize carbohydrates in other tissues and to a related hyperinsulinism (Rabinowitz and Zierler, 1962).

2. The increased adipose mass could be responsible for insulin resistance in other tissues. This could be caused by increased FFA levels, often found in obese patients (Randle et al., 1963).

3. Finally, the adipose tissue could be the site of the reduced utilization of carbohydrates and of the insulin resistance (Salans et al., 1968; Zinder et al., 1967). When the glucose uptake from adipose tissue represents a significant part of the global glucose uptake, variations in adipose tissue mass could influence insulin production.

In order to study the last two points, glucose tolerance and insulin sensitivity tests have been examined in lipectomized and control rats in which an increase in body weight was obtained by restriction of movement.

MATERIAL AND METHODS

Albino strain Wistar rats weighing 360-400 g underwent subcutaneous abdominal and epididymal adipose tissue ablation. The amount of tissue removed ranged from 7 to 11 g, equal to about 20-30% of the total body fat.

The rats were placed in small single cages and weekly measurements of body weight and food intake were made. The control rats were kept in the same conditions. Cell size determination was made on frozen-cut adipose tissue taken from the basal part of an epididymal fat pad and from the abdominal subcutaneous adipose tissue, following the method of Sjöström et al. (1971). At the fifth week, a glucose loading (1 g) and an insulin sensitivity test were carried out. Blood samples were taken from the femoral artery from overnight-fasted, anesthetized animals at 0, 30, 60 and 90 minutes for blood glucose, plasma FFA and plasma insulin (IRI) (Hales and Randle, 1963) determinations.

RESULTS

Rats kept in small single cages showed a rapid increase in body weight. At the fifth week the

* Supported by the National Research Council of Italy (Grant No. 7100804.04).

TABLE 1   Mean body weight variations in lipectomized and control rats

	Basal	1st week	2nd week	3rd week	4th week	5th week
Controls (No. = 10)	388 ± 16	419 ± 16	438 ± 16	446 ± 17	452 ± 16	456 ± 15
Lipectomized (No. = 10)	377 ± 21	399 ± 22	406 ± 20	410 ± 21	414 ± 21	418 ± 18

TABLE 2   Mean absolute weight increase in lipectomized and control rats

	1st week	2nd week	3rd week	4th week	5th week
Controls (No. = 10)	33.5 ± 4.3	52.4 ± 5.2	60.5 ± 6.1	66.1 ± 5.8	70.5 ± 5.3
Lipectomized (No. = 10)	21.7 ± 4.0	29.2 ± 4.4	33.4 ± 4.5	37.5 ± 5.3	41.0 ± 6.7
P	n.s.	< 0.005	< 0.005	< 0.005	< 0.005

TABLE 3   Weight of the viscera in lipectomized and control rats

	Controls		P	Lipectomized	P
	Before restriction of movement	At sacrifice			
Liver	13.20 ± 0.8	13.80 ± 1.0	n.s.	12.65 ± 0.6	n.s.
Kidney	1.35 ± 0.07	1.37 ± 0.08	n.s.	1.32 ± 0.05	n.s.
Heart	1.23 ± 0.06	1.21 ± 0.05	n.s.	1.24 ± 0.06	n.s.

TABLE 4   Food intake (g) during the period of restricted movement

	1st week	2nd week	3rd week	4th week	5th week	Total
Controls	236.9 ± 5.8	223.8 ± 6.3	232.2 ± 6.1	222.8 ± 8.3	206.6 ± 7.4	1122.5 ± 28
Lipectomized	224.1 ± 9.4	216.0 ± 4.1	221.4 ± 6.5	213.1 ± 7.3	198.4 ± 6.3	1063.4 ± 23
P	n.s.	n.s.	n.s.	n.s.	n.s.	n.s.

Food intake compared between 1st and 5th week: Controls $t = 3.1907$, $P < 0.01$; Lipectomized $t = 2.2711$, $P < 0.05$.

controls showed a body weight increase of 70.5 ± 5.3 g, significantly higher than that observed in the lipectomized rats (41.0 ± 6.7 g; $P < 0.005$) (Tables 1 and 2). The weight of the viscera in a group of rats sacrificed at the beginning of the experiment did not show any significant difference from the viscera weight in control and lipectomized animals (Table 3). This result enabled us to confirm that the increase of body weight depended almost entirely on an increase of body fat.

There was no significant difference between control and lipectomized rats regarding food intake (Table 4). The total intake in 5 weeks was lower in the lipectomized animals, but the difference was not statistically significant. However, there was a significant difference in food intake between the fifth and the first week in both groups.

TABLE 5 Mean cell diameter (μ) of subcutaneous and epididymal adipose tissue in control and lipectomized rats

	Lipectomized		P	Controls	P
	Before restriction of movement	At sacrifice		At sacrifice	
Epididymal (No. = 10)	89.1 ± 1.7	96.0 ± 1.5	< 0.01	98.4 ± 3.2	n.s.
Subcutaneous (No. = 10)	73.8 ± 2.9	98.2 ± 2.9	< 0.001	95.7 ± 3.8	n.s.

At the fifth week the size of subcutaneous and epididymal adipose tissue cells was significantly higher in comparison to the initial values (Table 5). This increase was more evident in subcutaneous adipose tissue, which would therefore seem to play an important role in body weight increase. The cell diameter of adipocytes both from subcutaneous and epididymal fat showed no significant differences in the control and lipectomized animals.

Basal values of blood glucose and plasma insulin, FFA and triglycerides are shown in Table 6. There were no significant differences in blood glucose and plasma insulin between the control and lipectomized groups. However, significantly lower levels of plasma FFA and triglycerides were found in the lipectomized group.

TABLE 6 Basal values of blood glucose, plasma insulin (IRI), FFA and triglycerides in lipectomized and control rats

	Glucose (mg/100 ml) (No. = 13)	Insulin (μU/ml) (No. = 8)	FFA (μmoles/l) (No. = 12)	Triglycerides (mg/100 ml) (No. = 15)
Controls	96.1 ± 7.3	18.3 ± 1.9	470.8 ± 49	90.8 ± 8.2
Lipectomized	93.5 ± 7.5	15.2 ± 1.6	345.7 ± 28	66.9 ± 6.6
P	n.s.	n.s.	< 0.05	<0.05

As can be seen in Table 7, the behaviour of blood glucose and plasma insulin was similar in both groups.

The hypoglycemic effect of a standard dose of insulin (1 U, i.m.) was similar in both groups. FFA values were consistently lower in the lipectomized group, although there were no significant statistical differences (Table 8).

DISCUSSION

The total body fat is determined by two factors, the number of the adipose cells and their size (Reh, 1953). An increase in body fat could depend both on hyperplasia and hypertrophy of adipose cells (Björntorp et al., 1966; Hirsh et al., 1966; Sailer et al., 1969). In most cases these factors were combined. In lipectomized rats the significant difference of adipose mass is entirely due to a reduction of the body cell mass.

The evaluation of body weight increase, enlargement of the adipose cell and food intake lead to the following conclusions.

Even if the caloric intake was not significantly different in both groups, the residual adipose tissue showed no compensatory hypertrophy in lipectomized rats. The difference in body weight increase in adult animals is then conditioned only by the cellularity of adipose tissue.

TABLE 7 Blood glucose and plasma insulin (IRI) values during oral glucose tolerance test in lipectomized and control rats (mean ± S.E.M.)

	Basal		30 minutes		60 minutes		90 minutes	
	Glucose (mg/100 ml)	IRI (μU/ml)	Glucose (mg/100 ml)	IRI (μU/ml)	Glucose (mg/100 ml)	IRI (μU/ml)	Glucose (mg/100 ml)	IRI (μU/ml)
Lipectomized (No. = 7)	96 ± 17	15.2 ± 1.6	124 ± 20	25.3 ± 3.0	138 ± 17	23.3 ± 5.0	173 ± 29	19.8 ± 2.0
Controls (No. = 7)	96 ± 16	18.3 ± 2.0	119 ± 22	29.4 ± 4.0	137 ± 20	28.7 ± 5.0	159 ± 26	27.1 ± 5.7
P		n.s.		n.s.		n.s.		n.s.
Controls, saline loading (No. = 5)	112 ± 11	15.0 ± 2.0	104 ± 10	13.5 ± 1.6	97 ± 12	13.9 ± 2.4	122 ± 22	15.5 ± 2.0

TABLE 8 Blood glucose and plasma FFA values during insulin sensitivity test in lipectomized and control rats

	Basal		30 minutes		60 minutes		90 minutes	
	Glucose (mg/100 ml)	FFA (μmoles/l)	Glucose (mg/100 ml)	FFA (μmoles/l)	Glucose (mg/100 ml)	FFA (μmoles/l)	Glucose (mg/100 ml)	FFA (μmoles/l)
Lipectomized (No. = 5)	83.4 ± 3.5	381 ± 61	33.6 ± 2.1	154 ± 51	22.0 ± 3.0	208 ± 71	19.6 ± 3.6	177 ± 60
Controls (No. = 5)	93.4 ± 1.0	462 ± 108	37.6 ± 2	220 ± 52	25.2 ± 0.7	262 ± 17	20.4 ± 1.8	314 ± 89
P	n.s.		n.s.		n.s.		n.s.	

Adipose tissue mass did not affect the caloric intake. The significant difference in food consumption between the first and the fifth week seems to indicate that the caloric intake is related to the growth rate.

The behaviour of blood glucose and plasma insulin during the oral glucose tolerance and insulin tests was similar in both groups. This could indicate that in normal animals fat mass does not significantly affect glucose homeostasis and insulin production during oral glucose tolerance test-induced hyperglycemia. The significantly lower basal FFA values in the lipectomized group could be related to reduced turnover of this substrate. The same mechanism could explain the significantly lower triglyceride levels in this group.

The results are in good agreement with in vivo and in vitro experiments, showing that lipid mobilization correlates with the number of adipose tissue cells, not with their size.

REFERENCES

Björntorp, P., Bergman, H. and Varnauskas, E. (1969b): Acta med. scand., 185, 351

Björntorp, P., Bergman, H., Varnauskas, E. and Lindholm, B. (1969a): Metabolism, 18, 840.

Björntorp, P., Hood, B. and Martinsson, A. (1966): Acta med. scand., 180, 117.

Björntorp, P. and Sjöström, L. (1971): Metabolism, 20, 703.

Dole, V. P. (1956): J. clin. Invest., 35, 150.

Felig, P., Marliss, E., Cahill Jr., G. F. (1969): New Engl. J. Med., 281, 811.

Gordon, E.S. (1960): Amer. J. clin. Nutr., 8, 740.

Hales, C.N. and Randle, P.J. (1963): Biochem. J., 88, 137.

Hirsh, J. and Knittle, J.L. (1970): Fed. Proc., 29, 1516.

Hirsh, J., Knittle, J.L. and Salans, L.B. (1966): J. clin. Invest., 45, 1023.

Issekutz Jr, B., Bortz, W.M., Miller, H.I. and Wroldsen, A. (1967): Metabolism, 16, 492.

Kalkhoff, R.K., Kim, H.J., Cerletty, J. and Ferrou, M.D. (1971): Diabetes, 20, 83.

Nestel, P.J. and Whyte, H.M. (1968): Metabolism, 17, 1122.

Opie, L.H. and Walfish, P.G. (1963): New Engl. J. Med., 268, 757.

Rabinowitz, D. and Zierler, K.L. (1962): J. clin. Invest., 41, 2173.

Randle, P.J., Garland, P.B., Hales, C.N. and Newsholme, E.A. (1963): Lancet, 1, 785.

Reh, M.I. (1953): Virchows Arch., 324, 234.

Sailer, S., Sandhofer, F. and Lish, M.J. (1969): Wien. Z. inn. Med., 50, 374.

Salans, L.B., Horton, E. and Sims, E. (1970): Clin. Res., 18, 463.

Salans, L.B., Knittle, J.L. and Hirsh, J. (1968): J. clin. Invest., 47, 153.

Stern, J.S., Batchelor, B.R., Hollander, N. and Hirsch, J. (1972): Lancet, 2, 945.

Sjöström, L., Björntorp, P. and Vrana, J. (1971): J. Lipid Res., 12, 521.

Zinder, O., Arad, R. and Shapiro, B. (1967): Israel J. med. Sci., 3, 787.

Nutritional, genetic and hormonal aspects of adipose tissue cellularity*

D. Lemonnier and **A. Alexiu**

Laboratoire de Nutrition Humaine, Institut Scientifique et Technique de l'Alimentation, CNAM, Paris, France

Little is known about the genetic, nutritional and hormonal control of adipose tissue cellularity in normal or obese human subjects. Adipose tissue obtained by needle biopsy (Hirsch et al., 1960) allows a good measure of fat cell size to be made from a determined subcutaneous site, but great site-to-site variability in the cell size of the subcutaneous adipose tissue occurs in normal and obese subjects. The relative contribution and cellularity of the main fat depots of the body to the total adipose tissue mass have not yet been defined. Thus, the calculation of the total number of fat cells from subcutaneous sites in human subjects remains a poor estimation. Conclusions about cell number must be made with caution (Bjurulf, 1959; Salans et al., 1971). It was thought useful to study the normal and pathological development of adipose tissue in animals.

Our procedures for the dissection of tissues and determination of adipose tissue cellularity have been previously described (Lemonnier, 1972).

ADIPOSE TISSUE GROWTH IN NORMAL ANIMALS

It is well established that the postnatal growth of rat or mouse adipose tissue results in an increase in both the size and number of the fat cells during the first 4 months of life. Later on, it has been claimed, the number is fixed and the tissue growth is due to adipocyte hypertrophy (Hellman et al., 1963; Hirsch and Han, 1969; Johnson et al., 1971). Most of these studies were carried out on the epididymal fat pads up to the age of 6 months. However, an increase in cell number has been described up to the 26th week in rat subcutaneous adipose tissue (Johnson et al., 1971).

Our investigations were extended to other adipose tissue sites and for a longer period. In the epididymal adipose tissue of normal rats fed on a control diet (FaT groups), adipocyte number did not change from the age of 2.5 to 12 months, and cell size increased (Figs. 5 and 6). Similar results were obtained at the same site in control female Swiss mice (Fig. 2). However, in the parametrial adipose tissue of control female rats, fat cell number showed a 2-fold increase between the ages of 2.5 and 12 months (Figs. 5 and 7).

As will be discussed below, adipocyte hyperplasia is frequently observed in the perirenal adipose tissue of the obese rat or mouse. Thus, to elucidate the possibility of fat cell multiplication in normal adult animals, we studied adipose tissue development at the perirenal site in normal male and female rats up to the age of 18 months. These animals were the lean littermates of genetically obese fa/fa rats. They were chosen because of the homogeneity of their adipose tissue and body weights. In our breeding stock none of the rats of this strain showed any tendency to spontaneous adiposity over a 2-year period. As shown in Figure 1, the increase in perirenal fat weight in the oldest male and female rats was due to an increase in cell number rather than in cell size. It seems that when a maximum in cell size is obtained, for a particular

* This work was supported by an INSERM research grant (Contrat libre No. 71.5.219.12).

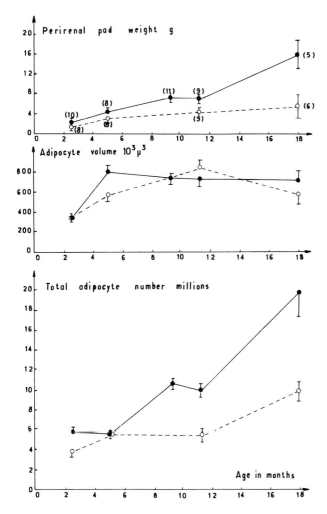

Fig. 1. Perirenal adipose tissue cellularity of control male (solid lines) and female (broken lines) rats fed a control diet at different ages. Vertical lines indicate mean ± S.E.M.

diet, the perirenal adipose tissue grows by fat cell hyperplasia. Our data strongly suggest that fat cell number is not fixed in every major adipose tissue depot in normal animals. They indicate that in aged animals fat cell hyperplasia might be the main or the exclusive method of adipose tissue growth in the perirenal site.

Differences between one adipose site and another were observed in the fat cell size of normal animals. The three sites studied in male mice showed significant site-to-site differences in fat cell size (Fig. 3). Such differences were also observed in the rat (Figs. 6 and 7). It is concluded that cell size in one fat depot is not representative of others.

CELLULARITY IN EXPERIMENTAL OBESITIES

Most of the studies concerning adipose tissue cellularity in obese rats or mice were carried out on the epididymal fat pad. This fat does not present any detectable modification in the number of fat cells in genetic (Hellman et al., 1962b; Lemonnier et al., 1971; Hellman et al., 1963; Lemonnier, 1971; Johnson et al., 1971; Johnson and Hirsch, 1972), hypothalamic (Hellman et al., 1962a: Hirsch and Han, 1969) or nutritional (Lemonnier and Alexiu, 1970; Lemonnier, 1972) obesity in rats and mice. Conversely, female animals made obese by a high-fat diet (Lemonnier, 1970) or genetically obese (Lemonnier, 1971) show fat cell hyperplasia of the genital fat pads.

Experiments were performed to determine:

a. the effect of a high-fat diet on cellularity during the first weeks of life,
b. whether the fat cell hyperplasia seen in the obese female is a site or a sex effect,
c. the effect of different kinds of edible fats on fat cell size and number,
d. the effect of the dietary level of B vitamins on adipose tissue cellularity,
e. the effect of under- and over-feeding in the neonate on adipose tissue cellularity in adults,
f. the adipose tissue cellularity in genetically obese animals,
g. the effect of a high-fat diet in genetically obese rats.

Obesity induced by high-fat diets

Effect of age, sex and site on the cellularity of adipose tissue in mice and rats rendered obese by a high-fat diet. Two diets were used: the control diet and the high-lard diet, containing 9 and 72% lipid as calories respectively. Both diets contained 22% protein as calories and were given ad lib (Lemonnier, 1972). The excess of body fat induced by the high-fat diet accounts for the differences in body weight (Lemonnier, 1969). The two diets were introduced to two groups of female Swiss mice during gestation and lactation, and the young were given the same diets as their mothers at weaning and throughout life.

As soon as they were 2 weeks old, the two groups of suckling mice differed in their body weight and in their embryonic fat pad weight (Table 1). The respective contributions of the undifferentiated and differentiated zones to pad weight were different in the two groups. The zone differentiated into fat cells was increased 6-fold in the young whose mothers ate the high-fat diet but their undifferentiated zone showed little change. In this group the considerable enlargement of the fat cells accounted for almost all the increased pad weight.

TABLE 1   Weight and cellularity of embryonic parametrial fat pads of 2-week-old mice whose mothers were fed a control or a high-fat diet (mean ± S.E.M.; 5 mice in each group)

Diet of the mothers	Body weight (g)	Embryonic fat pads weight (mg)	Undifferentiated volume (mm³)	Differentiated volume (mm³)	Adipose cell volume (10³μ³)	Number of adipocytes (10³)
Control	7.64 ± 0.220	23.6 ± 4.13	13.8 ± 0.722	6.95 ± 3.28	25.4 ± 6.10	259 ± 62.1
High-fat	9.16 ± 0.426**	51.5 ± 7.96**	18.6 ± 1.36*	32.7 ± 9.05**	144 ± 37.0**	257 ± 27.7 (NS)

Significant differences: * P < 0.05. ** P < 0.01, NS P > 0.05.

At this age the young do not take anything other than milk. Those whose mothers received the high-fat diet had doubly enlarged fat pads. There is a possibility that the observed changes in adipose tissue may reflect increased caloric intake of the mothers and/or changes in milk production and composition induced by the high-fat diet. It is significant that this excess weight is the effect of adipose cell enlargement and not of hyperplasia. This nutritional obesity, which appears very early, is not produced by fat cell hyperplasia in young animals since one cannot find any difference in the number of fat cells in the two groups until the age of 4 months.

Up to the age of 18 weeks hypertrophy of the adipose cells was only found in the high-fat diet group when a comparison was made with the control group of the same age (Fig. 2). Later on the number of fat cells continued to rise throughout the period of observation in the high-fat diet group. Therefore, in the obese group about 60% of the weight of the pads was due to new fat cells.

A similar figure was observed in the perirenal fat of the genetically obese Zucker (fa/fa) rat, in which hyperplasia appears after the 14th week (Johnson et al., 1971). In both cases there was a preceding fat cell enlargement followed by an increase in number, suggesting the formation of new fat cells in adults. However, the observed increase in fat cell number, induced genetically or by diet, does not exclude the possibility that preformed cells were laid down in early life. The idea of the existence in early-onset obesity of these preformed fat cells (Johnson et al., 1971; Hirsch and Han, 1969) has led to the categorization of human obesity according to the cellularity of the adipose tissue, with early-onset obesity characterized by hyperplasia and adult-onset obesity by hypertrophy (Hirsch and Knittle, 1970; Salans et al., 1971).

The current studies with mice cannot determine whether new fat cells arise or whether preexisting cells deposit lipid in adult obese mice. This hyperplasia in adults may have been induced by an increase in caloric intake during gestation and lactation. Thus, subsequent hypercellularity could have resulted from the filling up of fat cells formed at this time.

In order to discover whether adult animals are able to produce fat cell hyperplasia by eating a high-fat diet at that stage of life, the high-fat diet was introduced to adult animals. Lean control Zucker female and male rats were chosen for the reasons stated above. The diets were introduced at 5 months and the adipose tissue cellularity was measured 7 months later. As shown in Figures 6 and 7, the high-fat diet produced a slight increase in body weight but fat depots in the high-fat group were doubled in weight in the three sites studied: perigenital, perirenal and abdominal subcutaneous adipose tissue. In all cases there was an increase in cell size, but in the perirenal site the cellularity was increased by 50% in both sexes. These findings demonstrate that adult cell number can be affected by dietary manipulations. In male control rats on the control diet the cell number doubled (Fig. 1) during the time of administration of the high-fat diet to the parallel group. Thus, the further increase in cell number in the high-fat diet group seems to be an acceleration of the normal process observed at this site. In the corresponding control group of female rats the cell number was unchanged from 5 to 12 months of age. The formation of new fat cells in the perirenal site (Fig. 7) appeared to be triggered by feeding the high-fat diet. These results clearly suggest that a true hyperplasia was observed in animals rendered obese by the diet. However, radioactive methods are required to settle this important point. Indeed, the increase in cell number, as observed in rats, might be due to the development of preformed fat cells. This hypothesis would mean that preexisting fat cells exist not only in early-onset obesity but also in control animals, and that particular nutritional conditions, such as a high level of dietary fat, are able to trigger deposition of lipids in these preformed cells.

A marked increase in caloric intake and a marked adipocyte hyperplasia occurred in adult obese Swiss mice (Table 3). Therefore, one may suggest that the observed hyperplasia of adipose cells might be in part a consequence of the recent level of food intake induced by the early feeding pattern. However, hyperphagia does not necessarily induce fat cell hyperplasia. As shown by Hirsch and Han (1969), in hypothalamic obesity in rats there is no fat cell hyperplasia

TABLE 2   Caloric intake (mean ± S.E.M.) in 10-month-old mice fed a control diet ad lib or rendered obese by a high-fat diet

Diet	No. of mice	Body weight (g)	Caloric intake (cal/mouse/24 hr)
Control	6	43	14.22 ± 0.297
High-fat	6	60*	16.86 ± 0.942*
High-fat	2**	98*	23.35 ± 0.795*

* Significant (P < 0.01) with the control group of mice.
** The food intake of the 2 most obese mice was measured separately.

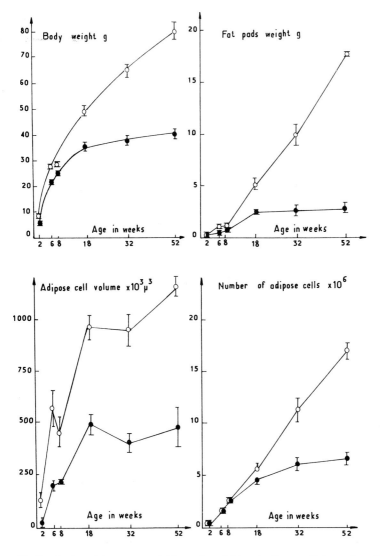

Fig. 2. Body weight, parametrial fat pads weight and adipose cell size and number in female mice fed a control (●) or a high-fat diet (o) at different ages. Vertical lines indicate mean ± S.E.M. (From D. Lemonnier, 1972, *J. clin. Invest.*; courtesy of the editors.)

although these animals are known to show the most striking hyperphagia. In contrast, rats fed a high-fat diet when adults showed a mild excess in their body weight and an apparently normal caloric intake. Caloric intake, measured daily for 11 days when the rats were 6.5 months old, did not show any significant differences between the control and high-fat diet groups (61.0 ± 1.22 vs. 64.3 ± 1.82 cal/24 hours respectively). However, adipose tissue weights were more than doubled and an increase in fat cell number was observed (Figs. 6 and 7). Thus the described hyperplasia was a specific effect of the level of fat in the diet.

Another aspect of this study was the demonstration of sex and site differences in adipocyte cellularity induced by dietary manipulations. In male Swiss mice the high-fat diet produced

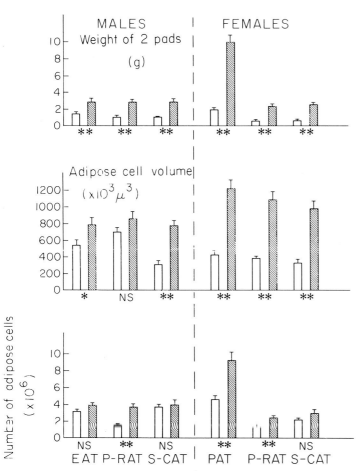

Fig. 3. Cellularity of adipose tissue of 32-week-old male and female mice fed a control (open bars) or a high-fat (shaded bars) diet. The mean body weight ± S.E.M. and the number of animals per group were: control males, 48.2 ± 1.81 (9); high-fat diet males, 67.7 ± 2.44 (10); control females, 34.6 ± 0.70 (9); high-fat diet females 64.4 ± 2.39 (11). EAT = epididymal adipose tissue, PAT = parametrial adipose tissue, P–RAT = perirenal adipose tissue, S–CAT = abdominal subcutaneous adipose tissue. Vertical lines indicate mean + S.E.M. *P < 0.05, **P < 0.01, NS P < 0.05. (From D. Lemonnier, 1972, *J. clin. Invest.*; courtesy of the editors.)

hyperplasia limited to the perirenal site and hypertrophy in the two other sites, so that cell size was the same for the three sites. In the females the high-fat diet resulted in more striking cell hypertrophy and hyperplasia (Fig. 3). This may explain why growth curves plateaued in males but not in females (Lemonnier, 1967). These findings also suggest that in certain specific conditions there might be a maximum cell size. In the control rats fat cell size was increased in the three sites in both sexes (Figs. 6 and 7). As for the rats, perirenal adipose cells were more numerous in the rats fed the high-fat diet but the number was unchanged in the two other sites. So, in each sex, adipose tissue sites in the high-fat diet obese rats and mice reacted to the diet in a site-specific way.

The regulation of the adipose tissue mass is determined by two processes: the formation of new fat cells and the regulation of fat cell size. It has been shown that a single factor, such as the diet fat content, governs both formation and maturation of new fat cells as well as regulation of fat cell lipid content. These two processes may occur at the same site or at different sites (Lemonnier, 1972).

Effect of dietary vitamin B level on cellularity in dietary fat-induced obesity in mice. Experiments have been reported using high-fat diets without any induction of obesity. Instead disturbed growth and low energy efficiency were observed (Tremolieres and Apfelbaum, 1963). Later the animals became obese when these diets were enriched with thiamine, riboflavin, pyridoxine, niacin and calcium pantothenate (Lemonnier, 1967). No effect of other vitamins on the onset of obesity could be seen (Lemonnier et al., 1968). These experiments were in agreement with the reported increased requirement for B-complex vitamins in animals fed a high-fat diet (Friedman and Olszyna-Marzys, 1966). However, fat storage was not evaluated and we have therefore reexamined the effects of various levels of B vitamins in a high-fat diet on the enlargement of the parametrial fat depots and adipocyte size and number in the Swiss mouse (Lemonnier et al., 1974).

Four-week-old female mice were fed ad lib on the control diet or the high-fat diet. In addition, four groups were fed ad lib on four high-fat diets, S/5, S/10, S/20 and S/40, containing all the components of the high-fat diet except that the amounts of thiamine, riboflavin, pyridoxine, calcium pantothenate and niacin were reduced to 1/5, 1/10, 1/20 and 1/40 respectively of the amounts present in the high-fat diet. Adipose tissue cellularity was studied when the mice were 7.5 months old.

The whole body and parametrial fat pad weights are shown in Table 3. The group fed the S/40 diet are not included because they died before the end of the experiment. The dietary B vitamins favoured an increase in adipocyte size and number in the parametrial adipose tissue. Obesity was totally prevented in the S/20 group, which was very similar to the control group as far as the weight of several organs (Lemonnier et al., 1974) and adipose tissue cellularity were concerned.

The amount of B vitamins in the basal control and high-fat diets used was about 5 times above the minimum recommended requirements (National Research Council, 1962). Thus, adipose cell hyperplasia was also demonstrated with low levels of B vitamins, i.e. under the minimum requirements (S/10 diet). The diets tested in this study (except the S/40 diet) did not lead to any symptoms of vitamin deficiency. Moreover, no significant differences in the daily energy intake between the groups were detected (Lemonnier et al., 1974).

These findings may explain why previous long-term feeding experiments with rats fed similar high-fat diets did not produce obese animals (Tremolieres and Apfelbaum, 1963). These diets were probably deficient in B-complex vitamins.

Effect of different edible fats on adipose tissue cellularity. It has been suggested that obesity can be easily induced with high-fat diets provided that the amount of polyunsaturated fatty acid in the diet is low (Dryden et al., 1956; Barboriak et al., 1958). To compare the effect of different edible fats on adipose tissue cellularity we chose corn oil, which is rich in polyunsaturated fatty acids, and hydrogenated palm oil, which contains a high percentage of saturated fatty acids.

Weaning rats were fed one of the following three diets: a standard laboratory diet (U.A.R.) containing about 5% of lipid, or this diet with either 15% (w/w) of hydrogenated palm oil or with the same amount of corn oil added. All rats were fed ad lib and killed by decapitation when 6 months old.

As shown in Table 4 the two high-fat diets did not induce obesity as judged by body weight; however, the weight of epididymal and perirenal adipose tissue was doubled. In the epididymal tissue there was only cell enlargement without significant cell hyperplasia. In the perirenal tissue both mechanisms were involved to produce the excess of pad weight: cell hypertrophy in the corn oil group and cell hyperplasia in the palm oil group. Polyunsaturated fat (sunflower oil), but not lard, has been shown to induce an increase in the DNA content of the epididymal fat pads (Raulin and Launay, 1967). From our data one cannot attribute this excess of DNA to an increase in fat cell number. There was no adipose cell hyperplasia in the corn oil-fed group. It is known that about 60% of the total DNA content of adipose tissue is due to stromal and vascular structures (Rodbell, 1964), and that this non-adipocyte DNA is highly sensitive to nutritional and hormonal status (Hollenberg et al., 1970). Furthermore, adipocytes may be tetraploid or binucleate (Mohr and Beneke, 1968).

Because our high-fat diets were very different in their glyceride:fatty acid ratio, one may assume that the nature of the fatty acids affects adipose tissue cellularity. Data obtained with

TABLE 3   Adipocyte size and number (mean ± S..M.) in the parametrial adipose tissue of mice fed a low-fat diet or one of four high-fat diets with varying levels of B vitamins for a 6.5 month period

	High-fat diets			
	S/20 (No. = 8)	S/10 (No. = 8)	S/5 (No. = 8)	S (No. = 7)
Body weight (g)	38.2 ± 1.20[a]	53.4 ± 2.14[b]	54.8 ± 2.09[b]	61.5 ± 3.33[c]
Parametrial fat pad weight (g)	2.27 ± 0.745[a]	7.10 ± 0.712[b]	7.71 ± 0.569[b]	10.28 ± 0.766[c]
Adipocyte volume ($10^3\mu^3$)	539 ± 97.6[a]	945 ± 46.6[b]	891 ± 78.1[b]	1291 ± 151[c]
Adipocyte number (10^6)	4.01 ± 0.698[a]	8.23 ± 0.838[b]	9.69 ± 0.727[b]	9.07 ± 0.844[b]

	Low-fat diet (No. = 7)
Body weight (g)	35.1 ± 1.41[a]
Parametrial fat pad weight (g)	2.23 ± 0.135[a]
Adipocyte volume ($10^3\mu^3$)	546 ± 51.8[a]
Adipocyte number (10^6)	4.56 ± 0.311[a]

Values indicated by a, b and c are significantly different (P < 0.05).

palm oil are very similar to those previously described in mice made obese by means of a diet including 40% (w/w) of lard. The main fatty acids of lard are palmitic, stearic and oleic acid, while those of palm oil have shorter chains. Thus it does not seem that the length of the fatty acid chain had an effect on the cellularity of adipose tissue. The large quantity of linoleic acid in corn oil may be concerned with adipose cell size. It has been shown that dietary fats rich in polyunsaturated fatty acids induce higher lipoprotein lipase activity (Pawar and Tidwell, 1968) in adipose tissue than diets rich in saturated and monosaturated fatty acids.

Effect of the level of food intake in the neonate on adipose tissue cellularity

Knittle and Hirsch (1968) have shown that rats bred in litters of 4 and 22 differed as to the size and number of epididymal adipose cells when 5 weeks old (2 weeks after weaning), and the differences became more marked with time. This method, when used with rats, produced low body weight in the large litters (Widdowson and McCance, 1960). Furthermore, it is known that animals undernourished in early life show striking reductions in cell number in many kinds of tissues (Winick and Noble, 1966). Thus these differences in fat cell number and size may be the effect of undernutrition.

At the age of 48 ± 12 hours, mice were randomly distributed to foster mothers in litters of 4, 9 or 20. After weaning at 28 days, all mice had free access to the same control diet. Although the

TABLE 4 Effect of two high-fat diets enriched with hydrogenated palm oil or with corn oil on adipose cell size and number in epididymal and perirenal adipose tissue

Diet	No. of rats	Body weight (g)	Epididymal adipose tissue			Perirenal adipose tissue		
			Tissue weight (g)	Adipocyte volume ($10^3\mu^3$)	Adipocyte number (10^6)	Tissue weight (g)	Adipocyte volume ($10^3\mu^3$)	Adipocyte number (10^6)
Control (1)	6	452 ± 11.3	7.8 ± 1.28	504 ± 48.7	16.5 ± 1.23	8.9 ± 1.89	837 ± 115	11.3 ± 0.96
Hydrogenated palm oil (15%) (2)	9	456 ± 13.6	14.3 ± 1.38	805 ± 76.9	19.6 ± 1.05	17.8 ± 2.25	878 ± 58.0	22.1 ± 2.85
Corn oil (15%) (3)	10	464 ± 10.0	12.8 ± 0.83	1,056 ± 88.2	13.9 ± 1.13	18.9 ± 1.23	1,355 ± 52.2	15.2 ± 0.78
Comparisons								
1 vs. 2		n.s.	**	*	n.s.	**	n.s.	**
1 vs. 3		n.s.	*	**	n.s.	**	**	n.s.
2 vs. 3		n.s.	n.s.	*	**	n.s.	**	*

*P < 0.05, **P < 0.01, n.s. P < 0.05.

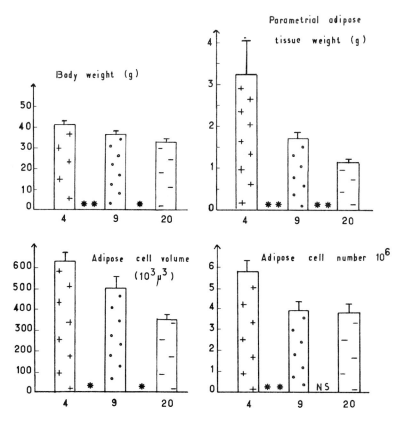

Fig. 4. Effect of early feeding pattern on parametrial adipose tissue cellularity of adult Swiss mice (4.5 months old) fed our control diet ad lib from weaning. Differences are established by comparison of the mice bred in litters of 4 and 20 with those bred in litters of 9. Vertical lines indicate mean + S.E.M. *P <0.05, **P <0.01, NS P> 0.05. (From R. Aubert et al., 1971, *C.R. Acad. Sci. (Paris)*; courtesy of the editors.)

mice were fed ad lib after weaning, differences in body lipids in the groups increased from weaning to the age of 18 weeks (Aubert et al., 1971). The females at 4.5 months showed significant differences in body fat which accounted for almost all the differences in body weight: an increase of 60% in body fat in those bred in litters of 4, and a decrease of 26% in those bred in litters of 20 when compared to mice bred in litters of 9 (Lemonnier et al., 1973). At this age parametrial fat cells were enlarged and more numerous in mice from small litters as compared to those raised in groups of 9. The third group raised in litters of 20 showed only a cell size reduction (Fig. 4).

Increasing litter size did not reduce the fat cell number in these mice. This result may be due to the less striking effect of this experimental device in mice than in rats. This study shows that early nutritional manipulations induce changes in adult adipose cell number, although it does not demonstrate whether these effects occur before weaning, i.e. at the time of the manipulations. However, because differences increased later on, these findings suggest that the changes observed in adults cannot be exclusively attributed to early nutrition but must also be due to later feeding patterns and behaviour, which may be modified by varying litter size (Seite, 1954). Thus it appears that mice which are overfed during the first 3 weeks of life become fat when adults, those underfed remaining thin.

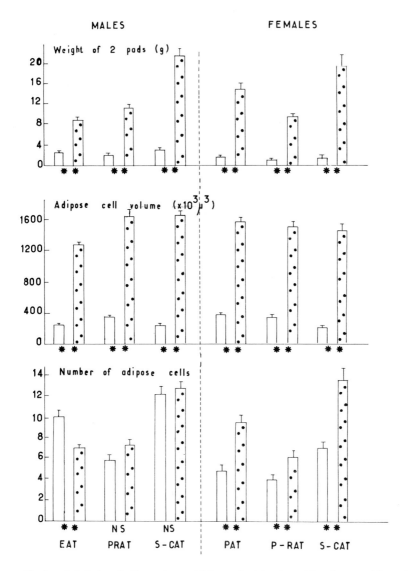

MALES   FEMALES

Fig. 5.   Cellularity of adipose tissue of 2.5-month-old male and female control (open bars) or genetically obese fa/fa rats (dotted bars). EAT = epididymal adipose tissue, PAT = parametrial adipose tissue, P-RAT = perirenal adipose tissue, S-CAT = abdominal subcutaneous adipose tissue. Vertical lines indicate mean ± S.E.M. *P < 0.05, **P < 0.01, NS P > 0.05. The mean body weight ± S.E.M. and the number of animals per group were: control males, 277 ± 10 (10); control females, 171 ± 4.2 (8); obese males, 365 ± 13.1 (8); obese females 311 ± 12.5 (8).

Adipose tissue cellularity in genetically obese rats

Adipose tissue cellularity was studied at three sites in genetically obese fa/fa Zucker rats and their lean control littermates at the ages of 2.5 and 12 months. They were fed ad lib on our control diet. At the age of 2.5 months (Fig. 5) the fat pad weights from the three sites were 3 to 7 times greater in the obese males and 8 to 11 times greater in the females than in lean rats of the same age. In both sexes adipose cell size was increased 4 to 6.5 times. Striking differences were observed in cell number: a 1.6 to 1.9 increase in female rats, a significant decrease in the epididymal tissue and no change at the two other sites in the male obese animals. These results confirm large sex differences at that age in fa/fa rats (Lemonnier, 1971).

At the age of 12 months, all the differences in cell number due to sex and obesity had disappeared in the perigenital sites (Figs. 6 and 7). Cell hyperplasia was only observed in the perirenal and subcutaneous fat pads in the obese animals. Cell size was unchanged in the male group from 2.5 to 12 months and slightly enlarged in the females. Our results are in good agreement with those of Johnson et al. (1971), who studied the cellularity of male fa/fa rats at different periods up to the 26th week. They showed that the fat cell number increases, in the perirenal and subcutaneous adipose tissue, throughout the experiment, suggesting the formation of new fat cells in adults.

Fat cell hyperplasia and hypertrophy was observed in genetically obese animals. In both sexes, each adipose tissue site reacted in a site-specific way. Thus mutant genes and diet produced similar effects on adipose tissue cellularity.

Effect of a high-fat diet in genetically obese rats

Genetically obese fa/fa rats and their lean control littermates were fed a high-lard diet ad lib between the ages of 5 and 12 months. As shown in Figures 6 and 7, the diet had no effect on the perigenital tissue weight in either sex, but a marked increase in weight was observed in the perirenal and subcutaneous fat of the male rats. This increase in weight was due to further fat cell hyperplasia and not to cell enlargement, as enlargement was already maximal in those animals fed the control diet. These results establish that the fat cell number is not genetically fixed in these fa/fa rats.

EFFECT OF CASTRATION AND ADIPOSE TISSUE ABLATION ON CELLULARITY

Whatever the kind of obesity, the epididymal tissue does not show any marked increase in cell number; conversely, hyperplasia is often observed in the parametrial fat pads. Thus it appeared of interest to castrate male animals. Subtotal ablation of the epididymal fat pad was also performed in some animals to study the effects on the cellularity of the perirenal adipose tissue. This site was chosen because it shows cell hyperplasia in obese animals. Six-week-old male Swiss mice were used and divided into four groups which were treated as follows: (1) sham operated, (2) ablation of the two epididymal fat pads, (3) castration, epididymal fat and epididymis not being removed, (4) castration and ablation of total epididymal fat and epididymis. They were fed our control diet ab lib up to the age of 18 weeks.

As shown in Table 5, the weight of the epididymal and perirenal fat was doubled in the castrated mice. This was the effect of a 2-fold increase in cell size. Cell number was unchanged. A similar result was obtained in the perirenal site when the epididymal pads were removed. In the case of castration plus ablation a further cell enlargement of the perirenal adipocytes was shown, corresponding exactly to the sum of the separate effects of castration and ablation. It is known that castration causes adiposity in animals; thus our data demonstrate that the increase in adipose tissue mass is due to fat cell enlargement.

Some authors have studied the effects of hormones on fat cell cellularity. No effect on the cell number could be demonstrated (Hollenberg and Vost, 1968; Vost and Hollenberg, 1970; Salans et al., 1972). In our experiments with ablation of the total epididymal fat pads a compensatory effect in the perirenal site was shown. This is in agreement with the theory of a global regulation of the adipose tissue mass (Liebelt, 1963).

CONCLUSIONS

The postnatal growth of adipose tissue in normal rats or mice results in an increase in both cell size and number during the first 4 months of life. Later on and up to 1 year of age the epididymal fat pads grow by cell enlargement. A contrasting phenomenon is demonstrated in the perirenal

MALES

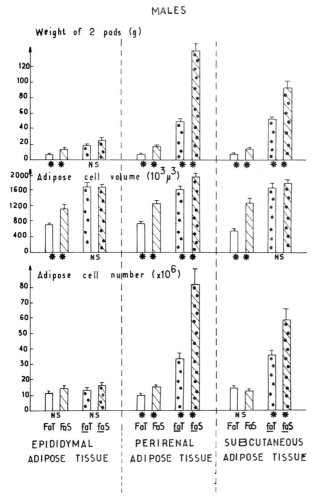

Fig.·6. Cellularity of adipose tissue of 12-month-old male control (open bars) or genetically obese fa/fa (dotted bars) rats fed a control (open bars) or a high-fat diet (shaded bars). The mean body weight + S.E.M. and the number of animals per group were: control rats on control diet (FaT), 532 ± 10.9 (9); control rats on high-fat diet (FaS), 561 ± 17.4 (8); genetically obese rats on control diet (faT), 671 ± 36.4 (6); genetically obese rats on high-fat diet (faS), 885 ± 31.0 (6). See legend to Figure 5.

site: cell size appears to be stable from 5 to 18 months and the observed increase in pad weight is due to an increase in fat cell number. It is concluded that fat cell hyperplasia may be of importance for fat storage in adult or old animals.

The fact that the number of fat cells in the perirenal fat is not fixed may explain the frequently observed adipocyte hyperplasia in that site in most experimental obesities. This is particularly the case in obesity induced by a high-fat diet or in genetically obese fa/fa rats. In both kinds of obesity and in each sex, each site reacts in a site-specific way with cell hypertrophy, hyperplasia or both. An increase in cell number can be obtained by means of a high-fat diet, when given to adults. The effect of the diet depends on the levels of dietary B vitamins and on the kind of fat added to the diet. The fat cell number is not fixed in the fa/fa rat; it increases when a high-fat diet is given. The fat cell hyperplasia induced by a high-lard diet is obtained in rats having a

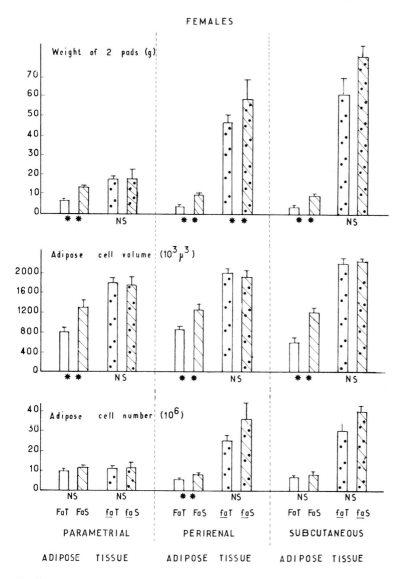

Fig. 7. Cellularity of adipose tissue of 12-month-old female control (open bars) or genetically obese fa/fa (dotted bars) rats fed a control (open bars) or high-fat (shaded bars) diet. The mean body weight +S.E.M. and the number of animals per group were: control rats on control diet (FaT), 286 ± 11.1 (9); control rats on high-fat diet (FaS), 344 ± 11.6 (9); genetically obese rats on control diet (faT), 596 ± 16.9 (7); genetically obese rats on high-fat diet (faS), 689 ± 30.5 (6). See legend to Figure 5.

normal caloric intake and thus appears to be a specific effect of the level of fat in the diet. However, an increase in cell size and number is obtained in adult mice when overfed in early life.

The absence of hyperplasia in the epididymal site is not related to the proximity of the testicle. Its ablation induces fat cell hypertrophy. A similar effect is obtained with ablation of the epididymal fat pads: a compensatory fat cell hypertrophy is observed in the perirenal fat. Age, site, sex, gene and, particularly, diet are important factors controlling the adipose cell number.

TABLE 5   Effect of castration and of epididymal adipose tissue ablation on adipose tissue cellularity in male Swiss mice

	Control	Ablation	Castration	Ablation + castration
Body weight (g)	37.8 ± 1.01[a]	38.7 ± 1.81[a]	39.9 ± 1.73[a]	41.3 ± 2.50[a]
No. of mice	11	10	9	9
Epididymal adipose tissue				
Weight of 2 fat pads (mg)	443 ± 81.6[a]	—	1130 ± 211[b]	—
Adipocyte volume ($10^3\mu^3$)	196 ± 24.9[a]	—	397 ± 61.0[b]	—
Adipocyte number (10^6)	2.37 ± 0.214[a]	—	2.78 ± 0.254[a]	—
Perirenal adipose tissue				
Weight of 2 fat pads (mg)	240 ± 48.1[a]	429 ± 10.9[b]	521 ± 113[b]	929 ± 213[c]
Adipocyte volume ($10^3\mu^3$)	243 ± 50.0[a]	309 ± 69.0[b]	469 ± 89.6[b]	693 ± 91.0[c]
Adipocyte number (10^6)	1.20 ± 0.147[a]	1.52 ± 0.195[a]	1.21 ± 0.086[a]	1.39 ± 0.189[a]

Values indicated by a, b and c are significantly different ($P < 0.05$).
Ablation and castration were performed in 6-week-old mice fed ad lib. Cellularity was studied at the age of 18 weeks.

REFERENCES

Aubert, R., Suquet, J.P. and Lemonnier, D. (1971): Effets à long terme de trois niveaux d'ingesta lactés sur les lipides corporels, la taille et le nombre de cellules adipeuses de la souris. C.R. Acad. Sci. (Paris), 273, 2636.

Barboriak, J.J., Krehl, W.A. and Cowgill, G.R. (1958): Influence of high-fat diets on growth and development of obesity in the albino rat. J. Nutr., 64, 241.

Bjurulf, P. (1959): Atherosclerosis and body-build with special reference to size and number of subcutaneous fat cells. Acta med. scand., 166, Suppl., 349.

Dryden, L.P., Foley, J.B., Gleis, P.F. and Hartman, A.M. (1956): Experiments on the comparative nutritive value of butter and vegetable fats. J. Nutr., 58, 189.

Friedman, L. and Olszyna-Marzys, A.F. (1966): Effect of dietary fat level on the B vitamin requirements of the rat. In: Proceedings, VII International Congress of Nutrition, Vol. I, p. 146. Editor: J. Kühnau. Pergamon Press, London.

Hellman, B., Taljedal, I.B. and Petersson, B. (1962a): Morphological characteristics of the epididymal adipose tissue in mice with obesity induced by gold-thioglucose. Med. exp. (Basel), 6, 402.

Hellman, B., Taljedal, I.B. and Westman, S. (1962b): Morphological characteristics of the epididymal adipose tissue in normal and obese-hyperglycemic mice. Acta morph. neerl.-scand., 5, 182.

Hellman, B., Thelander, L. and Taljedal, I.B. (1963): Post-natal growth of the epididymal adipose tissue in yellow obese mice. Acta anat., (Basel), 55, 286.

Hirsch, J., Farquhar, J.W., Ahrens, E.H., Peterson, M.L. and Stoffel, W. (1960): Studies of adipose tissue in man: a microtechnic for sampling and analysis. Amer. J. clin. Nutr., 8, 499.

Hirsch, J. and Han, P.W. (1969): Cellularity of rat adipose tissue: effects of growth, starvation and obesity. J. Lipid Res., 10, 77.

Hirsch, J. and Knittle, J.L. (1970): Cellularity of obese and nonobese human adipose tissue. Fed. Proc., 29, 1516.

Hollenberg, C.H. and Vost, A. (1968): Regulation of DNA synthesis in fat cells and stomal elements from rat adipose tissue. J. clin. Invest., 47, 2485.

Hollenberg, C.H., Vost, A. and Patten, R.L. (1970): Regulation of adipose mass control of fat cell development and lipid content. Rec. Progr. Hormone Res., 26, 463.

Johnson, P.R. and Hirsch, J. (1972): Cellularity of adipose tissue in six strains of genetically obese mice. J. Lipid Res., 13, 2.

Johnson, P.R., Zucker, L.M., Cruce, J.A.F. and Hirsch, J. (1971): The cellularity of adipose depots in the genetically obese Zucker rats. J. Lipid Res., 12, 706.

Knittle, J.L. and Hirsch, J. (1968): Effect of early nutrition on the development of rat epididymal fat pads: cellularity and metabolism. J. clin. Invest., 47, 2091.

Lemonnier, D. (1967): Obésité par des régimes hyperlipidiques chez le rat et la souris. Nutr. et Dieta (Basel), 9, 27.

Lemonnier, D. (1969): Experimental obesity in the rat by high-fat diets. In: Physiopathology of Adipose Tissue, p. 197. Editor: J. Vague. Excerpta Medica, Amsterdam.

Lemonnier, D. (1970): Augmentation du nombre et de la taille des cellules adipeuses dans l'obésité nutrition-nelle de la souris. Experientia (Basel), 26, 974.

Lemonnier, D. (1971): Sex difference in the number of adipose cells from genetically obese rats. Nature (Lond.), 231, 50.

Lemonnier, D. (1972): Effect of age, sex and site on the cellularity of the adipose tissue in mice and rats rendu obèse par un régime hyperlipidique. Arch. Anat. micr. Morph. exp., 59, 1.

Lemonnier, D. and Alexiu, A. (1970): Cellularité et caractères morphologiques du tissu adipeux du rat rendu obèse par un régime hyperlipidique. Arch. Anat. micr. Morph. exp., 59, 1.

Lemonnier, D., De Gasquet, P., Griglio, S., Naon, R., Reynouard, F. and Trémolières, J. (1974): Effect of dietary B vitamins level on fat storage, adipose tissue cellularity and energy expenditure. Nutr. Metabol., 16, 15.

Lemonnier, D., Manchon, P., Gradnauer-Griglio, S. and Sannier, J. (1968): Rôle de la composition de différents régimes hyperlipidiques donnés ad libitum et de la souche dans l'obésité alimentaire du rat. Ann. Nutr. Alim. 22, 107.

Lemonnier, D., Suquet, J-P., Aubert, R. and Rosselin, G. (1973): Long term effect of mouse neonate food intake on adult body composition, insulin and glucose serum levels. Hormone metab. Res., 5, 223.

Lemonnier, D., Winand, J., Furnelle, J. and Christophe, J. (1971): Effect of high-fat diet on obese-hypergly-cemic and non-obese Bar-Harbor mice. Diabetologia, 7, 328.

Liebelt, R.A. (1963): Response of adipose tissue in experimental obesity as influenced by genetic, hormonal and neurogenic factors. Ann. N.Y. Acad. Sci., 110, 723.

Mohr, W. and Beneke, B. (1968): Age dependance of nuclear DNA content of rat adipose tissue cells. Experientia (Basel), 24, 1052.

National Research Council (1962): Nutrient Requirements of Laboratory Animals. No. X. Publication 990. National Academy of Sciences, Washington, D.C.

Pawar, S.S. and Tidwell, H.C. (1968): Effect of ingestion of unsaturated fat on lipolytic activity of rat tissues. J. Lipid Res., 9, 334.

Raulin, J. and Launay, M. (1967): Enrichissement en ADN et ARN du tissu adipeux épididymaire du rat par administration de lipides insaturés. Nutr. et Dieta (Basel), 9, 208.

Rodbell, M. (1964): Localisation of lipoprotein lipase in fat cells of rat adipose tissue. J. biol. Chem., 239, 753.

Salans, L.B., Horton, E.S. and Sims, E.A. (1971): Experimental obesity in man: cellular character of the adipose tissue. J. clin. Invest., 50, 1005.

Salans, L.B., Zarnowski, M.J. and Segal, R. (1972): Effect of insulin upon the cellular character of rat adipose tissue. J. Lipid Res., 13, 616.

Seitz, P.F.D. (1954): The effect of infantile experiences upon adult behavior in animal subjects. I. Effects of litter size during infancy upon adult behavior in the rat. J. psychiat. Res., 110, 916.

Trémolières, J. and Apfelbaum, M. (1963): Modifications métaboliques produites par des régimes hyperlipi-diques athérogènes chez le rat. Monogr. Inst. nat. Hyg., 28, 37.

Vost, A. and Hollenberg, C.H. (1970): Effects of diabetes and insulin on DNA synthesis in rat adipose tissue. Endocrinology, 87, 606.

Widdowson, E.M. and McCance, R.A. (1960): Some effects of accelerating growth. I. General somatic development. Proc. roy. Soc. B, 152, 188.

Winick, M. and Noble, A. (1966): Cellular response in rats during malnutrition at various ages. J. Nutr., 89, 300.

DISCUSSION

Apfelbaum: Dr. Lemonnier did not find any compensatory hyperplasia in perirenal adipose tissue after bilateral removal of epididymal fat pads. We performed unilateral removal of the epididymal fat pad and did not find any increase in the number of adipocytes in the contralateral fat pad.

Observations on adipose tissue cellularity and development in rats and rabbits fed ad lib

M. DiGirolamo, L. Thurman and **J. Cullen**

Department of Medicine, Emory University School of Medicine, Atlanta, Ga., U.S.A.

The mechanisms regulating growth of the adipose tissue mass in the organism are largely unknown. Previous studies in man and in several mammalian species have shown that with age and ad lib feeding the adipose mass continues to expand even after maturity of the subjects, when most other organs have reached a stable weight. Although in the rat and in the hamster both fat cell number and fat cell size contribute, together with accumulation of water and stromal constituents, to the enlargement of the adipose mass (Benjamin et al., 1961; DiGirolamo and Mendlinger, 1971), studies by Hirsch and Han (1969) have shown that the contribution of the fat cell number is greater in the rat in the early stages of life (first 12-14 weeks) than in later stages, whereas the contribution of the fat cell size continues and predominates after maturity.

Exceptions to these patterns have been found in human obesity, in which fat cell number appears to be greater in obese than in lean mature subjects (Hirsch, 1972), and in the guinea pig (DiGirolamo and Mendlinger, 1971), in which the contribution of the fat cell number appears to be proportionally greater than that of the cell size in studies conducted in the first 11-13 months of life.

The present report summarizes the results of observations carried out in over 200 growing male Wistar rats and in 31 New Zealand rabbits during their first year of life while feeding ad lib on Purina laboratory chow or Purina rabbit chow, respectively. Both similarities and differences will be described in the comparison of adipose tissue development in the two species.

The basic procedures, described in detail elsewhere (DiGirolamo et al., 1971; DiGirolamo and Mendlinger, 1971), were as follows. Following sacrifice of the animals, adipose tissue was dissected in its entirety in the epididymal and perirenal regions and fragments of subcutaneous and mesenteric adipose tissue were removed in both rats and rabbits. Isolated fat cells were obtained by incubating 4 small slices (30-50 mg) of tissue removed from 4 separate points in each depot in Krebs-Ringer bicarbonate medium containing 4% defatted bovine serum albumin, 3 μmoles/ml glucose and 10 mg/g tissue collagenase from Cl. Histolyticum (Worthington Biochemical Corp.) at 37°C for 1 hour. The fatt cells were stained with methylene blue and the transverse diameter of 300 fat cells was measured by optical sizing. From the cell diameter the mean fat cell volume was calculated as previously described (DiGirolamo et al., 1971). Additional duplicate aliquots of the tissue (100-200 mg) were extracted in chloroform: methenol (2:1, v/v) according to the procedure described by Folch et al. (1975) and the triglyceride content of the tissue was determined as previously described (DiGirolamo et al., 1971). The number of fat cells contained in the tissue slices and in the entire epididymal or perirenal depot was calculated by dividing the triglyceride content of the tissue by the mean fat cell triglyceride content (fat cell volume × lipid density).

OBSERVATIONS IN THE RAT

Studies with epididymal adipose tissue

Figure 1 shows that as the body weight of the animals increases with age and ad lib feeding, the weight of the 2 epididymal fat pads continues to increase even after the animal body weight has

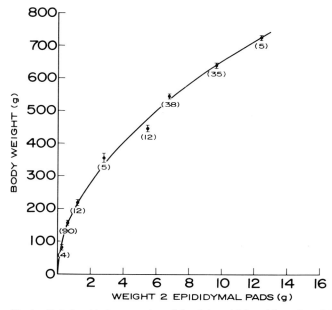

Fig. 1. Relationship between the weight of the epididymal fat pads and the body weight of 201 male Wistar rats. Values represent mean ± S.E.M.

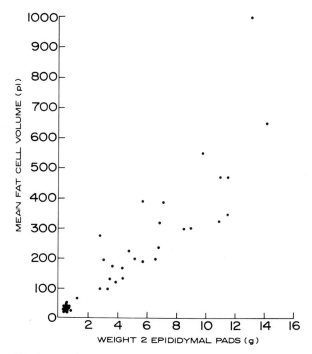

Fig. 2. Relationship between the weight of the epididymal fat pads and the mean fat cell volume (pl) calculated as described in the text.

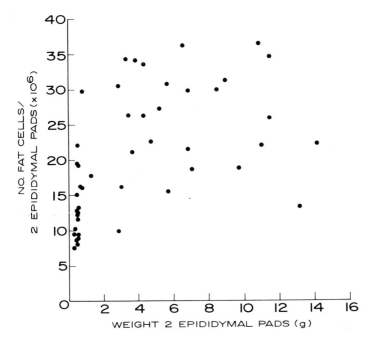

Fig. 3.   Number of fat cells per 2 epididymal fat pads in relation to the weight of the pads. Note the wide range of fat cell number observed in the pads of different animals that have reached maturity.

reached a 'plateau' level. Figure 2 shows that a progressive increase in fat cell volume is observed as the weight of the epididymal fat pads continues to increase. In Figure 3 the relationship between the weight of the 2 epididymal fat pads and the fat cell number is shown. It can be seen that in an initial phase (between 5 and 13-15 weeks) there is a progressive increase in fat cell number which however appears to 'plateau' after 13-15 weeks of age when the fat pads' weight is approximately 4 g. A wide range in fat cell number can be observed in different animals after maturity (range $15-36 \times 10^6$).

Observations by Hirsch and Han (1969) and in our laboratory (DiGirolamo et al., 1972) have shown that in the rat the fat cell size can be readily modified by varying the caloric nutrition of the animals, by hormonal stimulation (Salans et al., 1972) and by varying physical activity (Oscai et al., 1972), whereas the number of fat cells is relatively more stable and can only be modified in the rat by a quantitative variation in dietary intake of the animals in a very early stage of life (Knittle and Hirsch, 1968). The factors regulating the total number of adipose cells in a given depot or in a given animal species are not known. Hereditary factors as well as hormonal or environmental ones have been considered, but no definite clues have so far been found. In addition, precise estimation of cell number eludes accurate determination by available techniques when the fat cells of adipose tissue removed from animals at early stages of life are so small and contain so little lipid that either they do not float when separated from the stroma or are indistinguishable at microscopic observation from adjacent stromal tissue cells.

Interdepot variations in mean fat cell size

Comparison of morphological characteristics of the fat cells in 4 separate depots, of the rat (Table 1) has shown the following. At 8 weeks and at 1 year of age, considerable interdepot variations in mean fat cell volume are seen. Mesenteric fat cells appear to be smaller than epididymal, perirenal and subcutaneous fat cells. In 10-12-month-old rats the mesenteric fat cells were

TABLE 1   Interdepot variations in mean fat cell volume in 3 groups* of rats

Adipose depot	Average fat cell volume (pl)**		
	Group A	Group B	Group C
Mesenteric	21.4 ± 2.7	303.1 ± 56.0	99.6 ± 15.9
Subcutaneous	62.1 ± 5.5	336.2 ± 45.8	174.7 ± 21.2
Epididymal	57.0 ± 4.2	520.6 ± 73.7	264.4 ± 32.0
Perirenal	51.6 ± 6.9	581.7 ± 88.4	236.4 ± 27.5

* Group A (n=8)=8-week-old rats weighing 193 ± 2.5 g (mean ± S.E.M.) fed ad lib; Group B (n=8)=10-12-month-old rats weighing 596 ± 14.9 g fed ad lib; Group C (n=11)= rats similar in age and body weight to Group B but studied after a 20% body weight reduction was achieved by limitation of food intake to 10 g chow per day for 15-25 days.
** The average fat cell volume was calculated from the diameter determination of 300 isolated fat cells in the 4 adipose depots of each rat. The values shown represent the mean ± S.E.M. of n observations in each depot.

the smallest in size, epididymal and perirenal the largest, and the subcutaneous cells had an intermediate size.

When the capacity of the fat cells from the 4 depots of the rat to enlarge and reduce in size was studied during growth and following body weight reduction it was found that certain depots (viz. mesenteric and perirenal) were more capable of expanding and contracting in fat cell volume than others (viz. subcutaneous).

Heterogeneity in cell size of fat cell populations

The fat cell populations obtained by collagenase incubation present in each depot a natural distribution in size (diameter and volume) which is related to the mean cell size. Populations with a mean fat cell volume of less than 100 pl have shown greater heterogeneity in size (as determined by the coefficient of variation, CV – S.D./mean) than populations with a mean fat cell volume greater than 400 pl. The degree of heterogeneity in the size of the cells within a given population isolated from depots of animals fed ad lib has been found to be inversely related to the mean fat cell size (diameter or volume). This relationship has recently been confirmed in depots of animals which had been made to lose weight by undernutrition: when the mean cell volume was reduced, heterogeneity in fat cell size was found to be increased as compared to fat cells of obese animals fed ad lib.

OBSERVATIONS IN THE RABBIT

In the rabbit, the epididymal adipose tissue is located outside of the peritoneal cavity and is lodged in the inguinal canal. Figure 4 shows that, as the rabbit grows, both perirenal and epididymal adipose mass enlarge, but the epididymal pad weight is markedly smaller than that of the perirenal depot.

Interdepot variations in mean fat cell size

In the rabbit, as in the rat, marked interdepot variations in adipose cell size were observed at all stages of growth (Table 2). However, several differences were observed that clearly separated the pattern observed in the rabbit from that in the rat. In the rabbit, the epididymal fat cells were the smallest in volume of the 4 depots studied. This finding may be related to the location of the epididymal fat, which may be limited in growth within the inguinal canal. The mesenteric fat cells were surprisingly large when compared to the other depots, a finding in contrast to previous observations in the rat and in man. In addition, the subcutaneous depots of the rabbit

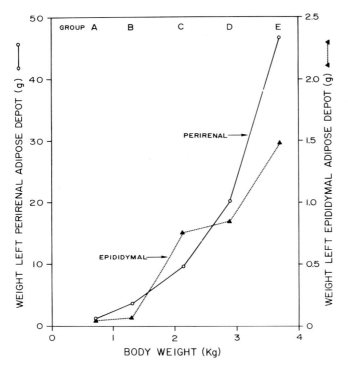

Fig. 4.   Weight of the left perirenal and epididymal adipose depots of rabbits in relation to the body weight. Values represent the mean of 5-7 animals in each group. Note the different scale for epididymal adipose tissue on the ordinate.

TABLE 2   Interdepot variations in mean fat cell volume in male rabbits* fed ad lib

Adipose depot	Average fat cell volume (pl) (mean ± S.E.M.)**				
	Group A	Group B	Group C	Group D	Group E
Mesenteric	87.59 ± 29.14	295.98 ± 44.61	308.35 ± 48.38	396.79 ± 51.30	678.40 ± 135.62
Subcutaneous	106.80 ± 38.74	316.51 ± 33.52	429.28 ± 61.05	575.71 ± 88.50	803.69 ± 133.19
Epididymal	70.08 ± 18.56	208.73 ± 26.85	194.66 ± 27.07	290.58 ± 22.71	331.17 ± 46.35
Perirenal	125.60 ± 34.41	328.74 ± 35.08	347.77 ± 36.62	481.61 ± 66.98	677.04 ± 96.22

* The 5 groups each consisted of 5-7 animals. The mean body weights of the animals in kg were 0.705, 1.315, 2.144, 2.884 and 3.699 respectively for groups A, B, C, D, and E.
** The average fat cell volume was calculated in each depot from the optical determination of the diameter of 300 isolated fat cells. The values for each group of rabbits were then averaged and are here shown for each depot studied.

(removed from the inguinal region) contained fat cells with the largest cell volume of any of the other depots studied. This finding was unexpected since in the rat subcutaneous fat cells (from the same inguinal location) had a mean volume intermediate between the mesenteric fat cells and the epididymal and perirenal fat cells. These observations, taken together, indicate that marked interdepot variations in fat cell size exist within the same species and also between species. They also cast some doubts on the concept of the 'adipose organ', a concept which implies uniformity of morphologic and dynamic characteristics in various locations of adipose tissue within the body.

Observations on fat cell size and number in epididymal and perirenal adipose tissue of the rabbit during growth and ad lib feeding

When the contribution of fat cell size and number to the adipose mass increase in the epididymal and perirenal adipose tissue depots was studied, certain interesting features were found. Groups of 5-7 animals of similar body weight were pooled and the changes in fat cell size and number were studied from one group to the next as the animals grew older and fatter (see Table 2 and Figure 5). In both the epididymal and perirenal depots, distinct changes in adipose cellularity were seen in the early stages of life. The first change was a marked increase in cell size with no evident change in number. Then the cells became more numerous with no apparent increase in size. These changes in the early stages of life were followed in later stages by a concomitant and parallel increase in both fat cell size and number.

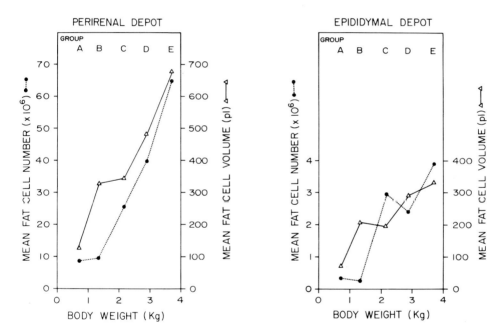

Fig. 5. Fat cell volume (pl) and fat cell number in the left epididymal and perirenal adipose depots of rabbits in relation to body weight. The values represent the mean of 5-7 animals in each group.

The precise mechanisms of this regulation in the early stages of life of the rabbit are not known. Among other explanations it is possible that hormonal influences in the early stages of growth may regulate the alternating increases in adipose cell number and cell size. The rabbit could therefore provide a model for studying what factors regulate the cell number increase in adipose tissue during the early life of the animals.

CONCLUSIONS

The present report summarizes observations on the spontaneous mode of development of adipose tissue in 4 separate locations in rats and rabbits fed ad lib. The interdepot differences in fat cell volume and the variable capacity of the different adipose depots to expand and contract in mean adipocyte volume have been emphasized.

A negative correlation between mean fat cell size and heterogeneity in size of the fat cells has been reported and extended to studies in which mean fat cell volume was reduced by undernutrition of the animals.

The alternating changes in fat cell volume and cell number of the rabbit in two adipose depots in the early stages of life are of interest and could provide a useful model for studying in vivo the factor(s) that regulates adipose mass growth in its two main parameters, adipose cell number and cell size.

REFERENCES

Benjamin, W., Gellhorn, A., Kundel, H. and Wagner, N. (1961): Effect of aging on lipid composition and metabolism in the adipose tissue of the rat. Amer. J. Physiol., 201, 540.

DiGirolamo, M. and Mendlinger, S. (1971): Role of fat cell size and number in enlargement of epididymal fat pads in three species. Amer. J. Physiol., 221, 859.

DiGirolamo, M., Mendlinger, S. and Fertig, J.W. (1971): A simple method to determine fat cell size and number in four mammalian species. Amer. J. Physiol., 221, 850.

DiGirolamo, M., Smith, J.E., Esposito, J. and Thurman, L. (1972): Interdepot variations in fat cell size, glucose metabolism and insulin response by adipose tissue of lean, obese and weight-reduced rats (abstract). Clin. Res., 20, 544.

Folch, J., Lees, M. and Sloane Stanley, G.H. (1957): A simple method for the isolation and purification of total lipids from animal tissues. J. biol. Chem., 226, 497.

Hirsch, J. (1972): Adipose cellularity in relation to human obesity. Advanc. intern. Med., 17, 289.

Hirsch, J. and Han, P.W. (1969): Cellularity of rat adipose tissue: effects of growth, starvation and obesity. J. Lipid Res., 10, 77.

Knittle, J.L. and Hirsch, J. (1968): Effect of early nutrition on the development of rat epididymal fat pads: cellularity and metabolism. J. clin. Invest., 47, 2091.

Oscai, L.B., Spirakis, C.N., Wolff, C.A. and Beck, R.J. (1972): Effects of exercise and of food restriction on adipose tissue cellularity. J. Lipid Res., 13, 588.

Salans, L.B., Zarnowski, M.J. and Segal, R. (1972): Effect of insulin upon the cellular character of rat adipose tissue. J. Lipid Res., 13, 616.

GENERAL DISCUSSION

Björntorp: Dr. Galton raised the question of the validity of counting fat cell numbers in human subjects when there is no method to count fat cells without fat. I agree that this is a difficulty which we have to take into consideration when we interpret our results from the physiological and clinical points of view. A classification of obesity in man is beginning to emerge from these fat cell data. In the groups so obtained one finds differences in the onset of obesity, which has recently been demonstrated by Salans et al. We have similar data and Drs. Apfelbaum and Guy-Grand also. Furthermore, there are metabolic associations with these fat cell classifications of obesity and the prognosis for weight reduction is poorer in the hyperplastic type of obesity. These are some of the clinical characteristics which seem to follow the fat cell classification system. These observations indicate to me that the fat cell measurements as we do them now are of value.

Adipose mass and obesity

Carbohydrate and lipid metabolism in human obesity*

P. Björntorp

Clinical Metabolic Laboratory, First Medical Service, Sahlgren's Hospital, University of Gothenburg, Gothenburg, Sweden

The rather close association between obesity and diabetes mellitus is well known among clinicians. Obese subjects have an increased risk of becoming diabetic. It is also clear that there is some statistical connection between obesity and hyperlipidemia. When it became possible to measure plasma insulin these associations were better defined. Insulin was secreted in excess of the transported glucose (Perley and Kipnis, 1966). It was then thought that the association with diabetes mellitus was caused by peripheral insulin ineffectiveness in obesity, leading to insulin overproduction, which placed an extra strain on the pancreatic insulin secretion mechanism.

It then became important to characterize the hyperinsulinemia of obesity further. Evidence was obtained which indicated a relation between plasma insulin and body fat (Bagdade, 1968; Bierman et al., 1968). The next step was to investigate which of the factors determining adipose tissue mass was associated with plasma insulin. Body fat depends on two factors, the triglyceride content of the fat cells containing the body fat, and the number of these fat cells. The triglyceride content is reflected in the size of the fat cells, because the major part of the cells is occupied by the stored triglyceride.

This then was the problem we wanted to study, and these studies will be briefly summarized in this paper.

CORRELATION BETWEEN PLASMA INSULIN AND FAT CELL SIZE

The first study was performed in 1968 when a sample of randomly selected men born in 1913 and living in Gothenburg was examined. Body fat correlated better with plasma insulin than did body weight. Of the body fat factors, fat cell size showed a stronger statistical correlation than body fat, while fat cell number even showed a negative correlation (Björntorp et al., 1970b). Men below the age of 55 who had survived a myocardial infarction also showed a correlation between fat cell size and plasma insulin. In these two groups of subjects there was also a significant positive correlation between fat cell size and the sum of insulin values during the glucose tolerance test (Berchtold et al., 1972).

Obese subjects were next examined. These were selected to be non-diabetic, non-dieting and weight-stable for at least several months. There was a rather weak, although statistically significant, correlation between fat cell size and plasma insulin in these subjects and a lack of correlation with body fat (Björntorp and Sjöström, 1971).

These findings have recently been confirmed by several groups. The Rockefeller University group also found that plasma insulin correlated better with fat cell size than with body fat in obese and non-obese subjects (Stern et al., 1972a). Similar findings were reported recently by the Stanford University group (Stern et al., 1972b). Brook and Lloyd (1973) have demonstrated a correlation between fat cell size and plasma insulin in obese children with an *r* value close to that found in obese adults (Björntorp and Sjöström, 1971).

This question then seemed to be settled, with general agreement that the factor in adipose

* This study was supported by grants from the Swedish Medical Research Council and the Swedish National Association Against Heart and Chest Diseases.

tissue which showed the strongest correlation with plasma insulin was the size, rather than the number, of fat cells. It is necessary, however, to interpret these results with great caution from at least two points of view. First, one does not find this correlation except under strictly defined conditions. Second, these and other observations indicate that plasma insulin levels are not necessarily related to adipose tissue mass, suggesting that fat cells do not affect insulin secretion. These two points are considered further below.

CONDITIONS IN WHICH THERE IS NO CORRELATION BETWEEN PLASMA INSULIN AND FAT CELL SIZE

The previously shown correlation between fat cell size and plasma insulin in obesity was found in obese subjects who were preselected to be non-diabetic, non-dieting and weight-stable. In an obese group without this preselection the correlation was not found (Sjöström and Björntorp, 1974). In diabetic obese subjects it is reasonable to believe that this is related to the abnormal insulin secretion. Weight increase or decrease influences plasma insulin. Obese subjects were investigated on two occasions separated by a 10-month interval. Body weight and adipose tissue composition were determined. Subjects who gained weight had higher insulin values than those of constant weight and those who lost weight (Sjöström and Björntorp, 1974). Weight stability is thus of importance for this question. It is also well known that dieting influences plasma insulin. Grey and Kipnis (1971) have convincingly shown this, particularly for carbohydrate intake.

Table 1 summarizes these points. The correlation is found in men and in hypertriglyceridemic men and women. This latter group has enlarged fat cells in comparison with controls (Björntorp et al., 1971a). The correlation is also found in men and women in another group with enlarged fat cells, namely obese subjects, but only under certain conditions. Dieting and diabetes remove the correlation, and so does physical activity, as will be discussed in more detail later. It is noticeable that the correlation was not found in non-obese women in this population (Björntorp et al., 1971b) nor in a larger population investigated later (Blohmé and Björntorp, unpublished data).

TABLE 1   Correlations (r values) between fat cell size and fasting plasma insulin values in various groups

Group	No.	Fat cell size vs. fasting insulin
Middle-aged men	49	0.51**
Men after a myocardial infarction	23	0.42*
Hypertriglyceridemic men and women	40	0.42*
Obese men and women	26	0.42*
Obese patients on restricted diet	16	n.s.
Obese patients after training	30	n.s.
Middle-aged women	23	n.s.
Maturity onset diabetes mellitus	24	n.s.

*$P < 0.05$, **$P < 0.01$.

Thus there is evidence for a sex difference in the correlation between insulin and fat cell size. In men a linear correlation seems to be present over the whole fat cell size range, while in women the correlation can only be demonstrated for larger fat cell sizes. The reasons for this have not been determined.

CORRELATION BETWEEN ADIPOSE TISSUE CELLULARITY AND PLASMA TRIGLYCERIDE

In randomly selected middle-aged men, adipose tissue cell size correlates with plasma insulin and there are correlations between insulin and glucose tolerance on the one hand, and plasma triglyceride on the other (Björntorp et al., 1970b). These latter correlations have also been demonstrated by Reaven et al. (1967) and Abrams et al. (1969) in randomized subjects. These findings in randomly selected men made it of interest to investigate endogenous hypertriglyceridemia, where glucose tolerance is decreased and plasma insulin and triglyceride are elevated. Body fat is also increased and this is due to enlarged fat cells rather than to an increased number of fat cells in comparison with controls (Björntorp et al., 1971a. This finding has also recently been confirmed by Stern et al. (see p. 316) and Walldius et al. (see p. 327).

To sum up, plasma insulin seems to be correlated with adipose tissue mass, primarily with fat cell size, under certain well defined conditions, in men, and possibly with enlarged fat cells in women. Dieting, weight decrease, diabetes and physical training disturb the correlation. Furthermore, under certain as yet poorly defined conditions the adipose tissue correlations are extended to include associations between plasma insulin and glucose tolerance on the one hand and plasma triglyceride on the other. It is noteworthy that fat cells are enlarged in endogenous hypertriglyceridemia.

What is the cause-effect relationship between plasma insulin concentration and fat cell size? It has been mentioned that the hyperinsulinemia of obesity might be a consequence of a peripheral insulin resistance. Such resistance has actually been suggested to be present in enlarged fat cells (Salans et al., 1968). It should be remembered, however, that there is also evidence for insulin resistance in muscle (Rabinowitz and Zierler, 1962; Felig et al., 1971) and in liver (Arky et al., 1967). The decreased insulin sensitivity might thus be a generalized phenomenon in hyperinsulinemic obesity. The fact that one can separate the insulin concentration dependence from adipose tissue by the factors mentioned suggests that adipose tissue is not an important regulator of insulin concentration, if it regulates this concentration at all. It seems more likely that insulin concentration in plasma exerts its well known effects on triglyceride assimilation and release mechanisms in adipose tissue as a primary event determining fat cell size. Large cells are then enlarged because of the elevated plasma insulin inhibiting lipid mobilization and increasing triglyceride synthesis.

Observations on the effect of physical training on the plasma insulin concentration in obesity seem to support this hypothesis rather well. The following part of this review will deal with these problems.

EFFECTS OF PHYSICAL TRAINING ON METABOLISM IN NON-OBESE SUBJECTS

The effects of physical training are usually followed in young subjects simply because they are easier to train physically. We were, however, lucky to be able to study a group of well trained, middle-aged men. They had been training regularly all their life, and are still active competitors in cross-country running or skiing. They train hard for at least an hour three times weekly and they frequently compete. In comparison with randomly selected men from the same city and of the same age they had less body fat due to smaller fat cells. They had a pronounced glucose tolerance and the insulin values during the glucose tolerance test were very low. They thus had a marked peripheral insulin sensitivity. Their blood lipids were also low (Björntorp et al., 1972a). Figure 1 shows that their fasting plasma insulin values were as low as those of well trained young athletes, here represented by a football team in Gothenburg. The football players had lower values than sedentary medical students (Björntorp et al., 1972b). There were no indications of any peculiarities in the diets of these athletes in the form of, for example, low carbohydrate intake or vegetarianism.

These studies can of course be criticized because of their comparison of different groups where the insulin differences might be due to other effects than those of physical training. We therefore also studied a group of subjects before and after training. These were men who had had a myocardial infarction before the age of 55 and who were trained physically for 9 months. They constituted the total number of patients surviving infarction of the age and sex in question during a limited period of time in Gothenburg, and every second subject was studied without training to serve as a control. These patients were not given any recommendations to change their diet except in a few cases of obvious dietary peculiarities. Body fat decreased during the training period. There was a marked decrease in plasma insulin, while glucose tolerance showed only minor improvements. This again indicated an increase in peripheral insulin sensitivity (Björntorp et al., 1972c). Fasting plasma insulin decreased but not to the low level of the athletes

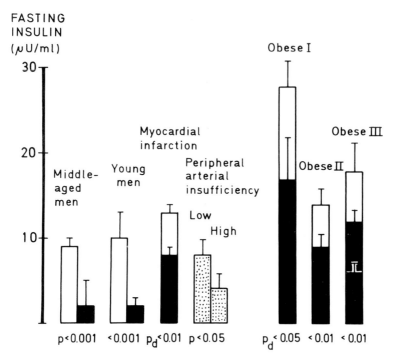

Fig. 1.   Effects of physical training on fasting plasma insulin concentration. From left to right: fasting plasma insulin values in sedentary (white column; Björntorp et al.. 1970b) or physically well trained (black column, Björntorp et al., 1972a) middle-aged men; in sedentary (white column; Björntorp et al., 1972b) or well trained (black column; Björntorp et al., 1972b) young men; in patients who had survived a myocardial infarction, before (white column) and after (black column) physical training (Björntorp et al., 1972c); in patients suffering from low or high peripheral arterial stenosis (stippled columns; Holm et al., 1973); and in obese subjects before (white columns) and after (black columns) physical training (in Obese III, after 3 and 6 months of training). (Obese I: Björntorp et al., 1970a, Obese II: Björntorp et al., 1973b, Obese III: Björntorp et al., 1974).

who had trained harder and for a longer time (Fig. 1). Figure 2 summarizes the results in all six groups examined. Apparently fat cell size, plasma insulin and plasma triglyceride are interrelated.

In contrast to these observations in physically well trained subjects are those found in patients who were immobilized because of disturbances of their muscle function. Patients with myotonic dystrophy without physical impairment showed some decrease in body cell mass and increase in body fat but no significant metabolic aberrations in comparison with controls, while patients with myotonic dystrophy who were disabled and more or less immobile had on average decreased glucose tolerance, hyperinsulinemia, and hyperlipidemia (Björntorp et al., 1973a). These results suggest that myotonic dystrophy by itself has little effect on carbohydrate and lipid metabolism; it is rather the immobilization in some of these patients which creates the metabolic disturbances.

POSSIBLE EXPLANATIONS OF THE INCREASED INSULIN SENSITIVITY AFTER PHYSICAL TRAINING

Physical training causes several changes in the circulatory system, perhaps in other endocrine functions in addition to insulin secretion, and in other functions of the body. Several of these

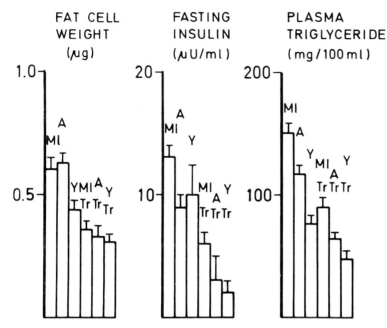

Fig. 2. Fat cell weight, fasting plasma insulin and triglyceride in: myocardial infarction patients before (MI) and after (MI, Tr) physical training (Björntorp et al., 1972c); randomly selected middle-aged men (A) (Björntorp et al., 1970b); physically well trained middle-aged men (A, Tr) (Björntorp et al., 1972a); young sedentary men (Y) (Björntorp et al., 1972b); and young well trained men (Y, Tr) (Björntorp et al., 1972b).

changes could theoretically be responsible for the increased insulin sensitivity after physical training. The effect could also be at the cellular level in the tissue which is primarily subjected to physical training, namely muscle. Muscle tissue shows several metabolic changes after physical training, as first shown by Holloszy (1967). Well trained middle-aged men had an increase both in the activity of succinic oxidase and in the metabolism of labelled glucose in their leg muscles as examined in vitro (Björntorp et al., 1972a). The question was then raised as to whether this change could be associated in some way with the increased general insulin sensitivity after physical training.

Another clinical group gave some evidence in this connection. Patients with intermittent claudication showed changes in the enzymes of their muscle tissue below the arterial stenosis which were similar to those shown in trained men. Succinic oxidase activity and glucose incorporation were both increased (Holm et al., 1972). It then became interesting to see whether or not this affected their insulin sensitivity. To test this it was necessary to exclude diabetic claudication patients, defined as those falling more than two S.D.s above controls in glucose tolerance tests. Furthermore, patients with signs of coronary artery disease were excluded because physical training was included in the program, and for this study we did not have available the alarm equipment necessary when exercising coronary patients. The remaining group, all men, were thus rather highly selected. By definition their glucose tolerance was not diabetic but they did not have as pronounced a glucose tolerance as the well trained men. Their insulin values were lower than those of the controls. Both the glucose and insulin curves were flatter than controls, which has also been noticed recently by Ghilchik et al. (1971). A quantitative aspect was introduced into the study by dividing the patients into those with high and those with low arterial stenosis. These two groups would then presumably have different amounts of adapted muscle tissue. The high stenosis patients had very low insulin values, in fact nearly as low as those of the well trained men. Finally, during training patients with low stenosis decreased their insulin values to those of the high stenosis patients (Holm et al., 1973).

These studies indicate that subjects with muscle enzyme adaptations similar to those of well trained men have a high peripheral insulin sensitivity like that of well trained men. This seems to increase the possibility that events in the muscle cell are of importance for insulin sensitivity. The intermittent claudication patients are not very active physically and probably lack several other adaptations in the body caused by physical training in addition to the muscle tissue changes. The quantitative analysis of studies of high and low stenosis seems to strengthen this possibility. We believe that these studies point to muscle as an important site for regulation of peripheral insulin sensitivity. However, they do not show how such an increased sensitivity can be brought about. Furthermore, they do not by any means exclude a significant role of the liver in this regulation.

PHYSICAL TRAINING IN OBESITY

These studies were all performed in a similar way. Severely obese subjects were selected, primarily those who had been obese since childhood and had a hyperplastic adipose tissue, and they were instructed not to change their ordinary diet but to eat freely during the period of physical training. Training was done using mainly the ergometer bicycle and similar exercises. In a first study (Björntorp et al., 1970a) the training period of 8 weeks caused an increase of aerobic power and muscle strength. Body mass did not increase. Furthermore, body fat in these subjects did not decrease, but rather showed a slight increase after training. Glucose tolerance did not change on average, while plasma insulin values became markedly lower. These values were then no longer above the levels found in non-trained controls. These findings showed that physical training was a quite powerful means of lowering plasma insulin in obesity in spite of the fact that body fat and fat cell size were maintained or even increased.

In the next study (Björntorp et al., 1973b) obese subjects with less glucose tolerance and less marked hyperinsulinemia were studied. None of these obese subjects was clearly diabetic, however. The training was not as hard in this group since the subjects were older and not as cooperative. The effects on plasma insulin were less marked but were seen in the intravenous glucose tolerance test and in fasting plasma insulin. The fact that the effect was found only in the intravenous test seems to indicate that the effect of physical training on insulin secretion is not mediated via enteric insulinogenic hormones.

In a third similar study (Björntorp et al., 1974) a strictly standardized training schedule was used which was maintained 3 times weekly for 6 months. The maximum work loads were based on each subject's maximal working capacity, and adjusted to remain 10-15 pulse beats below this level. Aerobic power increased in all these subjects after 3 and after 6 months. Plasma insulin decreased both during fasting (Fig. 1) and after an oral glucose tolerance test. In these subjects glucose tolerance was improved after 6 months, but body fat did not decrease on average.

One problem with these studies is that it may be difficult to separate acute effects of the last training session and more long-term adaptation of the metabolism after physical training. Acute work at 2/3 of the maximal rate, performed by obese subjects on the ergometer bicycle for 1 hour, did affect plasma insulin. This type of work comes close to leg exhaustion and almost empties muscle glycogen stores (Hultman, 1967). The insulinogenic index, as a crude index of insulin sensitivity of the periphery, apparently decreases for a couple of days after such work (Fahlén et al., 1972). In chronically trained obese subjects, however, with a lighter work load at each training session, the acute effect did not seem to be present 4 and 7 days after the last week of training because these two insulin curves were similar (Björntorp et al., 1974). We therefore believe that the effect of physical training on plasma insulin is at least partly an effect of chronic adaptation rather than an acute effect only.

It seems clear then that physical training exerts a lowering effect on plasma insulin in obesity. The effect on glucose tolerance is less striking and the insulin decrease is best detected in the intravenous test, indicating that the explanation of the effect lies in an increase of peripheral insulin sensitivity.

The question of whether or not normalization of plasma insulin occurs in these studies is difficult to answer. We have as yet no fully adequate control subjects with which to settle this point. Figure 1 compares the fasting insulin values of all the groups described. In the first study,

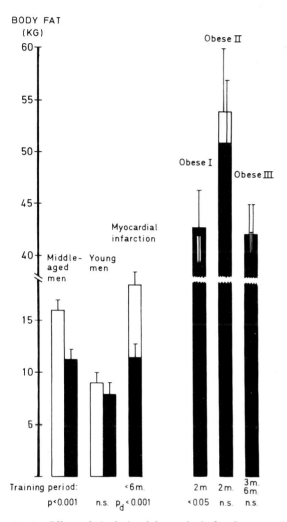

Fig. 3. Effects of physical training on body fat. Groups and symbols as in Figure 1.

including obese subjects, insulin was probably not normalized. In the second study insulin was not very high to start with. In the third study insulin values seem to have been lowered to the same region as the values in the trained myocardial infarction patients, who trained in an identical manner. It is not possible to state for certain that normalization is obtained, but a pronounced lowering undoubtedly occurs. Another striking observation is the fact that these obese subjects, characterized by hypercellularity of adipose tissue, increased body mass and early onset of obesity, fail to decrease in body fat during training, in contrast to non-obese training subjects (Fig. 3).

ACKNOWLEDGEMENTS

The results reported from the author's laboratory were obtained in collaboration with many investigators (see reference list) from the First and Second Medical Service, the Department of Physical Rehabilitation and Clinical Physiology and the Second Surgical Department of Sahlgren's Hospital, Gothenburg, Sweden.

REFERENCES

Abrams, M.E., Jarret, R.J., Keen, H., Boyns, D.R. and Crossley, J.N. (1969): Oral glucose tolerance and related factors in a normal population sample. II. Interrelationship of glycerides, cholesterol, and other factors with the glucose and insulin response. Brit. med. J., 1, 599.

Arky, R.A. and Freinkel, N. (1967): Alcohol hypoglycemia. V. Alcohol infusion to test gluconeogenesis in starvation, with special reference to obesity. New Engl. J. Med., 274, 426.

Bagdade, J.D. (1968): Basal insulin and obesity. Lancet, 2, 630.

Bierman, E.L., Bagdade, J.D. and Porte, D. (1968): Obesity and diabetes: The odd couple. Amer. J. clin. Nutr., 21, 1434.

Berchtold, P., Björntorp, P., Gustafson, A., Lindholm, B., Tibblin, G. and Wilhelmsen, L. (1972): Glucose tolerance, plasma insulin and lipids in relation to adipose tissue cellularity in men after myocardial infarction. Acta med. scand., 191, 35.

Björntorp, P., Bengtsson, C., Blohmé, G., Jonsson, A., Sjöström, L., Tibblin, E., Tibblin, G. and Wilhelmsen, L. (1971b): Adipose tissue fat cell size and number in relation to metabolism in randomly selected middle-aged men and women. Metabolism, 20, 927.

Björntorp. P., Berchtold, P., Grimby, G., Lindholm, B., Sanne, H., Tibblin, G. and Wilhelmsen, L. (1972c): Effects of physical training on glucose tolerance, plasma insulin and lipids and on body composition in men after myocardial infarction. Acta med. scand., 192, 439.

Björntorp, P., de Jounge, K., Krotkiewski, M., Sullivan, L., Sjöström, L. and Stenberg, J. (1974): Physical training in human obesity. III. Effects of long-term physical training on body composition. Metabolism, in press.

Björntorp, P., de Jounge, K., Sjöström, L. and Sullivan, L. (1970a): The effect of physical training on insulin production in obesity. Metabolism, 19, 631.

Björntorp, P., de Jounge, K., Sjöström, L. and Sullivan, L. (1973b): Physical training in human obesity. II. Effects on plasma insulin in glucose intolerant subjects without marked hyperinsulinemia. Scand. J. clin. Lab. Invest., 32, 41.

Björntorp, P., Fahlén, M., Grimby, G., Gustafson, A., Holm, J., Renström, P. and Scherstén, T. (1972a): Carbohydrate and lipid metabolism in middle-aged, physically well-trained men. Metabolism, 21, 1037.

Björntorp, P., Grimby, G., Sanne, H., Sjöström, L., Tibblin, G. and Wilhelmsen, L. (1972b): Adipose tissue fat cell size in relation to metabolism in weight-stable, physically active men. Hormone metab. Res., 4, 182.

Björntorp, P., Gustafson, A. and Persson, B. (1971a): Adipose tissue fat cell size and number in relation to metabolism in endogenous hypertriglyceridemia. Acta med. scand., 190, 363.

Björntorp, P., Gustafson, A. and Tibblin, G. (1970b): Relationships between adipose tissue cellularity and carbohydrate and lipid metabolism in a randomly selected population. In: Atherosclerosis, Proceedings, II International Symposium, p. 374. Editor: R. J. Jones. Springer, Berlin.

Björntorp, P., Schröder, G. and Örndahl, G. (1973a): Carbohydrate and lipid metabolism in relation to body composition in myotonic dystrophy. Diabetes, 22, 238.

Björntorp, P. and Sjöström, L. (1971): Number and size of adipose tissue fat cells in relation to metabolism in human obesity. Metabolism, 20, 703.

Brook, C.G.D. and Lloyd, J.K. (1973): Adipose cell size and glucose tolerance in obese children and effects of diet. Arch. Dis. Childh., 48, 301.

Fahlén, M., Stenberg, J. and Björntorp, P. (1972): Insulin secretion in obesity after exercise. Diabetologia, 8, 141.

Felig, P., Horton, E.S., Runge, C.F. and Sims, E.A.H. (1971): Experimental obesity in man: Hyperaminoacidemia and diminished effectiveness of insulin in regulating peripheral amino acid release. In: Program, 53rd Meeting of Endocrine Society, June, 1971, abstract 29.

Ghilchik, M.W. and Morris, A.S. (1971): Insulin response to glucose in patients with peripheral vascular disease, arteritis, and Raynaud's phenomenon. Lancet, 2, 1229.

Grey, N. and Kipnis, D.M. (1971): Effect of diet composition on the hyperinsulinemia of obesity. New Engl. J. Med., 285, 827.

Holloszy, J.O. (1967): Biochemical adaptations in muscle. Effects of exercise on mitochondrial oxygen uptake and respiratory enzyme activity in skeletal muscle. J. biol. Chem., 242, 2278.

Holm, J., Björntorp, P. and Scherstén, T. (1972): Metabolic activity in human skeletal muscle. Effect of peripheral arterial insufficiency. Europ. J. clin. Invest., 2, 321.

Holm, J., Dahllöf, A-C., Björntorp, P. and Scherstén, T. (1973): Glucose tolerance, plasma insulin and lipids in intermittent claudication with reference to muscle metabolism. Metabolism, 22, 1395.

Hultman, E. (1967): Studies on muscle metabolism of glycogen and active phosphate in man with special reference to exercise and diet. Scand. J. clin. Lab. Invest., 19, Suppl. 94.

Perley, M. and Kipnis, D.(1966): Plasma insulin responses to glucose and tolbutamide of normal weight and obese diabetic and non-diabetic subjects. Diabetes, 15, 867.

Rabinowitz, D. and Zierler, K.L. (1962): Forearm metabolism in obesity and its response to intraarterial insulin. Characterization of insulin resistance and evidence for adaptive hyperinsulinism. J. clin. Invest., 41, 2173.

Reaven, G.M., Lerner, R.L., Stern, M.P. and Farquhar, J.W. (1967): Role of insulin in endogenous hypertriglyceridemia. J. clin. Invest., 46, 1756.

Salans, L.B., Knittle, J.L. and Hirsch, J. (1968): The role of adipose cell size and adipose tissue insulin sensitivity in the carbohydrate intolerance of human obesity. J. clin. Invest., 47, 153.

Sjöström, L. and Björntorp, P.(1974): Body composition and adipose tissue cellularity in human obesity. Acta med. scand., in press.

Stern, J.S., Batchelor, B.R., Hollander, N., Cohn, C.K. and Hirsch, J. (1972a): Adipose cell size and immunoreactive insulin levels in obese and normal-weight adults. Lancet, 4, 948.

Stern, M., Olefsky, J., Farquhar, J. and Reaven, G. (1972b): Relationship between fat cell size and insulin resistance in vivo (abstract). Clin. Res., 20, 557.

DISCUSSION

Perlstein: What is the definition of obesity in this population?

Björntorp: These were all very obviously obese patients.

Perlstein: Was any attempt made to ascertain the proportion of carbohydrate in the diet of the obese training group?

Björntorp: The total calorie intake did not obviously decrease because some of the patients even increased their body fat. We have no hard data on the proportion of carbohydrate in the diet before and after training, but the impression was that there was no major change.

Perlstein: Was there any particular reason for using fasting insulin levels for reference rather than peak levels?

Björntorp: Fasting plasma insulin and the sum of the insulin values behave similarly. We have not analysed specifically the peak insulin values.

Breidhal: I, and I am sure many others, have been fascinated for some years by the beneficial effect of physical exercise on the control of diabetes mellitus and, like many others, have been unable to explain the insulin sensitivity induced by exercise. However, it is possible that the work of Bernstein may explain this. He has produced evidence for two functions of growth hormone with differing effects on carbohydrate metabolism, and one of these functions may produce the effects described by Dr. Björntorp. My question, therefore, is whether any measurements were made of growth hormone levels in any of the groups studied?

Björntorp: So far only incomplete, inconclusive measurements have been performed, but we would like to do this.

Kalkhoff: It is not clear to me how long it takes for the effects of exercise programs on plasma insulin to completely disappear after exercise is discontinued.

Björntorp: We believe that the acute effects after the last training session disappear after the first few days. There is probably also a more long-lasting effect still present 7 days after the last exercise (which is the longest period which we have so far studied) and it might well last longer. This is the subject of current studies.

Kalkhoff: Effects of exercise on plasma insulin in the obese are well described in your studies. However, you have never compared this group with a suitable healthy lean control population. Do you feel that relative hyperinsulinemia would still exist in the obese during regular exercise periods as compared to lean, age- and sex-matched control subjects?

Björntorp: As I stated in my presentation we have no adequate controls as yet, so I cannot answer this question.

Body constitution and blood glucose and serum insulin levels in a group of Tamil indians*

M. Fredman

Department of Anatomy, University of Cape Town Medical School, Cape Town, South Africa

The association of diabetes and obesity has been well documented clinically (Davidson and McLeod, 1971), as well as the relationship of diabetes to the specific distribution of fat (Vague et al., 1971; Feldman et al., 1969; Craig and Bayes, 1967). Recent studies (Fredman, 1972, 1974) suggest that the mesomorphy component as well as the endomorphy component is related to the diabetic state. Thirty-nine subjects, all Tamils, were investigated by the Endocrine Research Laboratory of the South African Medical Research Council and blood glucose and serum insulin levels in these patients were studied in relation to the somatotype constitution.

METHODS

Somatotype rating was arrived at by the anthropometric methods of Parnell (1958) and Heath and Carter (1969).

Blood glucose levels were determined by the Endocrine Research Laboratory on the Technicon Auto Analyzer using the modified method of Hoffman (1937), and serum insulin levels were determined using the Amersham kit by the method of Hales and Randle (1963). A normal subject was defined as one who presented with all glucose tolerance test values below normal limits, and a diabetic was a subject who had at least two values above normal limits.

RESULTS

The results were analyzed in two different ways.

TABLE 1 Significant differences of means between normal and diabetic Tamils

		Degrees of freedom	P
Blood sugar series			
Age	3.0785	29	< 0.005
Mesomorphy (Parnell)	2.4246	29	< 0.025
Mesomorphy (Heath-Carter)	2.4429	29	< 0.025
Ectomorphy (Heath-Carter)	1.4800	29	0.1
Fasting blood sugar	2.6463	29	< 0.01
30-minute blood sugar	3.0717	29	< 0.005
60-minute blood sugar	4.2673	29	< 0.005
120-minute blood sugar	3.1719	29	< 0.005
300-minute blood sugar	1.9434	20	< 0.05
Serum insulin series			
Age	4.8734	23	< 0.001
Mesomorphy (Parnell)	2.8223	23	< 0.005
Mesomorphy (Heath-Carter)	2.8058	23	< 0.005

* Supported by a grant from the Herman/Caporn Bequest Fund for Staff Research from the University of Cape Town.

TABLE 2 Correlation (r) of somatotypes and blood sugar levels in normal subjects (N) (n = 18) and diabetics (D) (n = 13)

	Blood sugar level									
	Fasting		30 minutes		60 minutes		120 minutes		300 minutes	
	N	D	N	D	N	D	N	D	N	D
Endomorphy (Parnell)	0.130	0.124	0.224	0.130	0.148	0.065	−0.057	−0.053	−0.222	0.446
Mesomorphy (Parnell)	0.430*	0.153	0.334	0.144	0.126	0.165	−0.356	0.119	−0.355	0.082
Endomorphy (Heath-Carter)	0.149	0.318	0.315	0.221	0.154	0.206	−0.196	0.105	−0.183	0.520*
Mesomorphy (Heath-Carter)	0.625**	0.322	0.295	0.298	−0.181	0.316	−0.433*	0.231	−0.260	0.337

*Significant correlation (P <0.01), **highly significant correlation (F <0.001).

Analysis of means

Means were obtained from the normal and diabetic subjects for the following variables: age; somatotype; subject's weight as a percentage of normal for corresponding height and age (Documenta Geigy, 1962); blood glucose and serum insulin values when fasting, and 30, 60, 120 and 300 minutes after a glucose tolerance test.

The mean values obtained for the normal and diabetic subjects were then tested for significant differences. The results of this analysis (Table 1) show that there was a statistically significant difference between the normal subjects and the diabetics for all the blood glucose determinations, as might have been anticipated. Significant differences between the means also emerged for age, and the mesomorphy component, estimated by both methods, with the diabetics having markedly higher ratings (5.04 to 3.67 on the Heath-Carter scale). There was no significant difference between the means of the serum insulin values or the means for the endomorphy and ectomorphy component estimated by either method, or the subject's weight as a percentage of normal.

Cross-correlation analysis

The individual values for each of the variables mentioned were subjected to a correlation analysis and the correlation values obtained for *r* are shown in Table 2, in respect of the endomorphy and mesomorphy ratings for normals and diabetics.

Significant correlations were noted between: (a) the fasting blood sugar and mesomorphy rating (by both methods) for normal subjects, (b) the 120-minute blood sugar and Heath-Carter mesomorphy rating in normal subjects, (c) the 300-minute blood sugar and the Heath-Carter endomorphy rating in diabetics and (d) fasting blood sugar and Heath-Carter ectomorphy rating of normal subjects.

Highly significant correlation values (not shown) were also obtained between: (a) the somatotype ratings by the methods of Parnell and Heath and Carter, with negative correlations for ectomorphy ($r = 0.77 - 0.95$), (b) the fasting blood sugar, and 30-, 60-, 120-, and 300- minute blood sugar levels in the diabetics ($r = 0.93 - 0.99$) (no correlations were found for these variables in the normal subjects), (c) the Parnell endomorphy/ectomorphy rating and all Heath-Carter component ratings and the subject's weight/normal weight ratio ($r = 0.61 - 0.88$).

DISCUSSION

Since the mean endomorphy ratings of the normals and diabetics do not differ significantly it can be assumed that either there is no significant difference between adipose tissue mass in these two groups or that the endomorphy rating correlates poorly with the adipose tissue mass. However, Wilmore (1970) has demonstrated a correlation between endomorphy and adipose tissue mass; thus the physical difference between normal subjects and diabetics is in fact not in adipose tissue mass.

This is further confirmed by the positive finding in this series of: (a) diabetes being associated with increased mesomorphy ratings, (b) the correlation between fasting blood sugar and both mesomorphy ratings in normal subjects, (c) the correlation between the 120-minute blood sugar level following a glucose tolerance test and the Heath-Carter mesomorphy rating in the normal subjects, and (d) the absence of significant correlations between endomorphy ratings and blood glucose and serum insulin values except for the 300-minute blood sugar reading in diabetics.

While adipose tissue mass may be a factor of great importance in the established diabetic, this study suggests that lean body mass is a more important factor in the development of the diabetic state, contrary to the generally accepted association of obesity and diabetes.

ACKNOWLEDGEMENTS

I wish to thank Professor W.P.U. Jackson for allowing me to investigate these patients and controls and for access to the records of the Endocrine Research Laboratory of the South African Medical Research Council.

REFERENCES

Craig, L.S. and Bayes, L.M. (1967): Androgynic phenotypes in obese women. Amer. J. phys. Anthropol., 26, 23.

Davidson, L.S.P. and McLeod, J. (1971): The Principles and Practice of Medicine, 10th ed. Churchill Livingstone, Edinburgh.

Documenta Geigy (1962): Scientific Tables, 6th ed. J.R. Geigy, Basle.

Feldman, R., Sender, A.J. and Siegelaub, A.B. (1969): Difference in diabetic and non-diabetic fat distribution patterns by skinfold measurements. Diabetes, 18, 478.

Fredman, M. (1972): Somatotypes in a group of Tamil diabetics. S. Afr. med. J., 46, 1836.

Fredman, M. (1974): Somatotypes of diabetic outpatients. S. Afr. J. med. Sci., in press.

Hales, C.N. and Randle, P.J. (1963): Immunoassay of insulin with insulin-antibody precipitate. Biochem. J., 88, 137.

Heath, B.H. and Carter, J.F.L. (1969): A modified somatotype method. Amer. J. phys. Anthropol., 27, 57.

Hoffman, W.S. (1937): A rapid photoelectric method for the determination of glucose in blood and urine. J. biol. Chem., 120, 51.

Parnell, R.W. (1958): Behaviour and Physique. Edward Arnold, London.

Vague, P., Vague, J. and Cloix, M.C. (1971): Relations entre le mode de repartition de la graisse et la diabète de la maturité chez les obèses. Acta diabet. lat., 8, 711.

Wilmore, J. (1970): Validation of the 1st and 2nd components of the Heath/Carter somatotyping method. Amer. J. phys. Anthropol., 32, 369.

Lipogenesis and frequency of feeding: enhanced serum insulin response and sparing of specific dynamic action by 2-meals-per-day feeding in man

S. Matsuki, K. Kataoka, Y. Suzuki and **Y. Takabayashi**

Department of Internal Medicine, Keio University School of Medicine, Tokyo, Japan

Tepperman and colleagues (Dickerson et al., 1943; Tepperman and Tepperman, 1958) demonstrated an increased rate of lipogenesis (conversion of carbohydrate into fat) in rats fed an adequate caloric diet when feeding was limited to 1 hour a day. Hollifield and Parson (1962) similarly found a 25-fold increase in lipogenesis in rats when feeding was limited to 2 hours a day and demonstrated, in addition, a marked increase in synthesis of enzymes in liver and adipose tissue of the pathways involved in lipogenesis. Cohn and Joseph (1960) found a remarkable increase in carcass fat and a fall in protein and water content in force-fed rats which were given food twice daily by stomach tube, as compared with control animals feeding ad lib. The experimental animals had utilised body protein to synthesize and store fat to produce a state that has been called 'non-obese obesity'.

It has been claimed that the once-daily feeding pattern exhibited by many Americans may be a significant factor in the high incidence of obesity in their population. The 'night-eating syndrome' described by Stunkard et al. (1955) is a pertinent observation. Human obesity may be an example of metabolic adaptation to a feeding pattern. Fabry et al. (1964) reported that excessive weight, increased serum cholesterol and diminished glucose tolerance were all significantly commoner among those who took meals less frequently.

The present study concerns observations on serum insulin response and oxygen consumption after meals in human subjects on 3-meals-per-day and 2-meals-per-day feeding regimes.

MATERIALS AND METHODS

The subjects were 5 males and 3 females, all healthy and non-obese. The prescribed diet comprised 360 g of bread, 360 g of milk, 150 g of egg, 75 g of corned beef, 42 g of cheese, 720 g of orange and 60 g of strawberry jam (2,240 calories, 93 g of protein, 60 g of fat and 330 g of carbohydrate), and was given in 2 or 3 meals. On the 3-meal regime food was given in 3 equal meals at 9:00 a.m., 1:00 p.m. and 7:00 p.m., while on the 2-meal regime 2 equal meals were given at 9:00 a.m. and 7:00 p.m. Each meal was eaten within 15 minutes.

Blood samples were drawn at 8:00, 10:00, 11:00 a.m., 1:00, 2:00, 3:00, 7:00, 8:00, 9:00 and 11:00 p.m., and analyzed for blood sugar (Auto-analyzer), serum insulin and growth hormone (double-antibody system) and plasma free fatty acid (Itaya-Ui's method). Oxygen consumption (closed circuit spirometry) was measured before and after morning and evening meals in 4 males and 2 females, after at least 30 minutes bed rest.

RESULTS AND DISCUSSION

Table 1 shows the results of blood analysis. Blood sugar and serum insulin after the evening meal were higher than after the morning meal on both regimes, and the increases after the evening meal on the 2-meal regime were significantly greater than those on the 3-meal regime (P < 0.05 and P < 0.01, respectively), although after the morning meal there was no difference between the two regimes. For serum growth hormone and plasma free fatty acid there were no

TABLE 1   Results (mean ± S.E.) of blood analysis in 8 subjects

	8:00 a.m.	10:00 a.m.	11:00 a.m.	1:00 p.m.	2:00 p.m.
3-meal regime					
Blood sugar (mg/dl)	80 ± 3	89 ± 11	82 ± 5	88 ± 3	100 ± 5
Insulin (μU/ml)	10 ± 2	44 ± 6	24 ± 3	12 ± 1	34 ± 4
Growth hormone (μg/ml)	2 ± 1	1 ± 1	3 ± 1	3 ± 1	2 ± 1
FFA (μmoles/l)	380 ± 30	276 ± 16	309 ± 15	422 ± 35	323 ± 26
2-meal regime					
Blood sugar (mg/dl)	82 ± 2	82 ± 7	82 ± 3	93 ± 3	95 ± 1
Insulin (μU/ml)	15 ± 7	47 ± 13	34 ± 6	25 ± 6	13 ± 2
Growth hormone (μg/ml)	2 ± 1	2 ± 1	2 ± 1	2 ± 1	8 ± 4
FFA (μmoles/l)	406 ± 46	339 ± 39	309 ± 20	319 ± 24	335 ± 28

	3:00 p.m.	7:00 p.m.	8:00 p.m.	9:00 p.m.	11:00 p.m.
3-meal regime					
Blood sugar (mg/dl)	96 ± 7	88 ± 3	125 ± 7	117 ± 9	80 ± 4
Insulin (μU/ml)	27 ± 4	13 ± 1	55 ± 9	52 ± 8	13 ± 1
Growth hormone (μg/ml)	2 ± 1	7 ± 3	2 ± 1	2 ± 1	13 ± 4
FFA (μmoles/l)	386 ± 36	548 ± 36	355 ± 28	379 ± 34	455 ± 43
2-meal regime					
Blood sugar (mg/dl)	95 ± 1	94 ± 12	169 ± 10	114 ± 5	93 ± 2
Insulin (μU/ml)	8 ± 3	8 ± 3	135 ± 26	110 ± 21	31 ± 7
Growth hormone (μg/ml)	12 ± 6	2 ± 1	2 ± 1	2 ± 1	11 ± 3
FFA (μmoles/l)	478 ± 46	806 ± 47	379 ± 54	356 ± 28	438 ± 49

significant differences between the two regimes except that growth hormone and free fatty acid increased before the evening meal on the 2-meal regime.

The results for oxygen consumption are presented in Tables 2 and 3, values being expressed in l/hour. The increase over 3 hours was calculated from the increased area up to 3 hours after a meal, the sum of the values at 1 hour and 2 hours, and 1/2 the value at 3 hours. The increase of oxygen consumption after the morning meal on the 2-meal regime was significantly greater than on the 3-meal regime. After the evening meal however there was no significant difference between the two regimes, except that the increases at 4 and 5 hours on the 2-meal regime were higher than those on the 3-meal regime. On the 2-meal regime the increase of oxygen consump-

TABLE 2   Increase of oxygen consumption (mean ± S.E.) after the morning meal (n = 6)

	Before	After 1 hour	After 2 hours	After 3 hours	Increase over 3 hours
3-meal regime	11.32 ± 0.57	2.18 ± 0.26	1.19 ± 0.56	1.86 ± 0.38	4.31 ± 0.60
2-meal regime	11.23 ± 0.66	3.29 ± 0.27	3.47 ± 0.46	3.35 ± 0.27	8.43 ± 0.45
Difference	n.s.	$P < 0.05$	$P < 0.01$	$P < 0.01$	$P < 0.01$

n.s. = not significant.

TABLE 3   Increase of oxygen consumption (mean ± S.E.) after the evening meal

	Before	After 1 hour	After 2 hours
3-meal regime	12.78 ± 0.52	2.53 ± 0.63	1.73 ± 0.29
2-meal regime	12.44 ± 0.77	2.14 ± 0.26	2.00 ± 0.24
Difference	n.s.	n.s.	n.s.

	After 3 hours	After 4 hours	After 5 hours	Increase over 3 hours
3-meal regime	1.00 ± 0.48	0.17 ± 0.27	0.09 ± 0.46	4.77 ± 1.05
2-meal regime	1.99 ± 0.20	1.19 ± 0.23	1.46 ± 0.37	5.14 ± 0.50
Difference	n.s.	$P < 0.05$	$P < 0.05$	n.s.

n.s. = not significant.

tion during the 3 hours after the evening meal was considerably smaller than that after the morning meal ($P < 0.01$), while on the 3-meal regime there was no difference between the increases after the morning and evening meals.

In the present study we observed that on the 2-meal regime the specific dynamic action (increase of oxygen consumption) (Bradfield and Jourdan, 1973; Duncan, 1964) of an evening meal was significantly spared, and that the insulin response to the evening meal was strikingly enhanced. If lipogenesis is increased by the 2-meal regime, a real increase in efficiency of food utilization may occur. Our findings suggest that the increased efficiency of food utilization is mainly due to the sparing of specific dynamic action, and that enhanced insulin response may by a causal factor in increased lipogenesis. In the night-eating syndrome, overeating at an evening or night meal instead of at a morning meal is necessary for increased lipogenesis.

REFERENCES

Bradfield, R.B. and Jourdan, M.H. (1973): Relative importance of specific dynamic action in weight-reduction diets. Lancet, 2, 640.
Cohn, C. and Joseph, D. (1960): Effects on metabolism produced by rate of ingestion of diet. Amer. J. clin. Nutr., 8, 682.
Dickerson, V.C., Tepperman, J. and Long, C.N.H. (1943): Role of liver in synthesis of fatty acids from carbohydrate. Yale J. Biol. Med., 15, 875.
Duncan, G.G. (1964): Diseases of Metabolism, 5th ed., p. 27. W.B. Saunders and Co., Philadelphia.
Fabry, P., Fodor, J., Hejl, Z., Braun, T. and Zvolankova, K. (1964): The frequency of meals: its relation to overweight, hypercholesterolemia, and decreased glucose tolerance. Lancet, 2, 614.
Hollifield, G. and Parson, W. (1962): Metabolic adaptation to stuff and starve program. I. Studies of adipose tissue and liver glycogen in rats limited to short daily feeding period. J. clin. Invest., 41, 245.
Stunkard, A.J., Grace, W.J. and Wolff, H.J. (1955): Night-eating syndrome, pattern of food intake among certain obese patients. Amer. J. Med., 19, 78.
Tepperman, J. and Tepperman, H.M. (1958): Effect of antecedent food intake pattern on hepatic lipogenesis. Amer. J. Physiol., 193, 55.

Epidemiology of adiposity

K. M. West

University of Oklahoma College of Medicine, Oklahoma City, Okla., U.S.A.

During the past 12 years standardized methods have been used in testing 13 populations in widely separated parts of the world. Among these populations conditions differed widely with respect to geography, culture, diet, race and economic status. In addition to the 10 national populations listed in Table 1, we studied 3 U.S. subpopulations (Cherokee Indians of North Carolina, whites of Bangor, Pennsylvania, and Plains Indians of Oklahoma). Organizations involved in these studies included the Interdepartmental Committee on Nutrition and the National Institutes of Health of the United States, the Institute for Nutrition of Central America and Panama and the Ministries of Health in 10 countries. Weight, height, skinfold thickness and other indices of nutritional status were related to factors such as diet, race, economic status, age, sex, parity, serum cholesterol, diabetes prevalence, etc. The major purpose of this report is to summarize and compare results in these populations with respect to their adiposity. Details concerning methods and workers have been published (West and Kalbfleisch, 1966, 1970, 1971).

ADIPOSITY AND DIABETES

Rates of diabetes in 12 age-matched groups varied more than 10-fold (West and Kalbfleisch, 1970). Rates of diabetes were also very closely linked with adiposity both among and within populations (West and Kalbfleisch, 1966, 1970, 1971). Although differences in rates of diabetes were very great among some of the races, differences were slight among racial or national groups when subjects were matched for adiposity. (West and Kalbfleisch, 1966, 1970, 1971; West, 1972).

EXTREMES OF ADIPOSITY

Because the conventional literature does not have much epidemiologic information on the extremes of massive obesity, I turned to to the *Guinness Book of World Records* (McWhirter and McWhirter, 1972). Apparently the heaviest person concerning whom reliable records are available was Robert E. Hughes, an American who weighed 1069 lb (486 kg)! His height was 6 feet and $\frac{1}{2}$ inch (184 cm). He must have had an adipose tissue mass of roughly 800 lb. Several persons had documented weights of over 800 lb, including a woman of 840 lb whose height was 5 feet 9 inches (175 cm). She was in show business and known as 'Baby Flo'. William J. Cobb, who was 6 feet tall, weighed a husky 232 lb in 1965, but this was after losing 570 lb from his previous weight of 802 lb. At birth Hughes weighed 11.25 lb, at age 6 203 lb, at age 10 546 lb, at age 18 693 lb, at age 25 895 lb, and at age 27 945 lb. This man died at 32 years (of 'uremia'), at which time his weight had declined from 1069 to 1035 lb. Baby Flo died at age 35. She weighed 10 lb at birth, 267 lb at age 11, and 621 lb at 25. We need more systematically collected data on the weight histories of the massively obese and similar data for their close relatives. Mr. Ben McCrory of North Carolina was the smallest of twins. In 1970 he weighed 640 lb, while his brother weighed 660 lb (McWhirter and McWhirter, 1972). A man who was well studied by

Bortz (1969) weighed 691 lb before losing 503 lb. He was only 69 inches (175 cm) tall. Calculations by Bortz suggested that only 5.8% of the 503 lb weight loss was body protein.

Successful marathon runners often have an adipose tissue mass equivalent to less than 5% of body weight (usually less than 3 kg) (Costill, 1972). In our studies in East Pakistan, 28% of the population weighed less than 70% of standard, and weights below 60% of standard were observed in 136 instances (2.1% of the general adult population)! The triceps skinfold thickness in these individuals was usually less than 3 mm (skin only?). The entire adipose tissue mass of such persons is probably less than 1 kg. Thus, the observed range of adipose tissue mass in the 'normal' population extends from less than 1 kg to more than 300 kg!

DIFFERENCES IN ADIPOSITY AMONG AND WITHIN POPULATIONS

Table 1 shows the mean percentage of standard weight for males and females in 10 groups, each of which is crudely representative of the general population of a country. The marked differences in the adiposity of some of these populations are striking. Table 2 compares in greater detail the fattest and the leanest of these populations. The leanest group we have tested were the men of rural East Pakistan. More than half of the adult men had a triceps skinfold thickness of less than 4 mm! Also, the majority of these men had a subscapular skinfold of less than 5 mm and

TABLE 1   Mean percentage of standard weight by age and sex in ten countries

Country	No. of subjects	Age group					
		20-24	25-34	35-44	45-54	55-64	>64
Costa Rica							
Rural							
Male	452	90.9	96.1	96.3	97.5	97.0	92.1
Female	649	98.4	100.4	110.4	111.6	108.8	95.6
Urban							
Male	45	96.8	99.4	100.8	106.8	97.5	98.0*
Female	Data lost						
El Salvador							
Rural							
Male	373	89.3	90.3	92.2	90.0	87.7	87.5
Female	570	97.3	95.6	97.9	99.4	93.1	89.0
Urban							
Male	49	91.5	92.8	105.2	102.5	92.5	109.2*
Female	111	106.3	106.5	112.2	114.4	118.6	96.5
Guatemala							
Rural							
Male	518	87.8	92.6	92.0	92.8	91.0	85.6
Female	642	93.9	98.8	102.1	99.0	94.0	93.9
Urban							
Male	109	87.5	97.2	96.8	100.8	100.8	99.2
Female	151	101.9	98.7	104.6	105.9	114.5	104.6
Honduras							
Rural							
Male	385	89.9	91.8	93.2	94.3	93.8	95.6
Female	539	94.3	98.3	103.0	102.0	97.8	91.5
Urban							
Male	75	97.1	98.1	98.7	104.4	103.3	80.0*
Female	125	98.5	103.2	110.1	111.7	104.4	99.8

TABLE 1   (continued)

Country	No. of subjects	Age group					
		20-24	25-34	35-44	45-54	55-64	>64
Nicaragua							
Rural							
Male	359	94.5	92.8	96.6	99.5	96.6	94.2
Female	595	98.0	105.3	105.7	106.7	104.8	99.8
Urban							
Male	46	96.8	104.5	112.5	93.9	131.3	99.2*
Female	103	98.8	105.3	121.7	116.8	111.7	122.5*
Panama							
Rural							
Male	403	91.0	93.6	93.3	92.0	89.7	90.5
Female	498	93.4	98.2	102.9	100.9	99.9	97.0
Urban							
Male	40	82.5	97.1	108.4	101.7	117.5	100.8*
Female	80	95.8	99.6	106.3	114.3	116.9	102.5*
Venezuela							
Males	340	95.4	102.6	107.1	105.8	101.6	98.2
Females	1299	97.2	103.5	110.3	112.5	102.6	100.6
Malaya							
Males	332	In Malaya no age-related data are available but in the general					
Females	234	population over 29 years of age the mean percentage of standard weight was 88 for men and 94 for women. Mean age of this group was 47.					
East Pakistan							
Rural							
Males	3148	76.0	76.6	76.3	75.7	75.8	73.1
Females	1465	80.6	78.8	78.0	77.2	74.5	72.9
Urban							
Males	1007	78.6	78.1	79.2	78.2	79.6	73.2
Females	273	79.6	82.6	85.1	87.0	82.6	85.0*
Uruguay							
Males	1010	96.5	109.3	110.8	114.2	111.0	110.0
Females	1701	101.5	112.0	122.7	129.8	127.5	126.0

* Less than 15 subjects.

weighed less than 77% of standard. In contrast, only 1% of the Cherokee Indians we studied in North Carolina were this thin, and about 1/3 of adult Cherokees weighed more than 129% of standard. Less than 1% of the males of rural East Pakistan weighed as much as 100% of standard. The heaviest population for which I have seen specific data is the group of Pima Indians studied in Arizona by Bennett et al. (1970). In a group of subjects between 35 and 54 years of age, 55% of the men and 89% of the women weighed more than 126% of 'desirable' standard. A majority of the Pima Indians in this age group have diabetes. Diabetes is roughly 30 times more common in Pima Indians than in age-matched residents of East Pakistan. The mean adipose tissue mass of middle-aged Pima Indians is roughly 10 times greater than the adipose tissue mass of the people in East Pakistan.

EFFECTS OF AGE AND SEX ON ADIPOSITY

We wanted to know whether the apparent effects of age and sex on adiposity previously observed in relatively affluent societies were peculiar to the cultural and social circumstances of these advanced societies, or whether they were biologic phenomena largely or wholly independent of environmental circumstances. The data in Table 1 show that in all 9 populations of Central America, South America and Asia, weight in relation to height declined in both sexes after the age of 64 (as it usually does in richer populations). As shown in Table 1, mean standard weight began to decline before the seventh decade in some of the populations. It should be kept in mind that even when matched for weight in relation to height, older subjects are usually somewhat fatter and less muscular than younger subjects. But skinfold measurements confirmed that subjects over 64 years of age had less adiposity. For example, in the men of Uruguay, subscapular skinfold thickness averaged 15 mm in men aged 55-59, but only 12 mm in men aged 60-64. In the poorest societies such as East Pakistan and Panama there was little or no difference between the weights or skinfold measurements of young adults as compared to middle-aged adults. Thus, increasing adiposity with age is not seen under all environmental circumstances. In most of the populations we studied, age-related increase in weight persisted to a higher age in females than in males (as it usually has in more affluent populations). Generally speaking, however, this particular difference among the sexes was less marked or absent in those populations that were very lean.

In most of the populations for which data are reported in Table 1, women seem to have gained more weight with age than men. This has usually been true in the more affluent societies. There are some interesting differences, however, in this respect. In rural Pakistan middle-aged women were not fatter than young women, and there was no difference between the sexes in age-related weight change. In contrast, adiposity was far greater in middle-aged black women of Oklahoma than in black men. The Indian women of Oklahoma also gained much more weight than men in the third, fourth, and fifth decades. It will be interesting to investigate further the causes of these marked differences among societies with respect to the effect of sex on age-related adiposity. These effects of age and sex are probably influenced by biologic, genetic, economic, and cultural factors.

In interpreting the data in Tables 1 and 2, it should be kept in mind that a woman of a standard weight of 100% usually has proportionately more fat than a man with a standard weight of 100%. In East Pakistan, although the mean standard weights of men and women were very similar, the mean thickness of the triceps skinfold was approximately twice as great in women. The thickness of the subscapular skinfold was only very slightly greater in women. In rural Central America there was very little difference between the weight of boys and girls 18 years of age, but the mean triceps skinfold of the girls was 14 mm, while that of the boys was only 7 mm. Evidence of this kind suggests that the greater adiposity of women as compared to men which has been observed previously in the more affluent groups, such as U.S. whites, is

TABLE 2 Comparison by age of skinfold measurements and percentage of standard weight in women of Uruguay and of rural East Pakistan

Age group	Mean width of triceps fold (mm)		Subscapular		Mean % of standard weight	
	Uruguay	E. Pakistan	Uruguay	E. Pakistan	Uruguay	E. Pakistan
25-34	19.6	6.8	17.8	6.7	114	79
35-44	21.7	7.2	18.1	6.9	124	78
45-54	24.5	7.0	22.9	6.9	132	77
55-64	23.0	6.5*	18.1	6.0*	129	74
>64	22.4		19.4		128	73

* Includes all subjects more than 54 years of age.

probably not mainly attributable to cultural, racial or environmental circumstances. Rather, it is a biologic phenomenon observed regularly under varying social, economic, cultural and racial circumstances. In contrast, the propensity of women to gain adiposity in the third, fourth and fifth decades at a rate faster than men (and for a longer period) may be strongly influenced by cultural, economic, social or racial factors. In Tecumseh, Michigan young women had triceps folds that were about 50% thicker than their subscapular folds (Montoye et al., 1965). But in rural East Pakistan these differences were very slight (Table 2). Although some evidence has been presented to suggest that distribution of fat is independent of adipose mass, the latter observations suggest that very lean women may have a smaller percentage of their fat in the triceps pad.

FACTORS ASSOCIATED WITH ADIPOSITY WHICH MAY BE ITS CAUSE

Dietary factors

In 12 populations we related the characteristics of diet to the degree of adiposity (West and Kalbfleisch, 1971). In 10 of these populations nutritional studies were quite detailed and in 2 populations the characteristics of the diet were evaluated only crudely. It was neither surprising nor exciting to find that people who ate more and exercised less were inclined to be fat. But our results with respect to the the characteristics of the diet may be of greater interest. These results have already been published in detail (West and Kalbfleisch, 1971; West, 1972). I will mention only a very few of our findings and conclusions. Although most fat populations consume a diet high in fat, high-fat diets are not invariably associated with a high prevalence of obesity. It may be noted, for example, that the diet of certain Masai tribesmen is high in fat, and they are quite lean.

In most populations that are fat, sugar consumption is high. This has led some to assign a very important etiologic role to sugar intake in the production of adiposity. In several populations including Bantus, Polynesians, Eskimos and North American Indians, rising sugar consumption has been associated with a rising prevalence of obesity (West, 1972). In our studies there was a fairly good correlation between sugar consumption and adiposity (West and Kalbfleisch, 1966, 1971; West, 1972). On the other hand, certain of our data suggested the possibility that this association might be partially or wholly coincidental. For example, sugar consumption was greater in Costa Rica than in Uruguay, but adiposity in Uruguay was much greater. Sugar consumption was about the same in Venezuela and Malaya, but adiposity in Venezuela was considerably greater than in Malaya. In widely separated parts of the world there are poor people who are sugar cane cutters. These groups have not been well studied, but for the most part they seem to be quite lean even though sugar consumption in these groups is usually high. It is difficult, however, to find an obese population in which sugar intake is low. But there are a few examples. I do not know the level of sugar consumption in the very fat Sumo wrestlers of Japan, but in general the Japanese diet is low in sugar. The 19th century kings of Karagwe (western shore of Lake Victoria) kept a harem of massively obese young women. They were fattened on milk, and had no refined sucrose. Thus, these epidemiologic observations neither confirm nor refute the hypothesis that sugar consumption is an important cause of adiposity.

Social, economic and cultural factors

In 13 populations we have studied there was a very consistent relationship between income and adiposity up to a level of about $700 per capita per year. At that point the regularity of this relationship terminates. For example, the people of Montevideo had approximately the same adiposity as those we tested in Bangor, Pennsylvania, but income levels were at least 3 times higher in Bangor. In the United States some poor subpopulations are leaner and some fatter than their more affluent countrymen. In Oklahoma poor blacks and poor Indians are much fatter than more affluent white, black or red people. In one U.S. study obesity was 7 times more prevalent in a group of poor young girls as compared to a more affluent group (Moore et al., 1962). It should be pointed out that these 'poor' subpopulations in the United States are surrounded by a very affluent society and that their income levels are relatively high by international standards even though they are considered to be 'deprived' in the United States. We do not yet know the extent to which economic, social or cultural factors are responsible for obesity in certain of the poor subpopulations of affluent societies. Probably a combination of factors is responsible. Our preliminary evaluations in the Plains Indians suggest, for example, that cultural attitudes play a strong role. In certain situations high calorie concentrated foods are cheaper than other dietary alternatives.

In rural Central America we compared the adiposity of 3 subpopulations who were classified on the basis of a determination of their socioeconomic status. Generally speaking, the subpopulations considered to be 'poor' had per capita incomes of approximately $200 per year, while the 'rich' had levels of income averaging about $900 per capita per year. The subgroup assigned a 'medium' status had an income level of intermediate degree. For men the mean percentages of standard weights were 89, 93 and 97, respectively, for the low, medium and high status subgroups. For women, the percentages of standard weights were 97, 99 and 107, respectively, for these 3 subgroups.

We noted above some possible differences in the distribution of subcutaneous fat among ethnic groups. Others have noted this also, but it is often difficult to determine the extent to which these differences are produced by racial or genetic factors, or by social, cultural, economic or other environmental factors (Robson et al., 1971).

Other factors

In our international studies we were unable to measure exercise levels with precision. But, generally speaking, it was evident that the lean populations exercised more than the fat ones. It is interesting that obesity is quite rare in wild animals.

We did not systematically study genetic factors, but we were impressed with the lack of evidence for any racial differences. For example, in Malaya adiposity was quite similar in Chinese, Malays and Indians. Blacks of Panama were far leaner than blacks of Oklahoma, and the fatness of the rural Panama blacks was the same as that of Panamanians of other races. Whites of Costa Rica were very lean, and whites of Montevideo were fat. American Indians of Guatemala were quite slender and American Indians of North Carolina were quite fat. Our experiences were consistent with the notion that adiposity is not strongly related to race in itself, but they suggested a strong role for social, economic and cultural factors.

In Uruguay and Venezuela, fatness was associated to a significant degree with parity. In 8 leaner populations this was not the case.

REFERENCES

Bennett, P.H., Burth, T.A. and Miller, M. (1970): The high prevalence of diabetes in the Pima Indians of Arizona, U.S.A. In: Diabetes Mellitus in Asia, p. 33. Editors: S. Tsuji and M. Wada. ICS 221. Excerpta Medica, Amsterdam.
Bortz, W.M. (1969): A 500 pound weight loss. Amer. J. med., 47, 325.
Costill, D.L. (1972): Physiology of marathon running. J. Amer. med. Ass., 221, 1024.
McWhirter, N. and McWhirter, R. (1972): Guinness Book of World Records, 17-18. Sterling Publishing Co., Inc., New York.
Montoye, H.J., Epstein, F.H. and Kjelsberg, M.O. (1965): The measurement of body fatness. A study in a total community. Amer. J. clin. Nutr., 16, 417.
Moore, M.E., Stunkard, A. and Srole, L. (1962): Obesity, social class, and mental illness. J. Amer. med. Ass., 181, 962.
Robson, J.R.K., Bazin, M. and Soderstrom, R. (1971): Ethnic differences in skin-fold thickness. Amer. J. clin. Nutr., 24, 864.
West, K.M. (1972): Epidemiologic evidence linking nutritional factors to the prevalence and manifestations of diabetes. Acta diabet. lat., 9, 405.
West, K.M. and Kalbfleisch, J.M. (1966): Glucose tolerance, nutrition, and diabetes in Uruguay, Venezuela, Malaya and East Pakistan. Diabetes, 15, 9.
West, K.M. and Kalbfleisch, J.M. (1970): Diabetes in Central America. Diabetes, 19, 656.
West, K.M. and Kalbfleisch, J.M. (1971): Influence of nutrition on prevalence of diabetes. Diabetes, 20, 99.

DISCUSSION

Angel: As you know obesity is associated with a marked increase in total body cholesterol stores (most of which is found in fat tissue) and an increase in endogenous cholesterol production. Do you have any population-based correlative data between obesity and diseases of cholesterol metabolism such as cholelithiasis or hypercholesterolemia?

West: We have published the mean cholesterol levels for men and women in the 10 populations (West and Kalbfleisch, 1966, 1970, 1971). There is a very good correlation between adiposity and cholesterol levels. But it should be kept in mind that this could be partly or wholly attributable to the fact that increasing affluence was also associated with increasing levels of dietary cholesterol and animal fat. In Central America we related adiposity to serum cholesterol levels. In subjects whose weights were 80-89% of standard the mean cholesterol values were 135 mg/100 ml; in subjects weighing more than 109% of standard the mean level was 151 mg/100 ml. It should be kept in mind, however, that most of those in the latter subgroup had only mild adiposity.

In the Plains Indians of Oklahoma, who are quite fat, cholelithiasis is extraordinarly common. Their serum cholesterol levels have not yet been systematically determined. However, it appears that they are not higher than in the general white population of the United States. We are now studying these Indians to confirm or refute the clinical impression of the local physicians that rates of artherosclerosis are low in the diabetics. Diabetes is present in about one third of middle-aged adults in this population. These Indians have an interesting peculiarity of their fat distribution. When compared with whites in the U.S.A. and Europe, they have a greater thickness of the subscapular pad in proportion to the triceps pad.

Epidemiological study of adipose mass in a population of men in their fifties: relationship to glucose, lipid, insulin, growth hormone and cortisol levels

E. Eschwege, G. Rosselin, J. Lellouch, A.J. Valleron, J.R. Claude, J.M. Warnet and **J.L. Richard**

Groupe d'Etude sur l'Epidémiologie de l'Athérosclérose (GREA)*, Paris, France

An oral glucose tolerance test was carried out at the second check-up in a prospective survey to measure, in a presumably healthy population, the frequency of cardiovascular disease caused by atherosclerosis, in relation to various clinical and biological parameters. In addition to a clinical examination, glucose, free fatty acid (FFA), triglyceride, insulin (IRI), growth hormone (HGH) and cortisol were assayed in each blood sample. The experimental details have been previously described (Eschwege et al., 1970; Rosselin et al., 1970, 1971). This communication will describe only the relationship between obesity and the hormonal and metabolic parameters measured during the oral glucose tolerance test. We will discuss these relationships firstly in all subjects, and secondly in normals compared to diabetics.

Ideal weight was obtained from the tables given by Documenta Geigy (1963), using the upper limit for the range for 'medium frame'. The degree of obesity was calculated as the difference between actual and ideal weight expressed as a percentage of ideal weight. Four weight categories were defined: non-obese = normal or below normal weight, plump = 1-10% above normal, moderately obese = 11-20% above normal, very obese = more than 20% above normal.

Among the 1828 subjects examined, 88 diagnosed diabetics, that is 5% of all subjects, and 22 with incomplete data were excluded. In the remainder, we defined: normal glycemic subjects with fasting plasma glucose < 110 mg/100 ml and 2-hour plasma glucose < 130 mg/100 ml, and newly diagnosed diabetics with 2-hour plasma glucose > 160 mg/100 ml.

RESULTS

All unknown diabetic subjects

The mean values of the parameters measured are indicated for the different weight categories in Table 1. In very obese subjects the fasting plasma IRI was 3-fold greater and the 2-hour plasma IRI 2-fold greater than in non-obese subjects. The plasma HGH and cortisol levels decreased as the degree of obesity increased, particularly for fasting plasma HGH and 2-hour plasma cortisol. Plasma glucose and serum triglyceride increased with increasing obesity. Fasting plasma FFA decreased with increasing obesity, except for the highest weight category, but the 2-hour values for FFA were similar in all categories. Correlation coefficients for these relationships are shown in Table 2. The correlations between obesity and fasting and 2-hour IRI, plasma glucose and triglyceride were strongly positive. Those between obesity and fasting and 2-hour HGH and cortisol were significantly negative. Because of the interrelationship of the different parameters, it is difficult to determine what is specifically due to obesity. Calculation of partial correlation coefficients suggests that obesity is separately linked to triglyceride, fasting and 2-hour IRI and fasting plasma glucose.

* GREA includes INSERM (Unité de Recherches Statistiques, Unité de Diabétologie, Section de Cardiologie), the Service de Dépistage Systématique de la Tuberculose et des Affections Cardiovasculaires, the Clinique Cardiologique, Hôpital Boucicaut and the Groupe d'Etudes sur la Fumée du Tabac.

TABLE 1   Mean fasting and 2-hour plasma IRI, HGH, cortisol, glucose, triglyceride and FFA levels by weight category in 1718 subjects

	Non-obese (No. = 313) (18%)	Plump (No. = 421) (25%)	Moderately obese (No. = 449) (26%)	Very obese (No. = 535) (31%)	P	Correlation coefficient*	P
IRI (μU/ml)							
Fasting	6	9	10	15	< 0.001	0.43	<0.001
2 hours	28	37	42	57	< 0.001	0.33	<0.001
HGH (ng/ml)							
Fasting	17	10	8	7	< 0.001	−0.18	<0.001
2 hours	8	6	5	5	< 0.05	−0.11	<0.001
Cortisol (ng/ml)							
Fasting	120	118	114	116	< 0.05	−0.05	
2 hours	94	89	83	84	< 0.001	−0.14	<0.001
Glucose (mg/100 ml)							
Fasting	97	100	102	105	< 0.01	0.29	<0.001
2 hours	96	99	101	109	< 0.001	0.16	<0.001
Triglyceride (mg/100 ml)	92	112	129	159	< 0.001	0.27	<0.001
FFA (mg/l)							
Fasting	101	97	95	104	< 0.01	0.04	
2 hours	43	42	41	42		0.01	

* Correlation coefficients were calculated for each parameter and the percentage of obesity.

TABLE 2   Correlation coefficients of the parameters measured in 1718 subjects

	IRI	HGH		Cortisol		Glucose		FFA		Triglyceride	% obesity
	2 hours	Fasting	2 hours	Fasting	2 hours	Fasting	2 hours	Fasting	2 hours		
IRI											
Fasting	0.48 ***	− 0.05	—	0.03	—	0.33 ***	—	0.09 ***	—	0.24 ***	0.43 ***
2 hours	—	—	− 0.05	—	0.05	—	0.49 ***	—	− 0.06*	0.20 ***	0.33 ***
HGH											
Fasting	—	—	0.34 ***	0.04	—	0.00	—	0.04	—	− 0.09 **	− 0.18 ***
2 hours	—	—	—	—	0.05	—	0.01	—	0.08 **	− 0.06 *	− 0.11 ***
Cortisol											
Fasting	—	—	—	—	0.47 ***	0.17 ***	—	0.25 ***	—	0.02	− 0.05
2 hours	—	—	—	—	—	—	0.15 ***	—	0.21 ***	− 0.07 *	− 0.14 ***
Plasma glucose											
Fasting	—	—	—	—	—	—	0.48 ***	0.19 ***	—	0.20 ***	0.29 ***
2 hours	—	—	—	—	—	—	—	—	0.11 ***	0.17 ***	0.16 ***
FFA											
Fasting	—	—	—	—	—	—	—	—	0.41 ***	0.21 ***	0.04
2 hours	—	—	—	—	—	—	—	—	—	0.19 ***	0.01
Triglyceride	—	—	—	—	—	—	—	—	—	—	0.27 ***

Significance: * P < 0.05, ** P < 0.01, *** P < 0.001.

TABLE 3   Mean fasting and 2-hour plasma IRI, HGH, cortisol, glucose, triglyceride and FFA by weight category in 95 newly diagnosed diabetics

	Non-obese (No. = 16 (17%))	Plump (No. = 19 (20%))	Moderately obese (No. = 18 (19%))	Very obese (No. = 42 (44%))	Significance	Correlation coefficient*
IRI (µU/ml)						
Fasting	8	10	11	24***	***	0.55***
2 hours	57***	65***	65***	112***	***	0.37**
HGH (ng/ml)						
Fasting	19	13	8	12*		−0.11
2 hours	13	4	5	7*		0.08
Cortisol (ng/ml)						
Fasting	135*	132*	142***	125**		−0.11
2 hours	122***	109**	109***	81	**	−0.39**
Glucose (mg/100 ml)						
Fasting	101***	115***	117***	123***	*	0.45***
2 hours	178***	189***	198***	200***		0.25*
Triglyceride (mg/100 ml)	128***	126	161	179**		0.25*
FFA (mg/l)						
Fasting	171***	142***	117***	152***	*	−0.08
2 hours	48	56***	49*	53***		0.01
Obesity %	−9	6	14	33**	***	

Significant differences from normal values: * P < 0.05, ** P < 0.01, *** P < 0.001.
* Correlation coefficients were calculated for each parameter and the percentage of obesity.

TABLE 4   Comparison of the parameters in normal and diabetic subjects for mean fasting and 2-hour values and for their relationship to the percentage of obesity

	Normals	Diabetics	Significance	Correlation coefficient (regression slopes) between % obesity and each parameter		Comparison of slopes
				Normals	Diabetics	
IRI (μU/ml)						
Fasting	9.5	15	***	0.39***	0.55***	***
2 hours	36	84.5	***	0.29***	0.37***	***
HGH (ng/ml)						
Fasting	9.5	12.3		-0.18***	-0.11	
2 hours	5.4	7.2		-0.07*	-0.08	
Cortisol (ng/ml)						
Fasting	113	131	***	-0.07*	-0.11	
2 hours	84	100	***	-0.15***	-0.39**	*
Glucose (mg/100 ml)						
Fasting	97	116	***	0.24***	0.45***	***
2 hours	90	189	***	0.09***	0.25*	***
Triglyceride (mg/100 ml)	120	155	***	0.27***	0.25*	
FFA (mg/l)						
Fasting	92	146	***	0.02	-0.08	
2 hours	40	52	***	-0.02	0.01	
Obesity %	12	17	***	—	—	

Significance: * $P < 0.05$, ** $P < 0.01$, *** $P < 0.001$.

Effect of diabetes on the relationship between obesity and hormonal or metabolic parameters

Table 3 shows the mean values of the different parameters for newly diagnosed diabetic subjects (2-hour plasma glucose > 160 mg/100 ml) according to the previously defined weight categories. As the degree of obesity increased, the mean values of fasting and 2-hour IRI and fasting plasma glucose increased, and the mean 2-hour cortisol value decreased. The fasting FFA values decreased with increasing obesity, except for the highest weight category, as in the normal subjects. Significant differences between subjects with normal and abnormal plasma glucose values in each weight category are also indicated in Table 3. The 2-hour IRI, fasting and 2-hour cortisol and fasting FFA were significantly higher in the diabetics. In the very obese group, all parameters differed significantly between diabetics and normal subjects except for the 2-hour FFA.

In Table 4 the differences between normals and diabetics are summarized without regard to weight. Also in this table is a comparison of the correlation coefficients between obesity and the different parameters in normals and diabetics. These comparisons were obtained by covariance analysis. The relationships between obesity and the plasma parameters were significantly different in diabetics compared to normals for fasting and 2-hour IRI and plasma glucose and 2-hour cortisol. The correlations between obesity and glucose or IRI were highly significant and positive in both groups and greater in diabetics than in normals. The regression slopes of cortisol on obesity and of obesity on cortisol were significantly different in normals compared to diabetics. The correlation was significantly negative in both groups but it was greater in diabetics than in normals.

DISCUSSION AND CONCLUSIONS

Difference in food intake has been shown to be highly correlated with the appearance of obesity (Strang, 1964). In this population the most important difference between obese and non-obese subjects was the higher alcohol calorie intake and the lower protein/total calories ratio (Pequignot et al., 1973). The relationship between chronic modification of diet of this type and changes of hormonal pattern is still unknown. However, it may be supposed that a higher IRI level is the link between this modification of the diet and both the increase in weight and the increase in triglyceride level. Our finding of an association of obesity with a high IRI level and a low HGH level is in accordance with previous work (Gajdos, 1971; Hales et al., 1968; Karam et al., 1965; Malaisse et al., 1968; Tchobroutsky et al., 1969; Vague et al., 1969; Yalow et al., 1965). We also found that obesity was negatively correlated with the cortisol level and it has recently been shown that plasma glucagon is low in obese subjects. Thus, it is difficult to attribute the resistance of the plasma glucose level to IRI in the obese to these hormonal changes.

The relationship between obesity and the different parameters was not greatly modified in diabetics. Thus, diabetic obesity does not appear to be characterised by a particular set of variables. However, diabetes makes still more apparent the link between obesity and parameters such as glucose, cortisol and IRI levels.

REFERENCES

Documenta Geigy (1963): Naturwissenschaftliche Tabellen, 6th ed., 634. J.R. Geigy AG, Basel.
Eschwege, E., Valleron, A.J., Rosselin, G., Claude, J.R., Warnet, J.M. and Richard, J.L. (1970): Diabetes and coronary heart disease epidemiological study. II. Multivariate statistical analysis of the glucose, lipid and hormonal levels systematically measured during 0-2 hours OGTT. In: Abstracts, VII Congress of the International Diabetes Federation, Buenos Aires, 1970, abstract no. 283. ICS 209. Excerpta Medica, Amsterdam.
Gajdos, A. (1971): Biochimie du tissu adipeux. In: Médecine et biochimie, problèmes d'actualité, 2nd series, p. 295. Masson et Cie, Paris.
Hales, C.N., Greenwood, F.C., Mitchell, F.L. and Straud, W.T. (1968): Blood glucose, plasma insulin and growth hormone concentration of individuals with minor abnormalities of glucose tolerance. Diabetologia, 4, 73.

Karam, J.H., Grodsky, G.M. and Forsham, P.H. (1965): The relationship of obesity and growth hormone to serum insulin levels. Ann. N.Y. Acad. Sci., 131, 374.

Malaisse, W., Malaisse-Lagae, F. and Coleman, D.L. (1968): Insulin secretion in experimental obesity. Metabolism, 17, 802.

Pequignot, G., Papoz, L., Cubeau, J. and Eschwege, E. (1973): Comportement alimentaire et anomalies 'limites' de l'hyperglycemie provoquée. Congrès International de diététique, Hanovre, 1973, p. 88.

Rosselin, G., Claude, J.R., Eschwege, E., Patois, E., Warnet, J.M. and Richard, J.L. (1971): Diabetes survey. Plasma insulin during 0-2 hour oral glucose tolerance tests systematically carried out in a professional group. I. Relationship with plasma glucose, free fatty acids, cholesterol, triglycerides and corpulence. Diabetologia, 7, 34.

Rosselin G., Claude, J.R., Eschwege, E., Warnet, J.M. and Richard J.L. (1970): Epidemiological study of diabetes and coronary heart disease. I. The implication of the 75 g 0-2 hour oral glucose test (0-2 h OGTT) in the detection and definition of diabetes. In: Abstracts, VII Congress of the International Diabetes Federation, Buenos Aires, 1970, abstract no. 309. ICS 209. Excerpta Medica, Amsterdam.

Strang, J.M. (1964): In: Obesity – Disease of metabolism, p. 735. Editors: G. Garfield and W.B. Durreau. Saunders Co., Philadelphia and London.

Tchobroutsky, G., Rosselin, G., Assan, R., Freychet, P. and Derot, M. (1969): Growth hormone secretion in obese subjects with and without diabetes. In: Physiopathology of Adipose Tissue, p. 269. Editor: J. Vague. Excerpta Medica, Amsterdam.

Vague, P., Boeuf, G., Depieds, R. and Vague, J. (1969): Plasma insulin levels in human obesity. In: Physiopathology of Adipose Tissue, p. 203. Editor: J. Vague. Excerpta Medica, Amsterdam.

Yalow, R.S., Glick, S.M., Roth, J. and Berson, S.A. (1965): Plasma insulin and growth hormone levels in obesity and diabetes. Ann. N.Y. Acad. Sci., 131, 357.

Relationship between the age of appearance of obesity and adipocyte diameter in 256 obese and 57 non-obese women

M. Apfelbaum, L. Brigant and **F. Duret**

Unité de Recherches Diététiques, Hôpital Bichat, Paris, France

Both the number and size of adipocytes are increased in obese subjects as compared to controls (Björntorp and Sjöström, 1971; Hirsch and Knittle, 1970), when these are considered as homogeneous groups. However, it may be questioned whether the obese group really is homogeneous or whether there exist subgroups, characterized by predominance of hyperplasia or hypertrophy. The existence of such subgroups would have two consequences: (1) as abnormal serum insulin is correlated with hypertrophy (Björntorp and Sjöström, 1971; Bray 1970) and not with cell number, metabolic disorders would be expected only if hypertrophy occurs; (2) hyperplasia could be prevented only in childhood, since the number of adipocytes seems to remain unchanged throughout adulthood. In adult men, weight loss in spontaneous (Hirsch and Knittle, 1970; Bray, 1970) or induced (Salans et al., 1971) obesity changes only adipocyte size. Bonnet et al. (1970), in a study of 22 normal children, showed that adipocyte size remained unchanged from the age of 5 to 15, indicating that cell number increases to account for the growth of adipose mass. Brook (1972) found the same results in a study of 64 normal and 54 obese children. He also showed that the mean adipocyte volume was smaller in obese adults with a history of obesity since childhood than in those who became obese in adulthood. Unfortunately this latter study was carried out on only 25 subjects.

Obesity appearing in childhood could therefore be due to hyperplasia and/or hypertrophy, whereas obesity appearing later in life could only be hypertrophic. What is the age limit for appearance of hyperplastic obesity in man? In the rat it is around 15 weeks (Hirsch and Han, 1969), which is about 1/7th of a rat's life or equivalent to 10 years in a man's life. However, puberty in the rat, which occurs at 6 weeks, does not stop adipocyte multiplication. Thus, extrapolating from rat to man seems particularly inappropriate in this case.

The present study compares mean adipocyte diameters in obese subjects classified according to the age of appearance of their obesity, with particular attention given to the frequency of hyperplasia when obesity appears in adolescence.

METHODS

Subjects were divided into 2 groups, 'obese' and 'non-obese'. The non-obese group included 26 subjects whose weight was between 95% and 105% of ideal weight (Metropolitan Life Insurance Co., *Statistical Bulletin,* 1959, Vol. 40) and had always been in that range, and 31 subjects who had consulted for obesity in the past, but whose weight was then between 90% and 110% of ideal. The obese group included 256 subjects whose weight exceeded 110% of the ideal. The subjects in the obese group were classed in 3 subgroups according to their answers to questions about the age of appearance of their obesity. Age categories were arbitrarily fixed as follows: childhood up to 13 (subgroup I), adolescence from 13 to 20 (subgroup II), and adulthood over 20 (subgroup III). Subjects who reported having become overweight after a pregnancy occurring before the age of 20 were classed in the adult group.

A subcutaneous abdominal biopsy was performed after Xylocain (lidocaine) anesthesia with a 6 mm needle connected to a vacuum pump. A 20 to 50 mg sample of adipose tissue was thus collected in a 25 ml container set up between the needle and the pump. The tissue was collected in 3 ml of a Krebs solution containing 3 mM of glucose and 4% bovine serum albumin (Pentex, Miles Lab., Kanakee, U.S.A.) with 5 mg of collagenase added (Worthington Biochem., New Jersey, U.S.A.). This preparation was poured into a plastic tube and incubated for 60 minutes at 37° C with gentle shaking. The infranatant was then removed with a plastic needle and the cells were washed twice with the Krebs solution to remove the collagenase.

The cells, without fixation, were placed on slides and photographed through a microscope. The adipocyte diameters were measured and distributed in frequency classes with a semi-automatic counter previously calibrated to a micrometric scale, which had also been photographed under the microscope, with a total magnification of 200. The validity of the method will be discussed elsewhere. For each subject, the images of at least 300 adipocytes were measured to determine the mean diameter, and the distribution of population was characterized by the skew and the kurtosis.

RESULTS

The data for mean adipocyte diameter and distribution in obese and non-obese groups are shown in Table 1. The non-obese group includes the 'control' subgroup and the subgroup of subjects who had consulted for obesity but were not overweight at the time of the study. No

TABLE 1   Relationship between the age of appearance of obesity and mean adipocyte diameters

Group	No. of subjects	Mean adipocyte diameter	% of hyper-plastic cases*
Non-obese	57	96.4 ± 1.56	0
Obese	256	109 ± 0.79	20
Subgroup I (obesity appeared in childhood)	80	107.1 ± 1.48	24
Subgroup II (obesity appeared in adolescence)	50	105.9 ± 1.76	31
Subgroup III (obesity appeared in adulthood)	126	110.8 ± 1.11	13

* Hyperplastic cases were defined as those with a mean diameter not greater than the 'non-obese' mean + 2 S.D.

Comparison of mean adipocyte diameter

		Variance ratio	Significance level of F	
Non-obese	vs. obese	57	0.001	(10.8)
Group I	vs. Group III	3.58	0.10	(2.71)
Group II	vs. Group III	4.67	0.05	(3.90)
Group I + II	vs. Group III	5.97	0.025	(5.02)

significant differences were found between mean adipocyte diameter and distribution parameters of these subgroups, which validates their inclusion in a single group.

When subgroups were formed according to the age of appearance of obesity it was found that the mean adipocyte diameter was significantly smaller in subgroups I and II than in subgroup III. Brook et al. (1972), studying a group of 21 obese females, found a mean lipid content per cell of 0.80 μg for subjects with onset of obesity in childhood, and a mean lipid content per cell of 1.01 μg for subjects with onset in adulthood. Knittle and Ginsberg-Fellner (1972) worked on 5 very obese adolescents and 5 very obese adults. Using their data we have calculated that in the adolescents the mean adipocyte weight was 0.60 μg and in the adults 0.857 μg (Knittle and Ginsberg-Fellner do not state the age of onset of obesity in their adult group but for their group of adolescents it was clearly not in adulthood).

The mean adipocyte diameter distribution was more homogeneous in subgroup III than in subgroups I and II. χ^2 was 12.17 in subgroup I (50%), 7.47 in subgroup II (75%) and 4.54 in subgroup III (97.5%). The greater homogeneity of subgroup III can be related to the lower frequency of purely hyperplastic cases (a purely hyperplastic case was defined by a mean adipocyte diameter no greater than that of the non-obese group + 2 S.D.). Thus, subjects who became obese in adulthood were characterized by a greater mean adipocyte diameter and a lower frequency of hyperplasia.

DISCUSSION

It is possible that the relationship between the age of appearance of obesity and hyperplastic predominance could have been explained by some differences in the history and present degree of obesity within the subgroups. For each subgroup the regression of present obesity against mean adipocyte diameter was computed (for subgroup I, $r = 0.341$; subgroup II, $r = 0.671$; subgroup III, $r = 0.490$). It appeared that there were no significant differences between subgroups.

There was a highly significant correlation ($r = 0.556$, $P < 0.001$) between mean adipocyte diameter and present obesity. This correlation might explain the correlation between mean adipocyte diameter and age of appearance of obesity. There are two arguments against this proposition: firstly, the mean present obesity was not significantly different between subgroups (% overweight: subgroup I, 36 + 2.9; subgroup II, 31 + 2,9; subgroup III, 36 + 2,0); secondly, the partial correlations presented in Figure 1 show that for a given age of appearance (cal-

Fig. 1   *Fig. 2.*

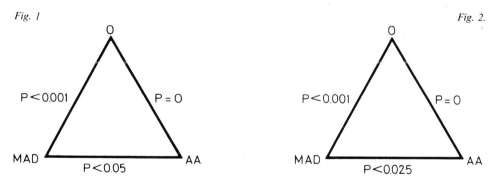

Fig. 1. Partial correlation between obesity (O), mean adipocyte diameter (MAD) and age of appearance of obesity (AA) in 256 obese and 57 control subjects.

Fig. 2. Partial correlation as in Figure 1, but excluding 78 subjects who had lost more than 5 kg compared to their maximum previous weight.

culated constant) there is a highly significant correlation between mean adipocyte diameter and obesity, and also that, for a given degree of obesity, there is a significant correlation between mean adipocyte diameter and age of appearance.

Thus, for a given degree of obesity, the mean adipocyte diameter is smaller in subgroups I and II than in subgroup III. So there appears to be no effect of present obesity, but what of previous obesity? For obvious reasons the mean age in subgroup III is higher than in the other subgroups (39.6 ± 1.0 compared with 28.3 ± 1.4 for subgroup I and 26.9 ± 1.4 for subgroup II). Figure 2 shows the partial correlations, excluding subjects who had lost more than 5 kg compared to their maximum previous weight, and shows that previous weight had not affected the relationship between age of appearance and mean adipocyte diameter.

Thus age of appearance and mean adipocyte diameter are directly related and this relationship is not dependent on the degree of present or previous obesity. Adipocyte hyperplasia frequently occurs in adolescents who, although not yet obese, are becoming so.

REFERENCES

Björntorp, P. and Sjöström, L. (1971): Number and size of adipose tissue fat cells in relation to metabolism in human obesity. Metabolism, 20/7, 703.

Bonnet, E., Gosselin, L., Chantraine, J. and Senterre, J. (1970): Adipose cell number and size in normal and obese children. Rev. europ. Étud. clin. biol., 15, 1101.

Bray, G.A. (1970): Measurement of subcutaneous fat cells from obese patients. Ann. intern. Med., 73, 565.

Brook, C.G.D. (1972): Evidence for a sensitive period in adipose cell replication in man. Lancet, 23, 624.

Brook, C.G.D., Lloyd, J.K. and Wolf, O.W. (1972): Relationship between age of onset of obesity and size and number of adipose cells. Brit. med. J., 2, 25.

Hirsch, J. and Han, P.W. (1969): Cellularity of rat adipose tissue: effects of growth, starvation and obesity. J. Lipid Res., 12, 706.

Hirsch, J. and Knittle, J. (1970): Cellularity of obese and nonobese human adipose tissue. Fed. Proc., 29/4, 1516.

Knittle, J.L. and Ginsberg-Fellner, F. (1972): Effect of weight reduction on in vitro adipose tissue lipolysis and cellularity in obese adolescents and adults. Diabetes, 21/6, 754.

Salans, L.B., Horton, E.S. and Sims, F.A. (1971): Experimental obesity in man: cellular character of the adipose tissue. J. clin. Invest., 50, 1005.

Fat and carbohydrate metabolism in athletes*

R. H. Johnson

Department of Neurology, University of Glasgow and Institute of Neurological Sciences, Southern General Hospital, Glasgow, U.K.

At the beginning of the present century studies of the physiology of exercise soon pointed to the important changes occurring with athletic fitness: the trained individual performs a task with a lower heart rate and has a higher maximum cardiac output, and maximum oxygen uptake ($\dot{V}O_2$max). Although the biochemical processes involved in muscle cell metabolism were extensively studied it was not until twenty years ago that corresponding alterations in biochemical processes were observed in relation to athletic fitness. The finding of lower concentrations of lactate and pyruvate in the blood of athletes during exercise, compared with untrained subjects, has been substantiated by several workers (Cobb and Johnson, 1963; Holmgren, 1956; Holmgren and Strom, 1959; Juchems and Kumper, 1968; Robinson and Harmon, 1941; Saltin and Karlsson, 1971b). This biochemical effect of athletic fitness might be secondary to physiological changes in blood flow or depend on alteration of enzymic systems. Thus athletes may produce less lactate, perhaps due to the greater blood flow found in their contracting muscles, compared to sedentary subjects (Elsner and Carlson, 1962). Alternatively, or in addition, they may dispose of it more rapidly, either by oxidation in muscle or by gluconeogenesis from lactate in the liver. As deep vein PO_2 barely changes during moderately heavy submaximal work (Jorfeldt and Wahren, 1970), it is probable that cytoplasmic glycolysis exceeds mitochondrial oxidative capacity in untrained subjects (Keul et al., 1967). With physical training there is a rise in the concentrations of mitochondrial respiratory enzymes which is associated with increased capacity to oxidize pyruvate (Holloszy, 1967), although other pathways are also involved (Molé et al., 1973). The lower lactate levels in athletes, therefore, appear to be related to biochemical changes (Hubbard, 1973). One would expect wider implications, with other substrates also being affected.

My interest in the effect of athletic training upon metabolism stemmed from comparing observations we made almost incidentally in two different investigations. The majority of a group of hill walkers who were fit, but not in athletic training, gave a positive test for acetoacetate in the urine after 3-5 hours of walking (Pugh, 1969). Another investigation was carried out on runners after they had competed in a marathon race (42 km), the winner of which took 2 hours 38 minutes. None of the competitors had ketonuria, even though the exercise was more strenuous (Pugh et al., 1967). A similar observation has been made on competitors after an 85 km championship ski race (Åstrand et al., 1963). These findings suggested that the phenomenon of post-exercise ketosis, first reported by Courtice and Douglas in 1936, might be different in athletes compared with non-athletes. Although changes in lactate and pyruvate with fitness have been extensively studied, post-exercise ketosis has received little attention and this observation deserved investigation.

POST-EXERCISE KETOSIS AND ATHLETIC FITNESS

We, therefore, studied middle and long distance athletes as well as non-athletes during and after running for 1½ hours (Johnson et al., 1969a). There were striking differences between the two

* I am grateful to the National Fund for Research into Crippling Diseases, the Scottish Hospital Endowments Research Trust, and the Muscular Dystrophy Group of Great Britain for financial support in this work.

groups. The athletes ran faster and at a steady rate throughout the period. The non-athletes ran progressively more slowly. Despite their greater activity the athletes had lower heart rates and had lower 'central' body temperatures associated with higher sweat rates (Pugh et al., 1967; Robinson, 1967). There were also major differences in metabolite concentrations in their blood. In addition to the expected finding of lower lactate and pyruvate concentrations during exercise they also had lower concentrations of free fatty acids (FFA) even though the amount of glycerol in the venous blood was similar. This suggests that in that study there was no major difference in fat mobilisation between the two groups. There was also marked development of post-exercise ketosis in the non-athletes, whereas the athletes had relatively low ketone body concentrations throughout (Fig. 1).

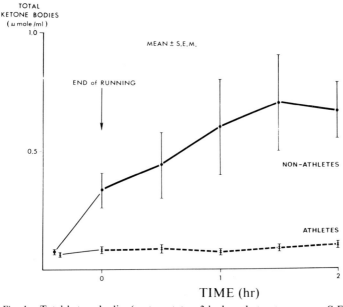

Fig. 1. Total ketone bodies (acetoacetate + 3-hydroxybutyrate, means ± S.E.M.) in 8 non-athletes and 5 trained athletes before and after running. The untrained subjects showed marked post-exercise ketosis. (From R.H. Johnson and J.L. Walton, 1972, *Quart. J. exp. Physiol.;* courtesy of the editors.)

These changes might be true metabolic differences between athletes and untrained subjects which could be innate or due to adaptation as a result of athletic training. Another explanation could be that the untrained subjects had worked relatively harder than the athletes by running to the limits of their capacity. The athletes, by running well within their limits, could have been subjected to smaller demands on their metabolic reserves. In order to investigate this we carried out a further investigation (Jennett et al., 1972; Rennie et al., 1974) in which heart rates and ventilatory studies indicated that competitive cyclists worked harder and nearer their maximal work capacity than untrained subjects (Astrand, 1964). In this investigation there was a large increase in the blood glucose concentrations in the cyclists, probably due to a decrease in the uptake of glucose by muscle, since plasma immunoreactive insulin fell to a greater extent in the cyclists during the exercise period. There was also a larger and more rapid rise in blood glycerol concentrations in the cyclists during the exercise, suggesting that they mobilised more fat than the noncyclists (Fig. 2). Nevertheless, the concentrations of plasma FFA and the rate of their increase were lower in the cyclists, suggesting that they utilized the FFA liberated during exercise more efficiently. These findings were in keeping with the conclusion that during exercise athletes show a reduction in the rate of carbohydrate oxidation compared with untrained subjects. The trained person relies more, even at comparable work loads, on a greater capacity to oxidize fat (Hermansen et al., 1967; Saltin and Karlsson, 1971a). After

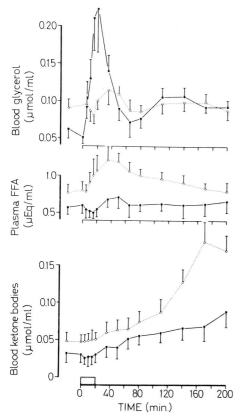

Fig. 2. Blood glycerol, plasma FFA and blood ketone bodies in 4 cyclists (●—●) and 5 untrained subjects (o----o) during and after 20 minutes of exercise (means ± S.E.M.). Although the cyclists worked relatively harder, and metabolised more fat, they had lower concentrations of ketone bodies after exercise (see also Fig. 5). (From M.J. Rennie et al., 1974, *Quart. J. exp. Physiol.*; courtesy of the editors.)

exercise, when muscle utilization of fuels virtually ceases, there was a rise in the ketone body concentrations in both groups (Fig. 2). The increase was much greater in the untrained subjects despite their lower level of lipolysis. These results indicated that the degree of development of post-exercise ketosis depends, not on the amount of exercise in relation to maximal work capacity, but on the athletic fitness of the individual.

The concentration of blood ketone bodies hardly alters during exercise, implying equilibrium between production and utilization. After strenuous exercise, however, the increase which we have found particularly marked in non-athletic individuals, could be due either to a higher rate of production or to a lower rate of utilization. If the rate of utilization is altered then the tolerance curves after ingestion of acetoacetate should differ. In order to investigate this possibility the tolerance to oral acetoacetate before, during and after exercise in a group of trained athletes has been compared with a similar series of investigations on untrained subjects (Fig. 3) (Johnson and Walton, 1972). No difference was found between the two groups in the rate of utilization of acetoacetate at rest. This is similar to the finding in obese individuals, for although the latter have a higher resistance to ketosis they metabolise 3-hydroxybutyrate at a similar rate to that observed in non-obese subjects (Kekwick et al., 1959). There was increased tolerance during exercise, suggesting that ketone bodies were being utilized as a fuel. Such a role for ketone bodies has been demonstrated previously in the whole animal (sheep) (Bergman et al., 1963) and in isolated tissues (Williamson and Krebs, 1961) and prevents a rise in ketone body

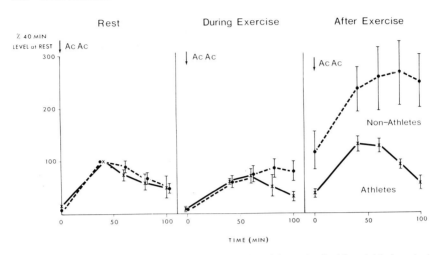

Fig. 3.   Acetoacetate tolerance in 3 trained athletes (**x**) and 6 untrained subjects (●) before, during and after exercise. Sodium acetoacetate (200 ml, 0.4 M) was ingested as indicated by the symbol AcAc. The results are expressed as percentages (mean ± S.E.M.) of the level of total ketone bodies 40 minutes after acetoacetate was ingested at rest. There was no difference in acetoacetate tolerance at rest, increased tolerance during exercise and reduced tolerance in the non-athletes after exercise. (From R.H. Johnson and J.L. Walton, 1972, *Quart. J. exp. Physiol.*; courtesy of the editors.)

concentrations during exercise (Drury et al., 1941). However, the change we observed in tolerance of acetoacetate during exercise was small, indicating that ketone bodies (as distinct from FFA) are of minor significance as a fuel for exercising muscle (Hagenfeldt and Wahren, 1971).

In the post-exercise period the untrained group had a much reduced tolerance to acetoacetate compared either to their own tolerance at rest or to that of the trained group after exercise. This observation suggests a reduction in the rate of utilization of ketone bodies after exercise by untrained subjects. This is different from the pre-exercise situation when the groups behaved

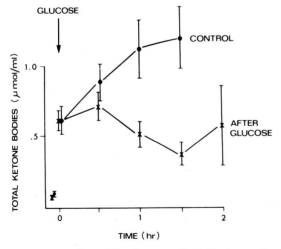

Fig. 4.   The effect of 50 g glucose orally 30-60 minutes after running for 1½ hours in 12 subjects compared with 9 otherwise similar subjects who were not given glucose. Administration of glucose diminished the post-exercise ketosis. (From R.H. Johnson et al., 1969b, *Lancet;* courtesy of the editors.)

similarly and is, therefore, also different from the situation in obesity. There may also be more continued production of ketone bodies in untrained subjects compared with the trained athletes as plasma FFA was much higher both during and after the exercise. In addition we have found that in untrained subjects the blood concentrations of FFA and the degree of post-exercise ketosis are depressed if glucose is ingested (Fig. 4) (Johnson et al., 1969b). These observations imply that post-exercise ketosis in untrained subjects develops due to a shortage of carbohydrate to meet energy needs and that in addition athletes have developed a better balance between utilization and production of ketone bodies.

There is a well substantiated negative relationship between fitness and obesity (Sloan, 1969; Welch et al., 1958), and we were concerned with the effect of obesity on the development of post-exercise ketosis. We found, however, that although there was a definite negative relationship between fitness and the development of post-exercise ketosis, there was no statistical relationship between its development and the degree of obesity in the group we studied (Johnson and Walton, 1971).

We have, therefore, accumulated much evidence that athletic training affects the susceptibility of an individual to the development of post-exercise ketosis, even though we have not so far examined individuals at different stages of a training programme. Our evidence suggests that there is a true alteration in metabolism and not an effect based on the work being a different proportion of an increased maximal work capacity.

HORMONAL STUDIES

The recognition of biochemical changes with athletic training has attracted attention to the possibility of facilitation of the enzymic pathways involved in muscle metabolism and to the alterations which develop in the hormonal status of the athlete. Dr. Holloszy and his colleagues (see p. 254) have been outstanding in developing our understanding of the effect of athletic training on cellular biochemical pathways. Evidence has been adduced that one of the enzymic mechanisms in which adaptation occurs with training is the transfer of pyruvate to alanine (Molé et al., 1973) and also the enzymic systems involved in ketone body oxidation (Winder et al., 1973). It is also possible for the pattern of metabolism with exercise to be altered in athletes by affecting concentrations of substrate stores. This may be achieved by altering their diet in the days prior to exercise (Bergström et al., 1967; Johnson and Rennie, 1973a).

The hormonal changes with athletic training have also attracted interest. Physical exercise depresses insulin release and stimulates the release of other hormones, including growth hormone (Hunter et al., 1965), cortisol, catecholamines (Von Euler, 1969) and androgens (Sutton et al., 1973). With most hormones there is evidence that the extent of the change is related to the intensity of the exercise.

Plasma insulin falls with exercise (Cochran et al., 1966). It has been suggested that the fall is smaller after an exercise programme (Devlin, 1963) but we have found that the greater fall occurs in trained subjects (Fig. 5). This was associated with a marked increase in blood glucose (Jennett et al., 1972; Rennie et al., 1974). Such a depression would be compatible with the reduced reliance by athletes on carbohydrate oxidation and their proportionally greater utilization of fat during exercise.

We have been particularly interested in growth hormone (HGH) for it has been suggested that it has a major role in the initial mobilization of fat (Hunter et al., 1965; Rabinowitz et al., 1965; Winkler et al., 1969). There is also evidence that athletically trained subjects who do not produce marked levels of post-exercise ketosis have smaller rises in HGH concentrations during exercise compared with untrained subjects (Fig. 5) (Sutton et al., 1968; Walton, 1970). We studied patients with hypopituitarism in whom HGH levels were low and did not vary with exercise (Johnson et al., 1971). Similar observations have also been made in normal subjects in whom the normal increase in HGH during exercise was abolished by fenfluramine (Ponderax, Servier Laboratories). This drug appears to have a direct effect on hypothalamic pathways controlling HGH release (Sulaiman and Johnson, 1973). In the study on patients with hypopituitarism at the end of exercise FFA and glycerol had increased more in the patients (Fig. 6) than in the controls, despite the patients' low HGH levels. The results provide evidence that HGH

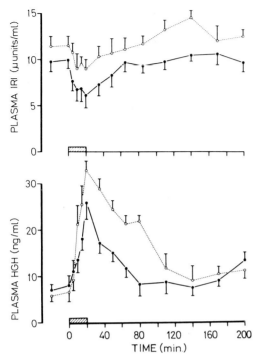

Fig. 5.   Plasma immunoreactive insulin (IRI) and plasma HGH in 4 cyclists (●—●) and 5 untrained subjects (o----o) during and after 20 minutes of exercise (see also Fig. 2). Although the cyclists worked relatively harder their hormone concentrations were lower. (From M.J. Rennie et al., 1974, *Quart. J. exp. Physiol.*; courtesy of the editors.)

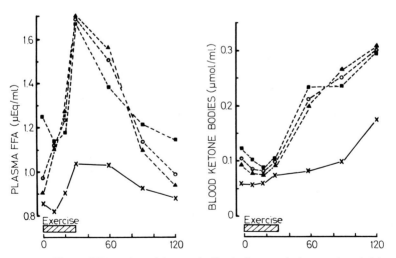

Fig. 6.   Plasma FFA and total ketone bodies in 8 controls (x:means) and 6 hypopituitary patients (o:means). The results in the patients have also been shown in 2 groups: (*a*) 2 patients on replacement therapy, 12.5 mg cortisone acetate twice daily and thyroxine 0.1 mg daily (■: means); and (*b*) 4 patients without replacement therapy (▲: means). The patients with hypopituitarism (no growth hormone) liberated FFA during exercise and ketone bodies rose after exercise more than in controls. This was independent of replacement therapy. (From R.H. Johnson et al., 1971, *Clin. Sci.;* courtesy of the editors.)

does not necessarily have a major role in fat mobilization or in the production of post-exercise ketosis (Basu et al., 1960; Hartog et al., 1967). Indeed it is probable that more than one hormonal system may activate lipolysis during exercise and when one is lost or blocked, others are available (Gollnick et al., 1970). We found that the rise in FFA and ketone bodies with exercise was actually higher in the patients with hypopituitarism, raising the possibility that HGH may depress FFA levels, via increased re-esterification, after the initial mobilization of fat with exercise (Winkler et al., 1969). It has been suggested that HGH may have a structure with two subunits acting antagonistically in this way (Bornstein et al., 1968).

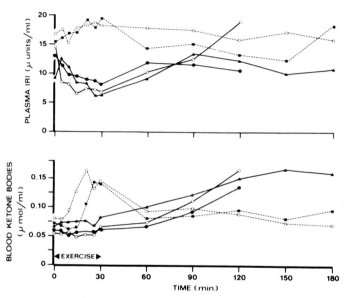

Fig. 7. Plasma immunoreactive insulin (IRI) and total ketone bodies (acetoacetate + 3-hydroxybutyrate) in 3 control subjects (—) and 2 acromegalic patients (----) during and after 30 minutes of exercise. The insulin rose in the patients instead of falling and the patients then showed depression, instead of elevation, of ketone bodies after exercise.

In a further study of patients with acromegaly, who have an excess of circulating HGH, we found that no increase in ketone body concentrations occurred after exercise (Fig. 7) (Johnson and Rennie, 1973*b*). Although one explanation could be that HGH has a dual action on fat metabolism, an alternative explanation is that the re-esterification of fat was facilitated by insulin (Bieberdorf et al., 1970), which was markedly raised in the patients compared with the controls. In our studies of exercise by athletes and non-athletic subjects insulin levels were lower in the athletes after exercise (Jennett et al., 1972; Rennie et al., 1974). The action of insulin in facilitating re-esterification of fat cannot, therefore, be implicated in the explanation of the athletes' lower concentrations of FFA and ketone bodies during and after exercise.

REFERENCES

Åstrand, P.O., Cuddy, T.E., Saltin, B. and Stenberg, J. (1964): Cardiac output during submaximal and maximal work. J. appl. Physiol., 19, 268.

Åstrand, P.O., Hallback, T., Hedman, R. and Saltin, B. (1963): Blood lactates after prolonged severe exercise. J. appl. Physiol., 18, 619.

Basu, A., Passmore, R. and Strong, J.A. (1960): The effect of exercise on the level of nonesterified fatty acids in the blood. Quart. J. exp. Physiol., 45, 312.

Bergman, E.N., Kon, K. and Katz, M.L. (1963): Quantitative measurement of acetoacetate metabolism and oxidation in sheep. Amer. J. Physiol., 205, 658.

Bergström, J., Hermansen, L., Hultman, E. and Saltin, B. (1967): Diet, glycogen and physical performance. Acta physiol. scand., 71, 140.

Bieberdorf, F.A., Chesnick, S.S. and Scow, R.O. (1970): Effect of insulin and acute diabetes on plasma FFA and ketone bodies in the fasting rat. J. clin. Invest., 49, 1685.

Bornstein, J., Krahl, M.E., Marshall, L.B., Gould, M.K. and Armstrong, J. McD. (1968): Pituitary peptides with direct action on the metabolism of carbohydrates and fatty acids (sheep). Biochim. biophys. Acta (Amst.), 156, 31.

Cobb, L.A. and Johnson, W.P. (1963): Hemodynamic relationships of anerobic metabolism and plasma free fatty acids during prolonged strenuous exercise in trained and untrained subjects. J. clin. Invest., 42, 800.

Cochran, B., Marbach, E.P., Poucher, R., Steinberg, T. and Gwinup, G. (1966): Effect of acute muscular exercise on serum immunoreactive insulin concentration. Diabetes, 15, 838.

Courtice, F.C. and Douglas C.G. (1936): The effects of prolonged muscular exercise on the metabolism. Proc. roy. Soc. B., 119, 381.

Devlin, J.G. (1963): The effect of training and acute physical exercise on plasma insulin-like activity. Irish J. med. Sci., 423.

Drury, D.R., Wick, A.N. and Mackay, E.M. (1941): The action of exercise on ketosis. Amer. J. Physiol., 134, 761.

Elsner, R.W. and Carlson, L.D. (1962): Post-exercise hyperemia in trained and untrained subjects. J. appl. Physiol., 17, 436.

Gollnick, P.D., Soule, R.G., Taylor, A.W., Williams, C. and Ianuzzo, C.D. (1970): Exercise-induced glycogenolysis and lipolysis in the rat: hormonal influence. Amer. J. Physiol., 219, 729.

Hagenfeldt, L. and Wahren, J. (1971): Metabolism of free fatty acids and ketone bodies in skeletal muscle. In: Muscle Metabolism During Exercise, p. 153. Editors: B. Pernow and B. Saltin. Plenum Press, New York.

Hartog, M., Havel, R., Capinschi, G., Earl, J. and Ritchie, B. (1967): The relationship between changes in serum levels of growth hormone and mobilisation of fat during exercise in man. Quart. J. exp. Physiol., 52, 86.

Hermansen, L., Hultman, E. and Saltin, B. (1967): Muscle glycogen during prolonged severe exercise. Acta physiol. scand., 71, 129.

Holmgren, A. (1956): Circulatory changes during muscular work in man. Scand. J. clin. Lab. Invest., 8, Suppl. 24, 1.

Holmgren, A. and Strom, G. (1959): Blood lactate concentration in relation to absolute and relative work load in normal man, and in mitral stenosis, atrial septal defect and vasoregulatory asthenia. Acta med. scand., 163, 185.

Holloszy, J.O. (1967): Biochemical adaptations in muscle. Effects of exercise on mitochondrial oxygen uptake and respiratory enzyme activity in skeletal muscle. J. biol. Chem., 242, 2278.

Hubbard, J.L. (1973): The effect of exercise on lactate metabolism. J. Physiol. (Lond.), 231, 1.

Hunter, W.M., Fonseka, C.C. and Passmore, R. (1965): Growth hormone; important role in muscular exercise in adults. Science, 150, 1051.

Jennett, S., Johnson, R.H. and Rennie, M.J. (1972): Ketosis in untrained subjects and racing cyclists after strenuous exercise. J. Physiol. (Lond.), 225, 47.P.

Johnson, R.H. and Rennie, M.J. (1973a): The effect of diet upon the metabolic changes with exercise in long-distance runners. J. Physiol. (Lond.), 232, 73P.

Johnson, R.H. and Rennie, M.J. (1973b): Changes in fat and carbohydrate metabolism caused by moderate exercise in patients with acromegaly. Clin. Sci., 44, 63.

Johnson, R.H., Rennie, M.J., Walton, J.L. and Webster, M.H.C. (1971): The effect of moderate exercise on blood metabolites in patients with hypopituitarism. Clin. Sci., 40, 127.

Johnson, R.H. and Walton, J.L. (1971): Fitness, fatness and post-exercise ketosis. Lancet, 1, 566.

Johnson, R.H. and Walton, J.L. (1972): The effect of exercise upon acetoacetate metabolism in athletes and non-athletes. Quart. J. exp. Physiol., 57, 73.

Johnson, R.H., Walton, J.L., Krebs, H.A. and Williamson, D.H. (1969a): Metabolic fuels during and after severe exercise in athletes and non-athletes. Lancet, 2, 452.

Johnson, R.H., Walton, J.L., Krebs, H.A. and Williamson, D.H. (1969b): Post-exercise ketosis. Lancet, 2, 1383.

Jorfeldt, L. and Wahren, J. (1970): Human forearm muscle metabolism during exercise. Scand. J. clin. Lab. Invest., 26, 73.

Juchems, R. and Kumper, E. (1968): Blood lactate response to exercise. New Engl. J. Med., 278, 912.

Kekwick, A., Pawan, G.L.S. and Chalmers, T.M. (1959): Resistance to ketosis in obese subjects. Lancet, 2, 1157.

Keul, J., Doll, E. and Keppler, D. (1967): The substrate supply of human skeletal muscle at rest, during and after work. Experientia (Basel), 23, 974.

Molé, P.A., Baldwin, K.M., Terjung, R.L. and Holloszy, J.O. (1973): Enzymic pathways of pyruvate metabolism in skeletal muscle: adaptations to exercise. Amer. J. Physiol., 224, 50.

Pugh, L.G.C.E. (1969): Thermal, metabolic, blood and circulatory adjustments in prolonged outdoor exercise. Brit. med. J., 2, 657.

Pugh, L.G.C.E., Corbett, J.L. and Johnson, R.H. (1967): Rectal temperatures, weight losses and sweat rates in marathon running. J. appl. Physiol., 23, 347.

Rabinowitz, D., Klassen, G. and Zierler, K. (1965): Effect of human growth hormone in muscle and adipose tissue metabolism in the forearm of man. J. clin. Invest., 44, 51.

Rennie, M.J., Jennett, S. and Johnson, R.H. (1974): The metabolic effects of strenuous exercise: a comparison between untrained subjects and racing cyclists. Quart. J. exp. Physiol., in press.

Robinson, S. (1967): Training, acclimatization and heat tolerance. Canad. med. Ass. J., 96, 795.

Robinson, S. and Harmon, P.M. (1941): The lactic acid mechanism and certain properties of the blood in relation to training. Amer. J. Physiol., 132, 757.

Saltin, B. and Karlsson (1971a): Muscle glycogen utilization during work of different intensities. In: Muscle Metabolism During Exercise, p. 289. Editors: B. Pernow and B. Saltin. Plenum Press, New York.

Saltin, B. and Karlsson, J. (1971b); Muscle ATP, CP and lactate during exercise after physical conditioning. In: Muscle Metabolism During Exercise, p. 395. Editors: B. Pernow and B. Saltin. Plenum Press, New York.

Sloan, A.W. (1969): Physical fitness and body build of young men and women. Ergonomics, 12, 25.

Sulaiman, W.R. and Johnson, R.H. (1973): Effect of fenfluramine on human growth hormone release. Brit. med. J., 2, 329.

Sutton, J.R., Coleman, M.J., Casey, J. and Lazarus, L. (1973): Androgen responses during physical exercise. Brit. med. J., 1, 520.

Sutton, J.R., Young, J.D., Lazarus, L., Hickie, J.B. and Maksvytis, J. (1968): Hormonal changes during exercise. Lancet, 2, 1304.

Von Euler, U.S. (1969): Sympatho-adrenal activity and physical exercise. In: Biochemistry of Exercise, p. 170. Editor: J.P. Poortmans. Karger, Basel.

Walton, J.L. (1970): Observations upon metabolic changes during and after exercise. M. Sc. Thesis, University of Glasgow.

Welch, B., Riendeau, R., Crisp, C. and Isenstein (1958): Relationship of maximal oxygen consumption to various components of body composition. J. appl. Physiol., 12, 395.

Williamson, J.R. and Krebs, H.A. (1961): Acetoacetate as a fuel of respiration in the perfused rat heart. Biochem. J., 80, 540.

Winder, W.W., Baldwin, K.M. and Holloszy, J.O. (1973): Adaptive increase in the capacity of skeletal muscle to oxidize B-hydroxybutyrate induced by chronic endurance exercise. Fed. Proc., 32, 889.

Winkler, B., Steele, R. and Altszuler, N. (1969): Effects of growth hormone administration on free fatty acid and glycerol turnover in the normal dog. Endocrinology, 85, 25.

Adipose mass and body composition

The relationship between total body potassium and body composition in the obese female

P. A. Delwaide and **E. J. Crenier**

Chimie Médicale, Toxicologie et Hygiène and Laboratoire de Biométrie humaine, University of Liège, Liège, Belgium

The total body potassium evaluated by isotopic dilution or whole body counting is in inverse proportion to weight and therefore reference tables contain body weight as a negative term in equations for the prediction, in mmoles/kg, of total potassium (Delwaide and Mersch, 1971). However, the value of this relationship could not be extended, without critical examination, to obese human subjects. For this reason we have determined body potassium in obese females who were not suffering from other disturbances and compared the results with those in control subjects. The determination of body composition is based on the dilution of tritiated water (Bernier and Vidon, 1965).

METHODS

Body potassium was measured by the 'whole body counting' technique using a counter with good geometry and high efficiency (Delwaide, 1969). Precise measurements up to 5% were obtained. The '4π' detector was not influenced by the counting geometry or the morphology of the subject. Nevertheless, in order to avoid any deviation due to the corpulence of the individuals, we used a series of polyethylene phantoms as references. The latter contained potassium solutions in different volumes and concentrations adapted, in each case under study, to the somatotypes of the individual. Individualized reference values based on age and weight were set up for this type of instrument and for the local population only after carrying out a series of studies on control subjects (Delwaide and Mersch, 1971). Total water measurements within a 5% precision limit were obtained by the dilution of tritiated water. By using the hydration coefficient 0.72 and the classic formulae we were able to find the amount of total body fat and lean body mass for each subject. This method as a basis for the study of body composition has been considered doubtful under various conditions (François, 1969). Further studies based on an independent method of lean mass determination by anthropometric measurements (Crenier, 1966) allowed us to check the dilution method on control subjects. The body fat values obtained by both methods were shown to be significantly correlated ($r = 0.8992$, P < 0.001 in 59 subjects) if we take into consideration the fact that the determinations were indirect and therefore amplify the experimental errors of direct measurements (Delwaide and Crenier, 1973). We evaluated the amount of total protein by calculating the difference between body weight and the combination of water, body fat and bone, the latter being taken from the tables of Bernier and Vidon (1965). Body surface was calculated by Dubois' formula. Total potassium can thus be related to the following parameters: body weight, body surface, lean body mass and total proteins.

RESULTS

The measurements of total potassium and body composition as based on the dilution of tritiated water were carried out on 59 control female subjects and on 26 obese female subjects. The overweight in the latter group averaged 40% more than the theoretical weight obtained from Lorentz's formula (Vandervael, 1964), but these subjects showed no signs of associated illnesses. All measurements were made before beginning any therapy, such as fasting or administration of diuretics, which could influence potassium levels. Table 1 gives the average values of the various parameters studied in both the control and obese subjects. The reference norms for water and the derived variables were taken from the tables of Bernier and Vidon (1965); those for potassium come from our own reference tables (Delwaide and Mersch, 1971). The average

TABLE 1   Comparison of body composition and potassium values in female subjects (mean ± S.D.)

Variable	Control (n = 59)	Obese subjects (n = 26)	Student's t test
Weight	56.9 ± 5.8	93.2 ± 12.5	
Body surface (m²)	1.617 ± 0.096	1.900 ± 0.260	7.22
Total water (l)	31.39 ± 3.67	36.79 ± 5.73	5.15
% body weight	55.2 ± 4.88	39.5 ± 6.07	12.50
Body fat (kg)	13.3 ± 4.34	42.2 ± 13.5	
% body weight	23.2 ± 6.78	44.4 ± 9.50	
Lean body mass (kg)	43.6 ± 5.10	51.5 ± 7.74	5.24
% body weight	76.8 ± 6.78	55.6 ± 9.50	11.50
Total protein (kg)	8.21 ± 1.43	10.19 ± 2.10	5.79
% body weight	14.4 ± 1.86	11.2 ± 2.43	6.60
Potassium			
Total (mmoles)	2615 ± 395	2753 ± 434	(1.42)
mmoles/kg body weight	45.9 ± 6.15	29.8 ± 4.34	12.0
as % of reference values	—	84.4	
mmoles/m² body surface	1620 ± 211	1487 ± 535	(1.62)
mmoles/kg lean body mass	56.9 ± 10.5	53.7 ± 9.54	(1.32)
mmoles/kg total protein	308 ± 70	270 ± 58	2.39

values of total water obtained from the obese subjects agreed with those of the tables (41% of body weight). As a result our subjects did not present any associated hyperhydration and our calculations of body composition remain valid. The total potassium (in mmoles/kg) as compared to our norms showed a net decrease. On the basis of unit body weight and age it was found that the obese subjects had a definite potassium deficit. It should be emphasized that weight is a negative term in the prediction equation; the amount of potassium in obese subjects is distinctly lower than that demonstrated by the normal drift in relation to weight. However, the total potassium referred to other composition indices showed less distinct differences in these cases of obesity when compared to those of control cases. When expressed as a function of the lean body mass (as calculated in reference to total water) the potassium did not show any significant inverse correlation with overweight. On the other hand, overweight showed significant negative correlations with the following parameters: potassium in mmoles/kg($r = -0.5797$), total water as a percentage of body weight ($r = -0.6608$), total protein as a percentage of body weight ($r = -0.4590$).

DISCUSSION AND CONCLUSION

Direct determination of total potassium from γ-rays of ^{40}K eliminates errors caused by incomplete mixing of ^{42}K (Linquette et al., 1969). It is indeed reasonable to assume that we can know the body potassium store which can be assimilated with intracellular potassium. The relationship between potassium store and indices of body composition established from the dilution of tritiated water is of great importance in spite of the simplicity of the methods. Our values for the potassium/lean mass ratio are nevertheless lower than the classic values, but in line with more recent data (François, 1969; Womersley et al., 1972). In the cases we examined overweight was not accompanied by excessive water retention. Obesity was accompanied by a 20% increase in the absolute values of lean mass and protein for an average 40% overaccumulation of fat (Quoidbach et al., 1969; Linquette et al., 1969). There was also a slight, insignificant increase in the absolute value (mmoles) of total potassium. The interpretation of potassium value in function of lean mass and proteins is difficult to define; the figures obtained are low, especially the potassium/protein ratio. However, low potassium is distinct when expressed as a function of gross weight, but is attenuated when expressed as a function of a more physiological index. The present state of our research does not allow us to explain this phenomenon, but it does confirm the notion that the measurement of total potassium alone is not an indication for the estimation of fat accumulation when using coefficients from control subjects. The ^{40}K technique would tend to overestimate the fat mass (Myhre and Kessler, 1966). Therefore, it is necessary to use two independent systems of measurement.

REFERENCES

Bernier, E. and Vidon, J. (1965): Valeurs normales de l'eau et des électrolytes. Métab. Eau Électrol., 7, 5.

Crenier, E.J. (1966): La prédiction du poids corporel 'normal'. Biomét. Hum., 1, 10.

Delwaide, P.A. (1969): A 4π plastifluor body counter for clinical use: calibration. Int. J. appl. Radiat., 20, 623.

Delwaide, P.A. and Crenier, E.J. (1973): Body potassium as related to lean body mass measured by total water determination and by anthropometric method. Human Biol., 43, 509.

Delwaide, P.A. and Mersch, G. (1971): Potassium corporel total de sujets adultes en bonne santé. Étude statistique. Biomét. Hum., 6, 1.

François, B. (1969): Makeshift determination of total body fat. In: Physiopathology of Adipose Tissue, p. 349. Editor: J. Vague. Excerpta Medica, Amsterdam.

Linquette, M., Fossatti, D., Lefebvre, J. and Chechan, C. (1969): The measurement of total water, exchangeable sodium and potassium in obese persons. In: Physiopathology of Adipose Tissue, p. 320. Editor: J. Vague, Excerpta Medica, Amsterdam.

Myhre, L.G. and Kessler, W.V. (1966): Body density and potassium-40 measurements of body composition as related to age. J. appl. Physiol., 21, 1251.

Quoidbach, A., Busset, R., Dayer, A. and Mach, R.S. (1969): Evaluation of body composition in obesity using isotopic dilution techniques. In: Physiopathology of Adipose Tissue, p. 302. Editor: J. Vague. Excerpta Medica, Amsterdam.

Vandervael, F. (1964): Biométrie Humaine. Desoer, Liège.

Womersley, J., Boddy, K., King, P.C. and Durnin, J.V. (1972): A comparison of the fat-free mass of young adults estimated by anthropometry, body density and total potassium content. Clin. Sci., 43, 469.

The assessment of obesity from measurements of skinfold thickness, limb circumferences, height and weight

J. Womersley and **J.V.G.A. Durnin**

Institute of Physiology, University of Glasgow, Glasgow, U.K.

Measurements of skinfold thickness, limb circumferences, height and weight were made on 209 male and 272 female subjects between the ages of 16 and 72 years. The measurements of skinfold thickness were made at the biceps, triceps, subscapular and suprailiac sites; the circumferences were measured at the upper arm, the calf and the thigh. The exact sites and procedures used were as described by Durnin and Rahaman (1967).

The body density of each subject was determined by the technique of underwater weighing, and the residual volume of air present in the lungs at the moment of weighing under water was estimated by nitrogen washout (Rahn et al., 1949). Each measurement of body density was carried out in triplicate and the results were calculated on an Olivetti desk computer while the subject remained in the water; repeat measurements could therefore easily be made if the first 3 results were not in good agreement.

The fat content of each subject was calculated according to the equation of Siri:

$$\% \text{ fat} = \left(\frac{4.95}{\text{density}} - 4.50 \right). 100$$

Table 1 gives the number of subjects in each age group together with the means and standard deviations of their height, weight and fat content. The subjects in the youngest age group of both sexes were selected from the population of normal Glasgow schoolchildren. The older subjects were partly selected on a non-random basis from business and professional colleagues. But in order to include a larger number of particularly obese and particularly lean individuals volunteers were obtained from an obesity clinic, in response to a newspaper advertisement and from health clubs, the University gymnasium and a ballet company.

TABLE 1   Basic data on the subjects

Age group	No.	Mean height (cm) ± S.D.	Mean weight (kg) ± S.D.	Mean fat content (% body weight) ± S.D.
Males				
17-19	24	177 ± 9	73.1 ± 16.1	15 ± 7
20-29	92	177 ± 7	70.1 ± 12.2	15 ± 7
30-39	34	176 ± 6	79.8 ± 10.4	23 ± 5
40-49	35	175 ± 7	76.9 ± 9.2	25 ± 7
50-72	24	172 ± 8	80.4 ± 11.7	28 ± 9
Females				
16-19	29	163 ± 6	57.8 ± 10.9	26 ± 8
20-29	100	163 ± 6	63.2 ± 14.4	29 ± 10
30-39	58	162 ± 5	68.1 ± 16.1	33 ± 9
40-49	48	162 ± 7	68.1 ± 13.8	35 ± 8
50-68	37	161 ± 6	69.1 ± 16.1	39 ± 8

We have shown (Durnin and Womersley, 1974) that the most reliable estimate of body fat from the 4 skinfolds we measured is obtained if a combination of 3 or of all 4 sites is used, and the most convenient method is simply to sum the measurements. In the present paper the term 'total skinfold' therefore refers to the sum of the skinfold thicknesses at the 4 sites. No logarithmic or other transformations have been carried out since our investigations have shown that these manipulations only slightly improve the accuracy of prediction of body density and body fat content.

Separate correlation coefficients were calculated between body density and (1) the 'total skinfold', (2) each of the 3 circumferences, (3) height and (4) weight. The values for these correlation coefficients are given in Table 2. Except in the group of 40 to 49-year-old women, the 'total skinfold' correlates at least as well with body density as do any of the 3 measurements of circumference. And, although it is not shown in the Table, in almost all cases each of the 4 individual skinfold measurements show a higher correlation with body density than do any of the 3 individual circumferences. The arm and thigh circumferences appear fairly consistently to correlate more highly with body density than does the calf circumference.

TABLE 2 Correlation coefficients for the relationship between body density and different anthropometric measurements

Age group	Height	Weight	Total skinfold	Arm circumference	Calf circumference	Thigh circumference
Males						
17–19	−0.11	−0.82	−0.88	−0.81	−0.86	−0.88
20–29	0.19	−0.64	−0.81	−0.47	−0.38	−0.48
30–39	0.21	−0.49	−0.69	−0.40	−0.34	−0.42
40–49	0.26	−0.51	−0.80	−0.50	−0.52	−0.57
50–72	0.60	−0.40	−0.87	−0.45	−0.15	−0.33
17–72	0.28	−0.63	−0.79	−0.59	−0.33	−0.48
Females						
16–19	−0.06	−0.78	−0.86	−0.66	−0.62	−0.71
20–29	0.07	−0.76	−0.85	−0.78	−0.54	−0.73
30–39	0.17	−0.73	−0.76	−0.72	−0.57	−0.63
40–49	0.24	−0.79	−0.76	−0.78	−0.68	−0.69
50–68	0.09	−0.79	−0.84	−0.51	−0.19	−0.73
17–68	0.15	−0.74	−0.80	−0.75	−0.49	−0.62

It can also be seen from Table 2 that in the women body weight is highly correlated with body density.

Table 3 gives the coefficients of multiple correlation between body density and the variables height, weight, age, total skinfold and the 3 measurements of circumference; the standard errors of the estimates of body density from these variables are also shown. In the younger men and in the older women the use of several variables rather than skinfold thickness alone brings about a significant increase in the accuracy of predictions of body density. The variables which have the greatest predictive value vary considerably between the different groups, however, and this may mean that the particular selection of variables for a specific sex and age group may not be appropriate for the same sex and age group of a different population.

Table 4 gives the correlation coefficients for the relationships between body density and various indices that have been used by different workers for the estimation of obesity from measurements of height and weight. In both sexes there appears to be little difference between the various indices. In the men the correlation between skinfold thickness and body density is usually considerably higher than that between body density and any of the different indices of obesity. In the older women, however, the correlations between certain height-weight indices and body density are sometimes better than between body density and the total skinfold.

TABLE 3   Coefficients of multiple correlation (R) and standard errors of the estimate (S.E.E.) of body density using multiple regression analysis with 7 different variables

	Males							
	17–19 years			20–29 years			30–39 years		
Variable	R	S.E.E.	Variable	R	S.E.E.	Variable	R	S.E.E.
Total skinfold	0.88	0.0078	Total skinfold	0.83	0.0092	Total skinfold	0.77	0.0078
Calf circumference	0.96	0.0049	Height	0.84	0.0090	Age	0.81	0.0073
Age	0.97	0.0046	Age	0.85	0.0088	Calf circumference	0.86	0.0066
Weight	0.97	0.0045	Thigh circumference	0.86	0.0087	Height	0.87	0.0065
Thigh circumference	0.98	0.0038						
	40–49 years			50–72 years			17–72 years		
Total skinfold	0.83	0.0084	Total skinfold	0.87	0.0099	Total skinfold	0.82	0.0114
Arm Circumference	0.84	0.0084	Thigh circumference	0.88	0.0099	Age	0.89	0.0089
			Height	0.88	0.0101	Thigh circumference	0.90	0.0088
						Height	0.90	0.0086

	Females							
	16–19 years			20–29 years			30–39 years		
Variable	R	S.E.E.	Variable	R	S.E.E.	Variable	R	S.E.E.
Total skinfold	0.85	0.0089	Total skinfold	0.85	0.011	Arm circumference	0.73	0.014
Thigh circumference	0.86	0.0089	Arm circumference	0.87	0.011	Age	0.77	0.013
Arm circumference	0.86	0.0090	Age	0.87	0.011	Height	0.79	0.013
			Height	0.87	0.011	Weight	0.80	0.013
	40–49 years			50–68 years			16–68 years		
Weight	0.80	0.0098	Total skinfold	0.81	0.0093	Total skinfold	0.80	0.013
Height	0.86	0.0085	Thigh circumference	0.87	0.0079	Age	0.84	0.011
Age	0.87	0.0084	Age	0.88	0.0078	Thigh circumference	0.85	0.011
Arm circumference	0.87	0.0083	Weight	0.89	0.0078	Height	0.86	0.011
Total skinfold	0.88	0.0083						

The first variable listed in each group is the one which most accurately predicts body density. The variable which, when added to this first variable in a multiple regression equation, most improves the accuracy of prediction is listed next, and the multiple correlation coefficient for these two variables with body density and the standard error of the estimate is given. Other variables are added according to the degree to which they improve the prediction of body density, until the inclusion of further variables has no further practical value.

TABLE 4 Correlation coefficients for the relationships between body density and various indices of obesity

Age group	Weight-height index					
	$\dfrac{W}{H}$	$\dfrac{W}{H^{1.8}}$	$\dfrac{W}{H^{1.67}}$	$\dfrac{W}{H^2}$	$\dfrac{W}{H^3}$	$\dfrac{H}{W}^{0.33}$
Males						
17-19	-0.87	-0.89	-0.89	-0.88	-0.85	0.83
20-29	-0.72	-0.76	-0.75	-0.76	-0.77	0.76
30-39	-0.52	-0.52	-0.52	-0.52	-0.51	0.51
40-49	-0.59	-0.62	-0.62	-0.62	-0.60	0.59
50-72	-0.59	-0.70	-0.69	-0.72	-0.77	0.77
17-72	-0.72	-0.76	-0.75	-0.76	-0.76	0.78
Females						
16-19	-0.81	-0.81	-0.81	-0.81	-0.79	0.80
20-29	-0.80	-0.81	-0.81	-0.81	-0.81	0.81
30-39	-0.76	-0.77	-0.77	-0.78	-0.78	0.79
40-49	-0.83	-0.85	-0.84	-0.84	-0.83	0.85
50-68	-0.82	-0.82	-0.82	-0.82	-0.81	0.86
16-68	-0.78	-0.79	-0.79	-0.79	-0.78	0.82

REFERENCES

Durnin, J.V.G.A. and Rahaman, M.M. (1967): The assessment of the amount of fat in the human body from measurements of skinfold thickness. Brit. J. Nutr., 21, 681.

Durnin, J.V.G.A. and Womersley, J. (1974): Body fat assessed from total body density and its estimation from skinfold thickness: measurements on 481 men and women aged from 16 to 72 years. Brit. J. Nutr., in press.

Rahn, H., Fenn, W.O. and Otis, A.B. (1949): Daily variations of vital capacity, residual air and expiratory reserve, including a study of the residual air method. J. appl. Physiol., 1, 725.

Influence of diet, strain, age and sex on fat depot mass and body composition of the nutritionally obese rat*

R. Schemmel and **O. Mickelsen**

Department of Food Science and Human Nutrition, Michigan State University, East Lansing, Mich., U.S.A.

Fundamentally, all obesity can be considered to be of nutritional or dietary origin. This is apparent when it is realized that an increase in body weight occurs only when the intake of calories exceeds the caloric expenditure or requirement. The magnitude of the excess caloric intake required to produce obesity depends upon a variety of factors. A growing child must be in positive caloric balance, otherwise growth will not occur. However, if caloric intake exceeds the demands for growth and if the other physiological demands of the child are normal, obesity will likely develop. There is, however, some question as to whether obesity in early childhood predestines the individual to a life-long obese state.

In animals, obesity has been produced by a variety of techniques. Rats and mice have been used most extensively for such studies. In the rat, hyperphagia can be produced by lesioning the ventromedial nucleus. Although the primary result of such a lesion is a ravenous appetite and resulting development of obesity, there are other effects which have been explored only in the last decade or so. One that occurs in the immature rat is the cessation of skeletal growth. An explanation for the skeletal abnormality has been proposed by Han and Liu (1966), who suggested that there are two centers which are lesioned whenever the 'satiety' center is destroyed. The other center monitors the release of growth hormone from the pituitary. Destruction of the center regulating the release of growth hormone results in a cessation of skeletal growth.

Lesions in the ventromedial nucleus result in an increased size of the islet cells in the pancreas (Kennedy and Parker, 1963). Islet cell enlargement is associated with increased insulin release, which occurs as early as the fourth day after the rats have been lesioned (Frohman et al., 1969).

Other methods are available for the production of obesity in experimental animals. Most of these entail problems which limit their application to other forms of obesity, especially to that which occurs with increasing frequency among human subjects. For these reasons, we decided to see if we could produce obesity by an ad lib dietary means without resorting to forced feeding, or making the animal eat whenever it was thirsty (Ingle, 1949).

Our approach was based on the observation that when normal young men were suddenly changed from a typical American diet (fairly high in fat) to one that was very low in both calories and fat it required a number of days before the subjects became conscious of the caloric restriction (Keys et al., 1950). The reverse of this procedure was explored as a method of producing obesity in experimental animals. It was hypothesized that abruptly switching rats from a low-fat ration composed primarily of natural grains to one containing a high concentration of fat and very little roughage would result in their having difficulty in adjusting their caloric intake of the new ration to exactly meet their requirements.

Our first test of this hypothesis was with Sprague-Dawley rats which had been fed, from weaning, a grain ration providing 3.4 kcal per g. When male rats weighing about 300 g were switched to a semi-purified, high-fat ration, they reduced the weight of feed consumed. This reduction occurred during the first 24 hours. It probably stemmed from the fact that the high-fat ration provided 6.6 kcal per g. Although the rats very rapidly reduced their intake when fed the

* Published as Article No. 4654 from the Michigan Agricultural Experiment Station. This work was supported in part by Grants AM 08496 and HD 04502 from the National Institutes of Health.

high-fat ration, complete caloric equilibration was not achieved. The caloric intake of the high-fat ration remained about 10% higher than it had been when the grain ration was fed. This excess of caloric intake, perhaps combined with other factors, accelerated the gain in body weight. Although the rats fed this high-fat ration after a number of months were 20% heavier than those fed the grain ration, the resulting state of obesity was not as great as desired. To overcome that limitation, adult rats were force-fed the high-fat ration. This resulted in rapid increases in body weights which were twice those of controls fed a commercial stock diet (Mickelsen et al., 1955). However, the increased labor involved and the occasional failure to get the ration into the rat's stomach during the injection of the feed led to the exploration of other techniques.

By feeding the high-fat ration immediately after the rats were weaned, it was possible to increase the rate of body weight gain, the maximum body weight attained and the percentage of fat in the rat's body. Since the Sprague-Dawley rats used in the original work were prone to middle ear and upper respiratory infections, when maintained over long periods of time, Osborne-Mendel rats were tested for their susceptibility to obesity. The latter rats did develop upper respiratory infections but to a lesser extent than the Sprague-Dawley rats and only very infrequently showed signs of acute middle ear infections. For these reasons, Osborne-Mendel rats were used for most of our long-term obesity studies.

Rates of body weight gain approaching 10 g per day were maintained in some Osborne-Mendel male rats fed the high-fat ration for the first 7 weeks following weaning. A few of these rats were continued on the high-fat regimen until they attained weights exceeding 1700 g. The body fat content of the latter rats approximated 50% (Mickelsen et al., 1955; Schemmel et al., 1969b).

The high-fat ration contained 60% Crisco (a hydrogenated vegetable shortening manufactured by Proctor and Gamble, Cincinnati, Ohio), 25% high-protein casein, 5% of the Wesson modification of the Osborne-Mendel salt mixture, 2.2% of a vitamin mixture, 2.0% of a cellulose-type non-nutritive fiber, 0.01% Aureomycin (chlortetracycline; generously donated by the American Cyanamid Co., Princeton, N.J.), 2% liver powder, 0.25% Dl-methionine and 3.54% sucrose (Schemmel et al., 1969b). The control rats, which never became obese, were fed a natural grain ration (Campbell et al., 1966). Because of the extremes in fat content, these rations differed very markedly in caloric density. The high-fat ration had 660 kcal of digestible energy per 100 g, whereas the grain ration, which contained only 3% fat, had 335 kcal.

Modifications of the high-fat ration have not proved satisfactory in producing obesity when oils or fats softer than Crisco have been used at levels of 50% or more (Barboriak et al., 1958). When large amounts of an oil are used in a ration, casein is not the ideal source of protein since it does not facilitate the formation of a stable emulsion. If casein is used, the oil rapidly separates and layers at the top of the feeding cup. As soon as that happens, the rat ceases to eat and fails to gain weight or does so very slowly and never becomes obese, despite any genetic propensity in that direction. Recent work in our laboratory indicates that a whey protein concentrate (Protolac-Calcium, Borden, Inc., New Ulm, Minn.) permits incorporation of liquid fats into the ration at high levels (Fino et al., 1973).

Initially, one of the major criticisms of the high-fat ration was that it was so low in some essential nutrient that the rats ate extra amounts of feed to overcome the dietary deficiency. These critics have usually referred to the work of Johnson et al. (1936), which indicated that when rats were fed a low-protein ration, they had a higher percentage of fat than rats fed a ration containing an adequate amount of protein. These investigators reported that rats fed a ration containing 7.5% of protein deposited 5.2 g of fat in their bodies over a 2-week period, whereas the control rats fed a similar ration containing 27% protein formed only 2.8 g of body fat. Such small variations in body fat content become of minor significance compared to the marked differences between our normal and obese rats.

That the obesity produced in the Osborne-Mendel rats by feeding them the high-fat ration on an ad lib basis is not due to a protein or some other nutritional deficiency is further supported by our attempts to overcome such presumed deficiency. This we did by adding supplements that are recognized as being very good nutritionally. Liver powder was one of the first so added. If the obesity in our animals had been due to a dietary deficiency, it would be anticipated that the addition to the ration of the nutrient present at an inadequate level would decrease the percentage of fat in our rats and at the same time reduce their rate of weight gain. Such was not the case. Furthermore, doubling the dietary level of the water-soluble vitamins and adding an antibiotic such as Aureomycin, which is recognized as partially effective in overcoming minor nutritional deficiencies (Mickelsen, 1956), did not reduce the degree of obesity. If anything, these nutritional supplements increased the rate of weight gain and accentuated the degree of obesity.

When this work was undertaken, it was desired to secure some information as to whether heredity played any role in this nutritional type of obesity and, if so, to what extent the genetic propensity could be influenced by various dietary components. Another objective was to evaluate the extent to which the obese state was associated with an increase in muscle and skeletal mass. One other facet of our work that we shall allude to in this presentation is the influence of nutritional obesity imposed on rats prior to weaning on the body weight and body fat content ultimately attained by these animals.

THE ROLE OF GENETICS IN BODY FAT DEPOSITION

Attempts to evaluate the influence of environmental and/or genetic factors on human obesity have been plagued by the difficulty in identifying the degree to which these two factors interact (Tepperman, 1957; Mayer, 1965; Newman et al., 1937; Angel, 1949; Gurney, 1936; Ellis and Tallerman, 1934). Furthermore, it is difficult to follow alterations in body composition and fat depot development in human subjects. There are a number of reasons for this, including the difficulty of studying a group of human subjects over a long period of time and the problems associated with ascertaining (1) the rate at which various fat pads in the body increase in size and (2) the distribution and amount of fat in the abdominal area.

The paucity of genealogical information, including such simple anthropometric measurements as heights and weights, makes it impossible to go back more than one or two generations in any study involving human subjects. Even where a number of generations are available for an elementary evaluation of general fatness or leanness, it is impossible to separate the genetic and environmental factors that influence the development of obesity. Such data would be limited by the criticism that the children of obese parents are exposed, early in life, to individuals who always eat large meals. As a result of such exposure, the children establish habits of overeating.

The use of an experimental model in which obesity can be developed by purely dietary or nutritional means obviates many of the problems associated with human studies. Information secured by that kind of investigation should provide incontrovertible evidence on the role of genetics and diet in the development of obesity.

To determine the propensity to develop obesity, we used the following strains of rats: (1) Osborne-Mendel, (2) Sprague-Dawley, (3) Hoppert, (4) Wistar-Lewis, (5) Hooded, (6) M.S.U. Gray and (7) S 5B/Pl. After establishing the tendency of these rats to become obese, more extensive studies were carried out with the strains that were at the two extremes of the spectrum. These were the Osborne-Mendel and S 5B/Pl rats. In both of these strains, the males and females behaved similarly, with all Osborne-Mendel rats readily becoming obese when fed the high-fat ration, whereas none of them even approached that condition when fed a grain ration. On the other hand, the S 5B/Pl rats, when fed the high-fat ration, showed practically no greater body weight gains than their littermates fed the grain ration.

The rats were housed in a room maintained at $23 \pm 1°C$ with alternating 12 hour periods of light and darkness. Tap water and diet were always available to the animals.

Littermate, weaning rats of all but the Sprague-Dawley and Hoppert strains were randomly assigned to either the high-fat or the grain ration described above. Since the Sprague-Dawley and Hoppert rats were secured from ouside the laboratory, they were paired on the basis of body weights and assigned to the two diets in a random manner. The susceptibility of the different strains of rats to obesity was evaluated on the basis of body weight gains, weight of fat depots, body fat content and efficiency of dietary energy utilization. For these parameters, the significance of the differences was determined by the analysis of variance or Student's t test. Carcasses were prepared for moisture, fat, nitrogen and ash determinations as described by Mickelsen and Anderson (1959). Individual fat depots were removed according to the procedures of Schemmel et al. (1970b).

TABLE 1   Body weight and fat (mean ± S.E.M.) in male and female rats of 7 strains at 23 weeks of age fed either a low-fat grain or high-fat ration for 20 weeks*

	Body weight (g)		Body fat (%)	
	Grain	High fat	Grain	High fat
Males				
Osborne-Mendel	445.3 ± 17.1	692.6 ± 26.6	13.0 ± 1.2	39.0 ± 3.1
Sprague-Dawley	423.0 ± 22.5	592.9 ± 19.5	14.1 ± 1.1	30.4 ± 1.9**
Hoppert	402.5 ± 12.1(4)***	495.9 ± 69.3(4)	16.4 ± 1.7(4)	31.1 ± 5.4(4)
Wistar-Lewis	363.7 ± 12.9	515.4 ± 8.0	13.6 ± 1.1	33.9 ± 3.0
Hooded	392.4 ± 14.4(4)	557.9 ± 25.8	12.3 ± 1.1(4)	32.6 ± 1.6
Gray	378.7 ± 8.9	493.5 ± 20.2	14.6 ± 1.5	27.5 ± 2.3
S 5B/Pl	303.8 ± 6.1	345.7 ± 15.8	9.9 ± 0.5	14.1 ± 0.8
Females				
Osborne-Mendel	301.0 ± 6.2	452.3 ± 35.5	14.9 ± 0.8	40.4 ± 3.1
Sprague-Dawley	279.3 ± 10.0	331.0 ± 11.0	14.8 ± 1.5	30.4 ± 0.9
Hoppert	243.8 ± 11.1	296.2 ± 38.2	13.7 ± 1.8	28.2 ± 5.1
Wistar-Lewis	204.0 ± 7.7	309.7 ± 14.7	12.5 ± 1.5	35.3 ± 2.2
Hooded	216.9 ± 9.7	269.6 ± 11.9	11.1 ± 0.4	30.7 ± 1.4
Gray	203.2 ± 2.5	249.0 ± 10.4	12.0 ± 1.1	26.1 ± 1.2
S 5B/Pl	162.8 ± 6.2(4)	170.0 ± 4.7(4)	10.9 ± 0.5(4)	13.7 ± 0.8(4)

* Adapted from Schemmel et al. (1970a).
** By Dunnett's test, body weight (g) and body fat (g/100 g body weight) were significantly different from Osborne-Mendel rats within sex and diet groups: $P < 0.01$.
*** There were 5 rats per group, except for where 4 appears in parentheses, when there were 4 rats per group.

Each group of rats was fed its diet continuously for either 10 or 20 weeks. At both these times, the average body weights of the 10 rats in each group were greater for those fed the high-fat than for those fed the grain ration (Table 1). The differences in all parameters were greater the longer the animals were fed their respective rations. The Osborne-Mendel rats showed the greatest differences for the animals of one strain fed the two rations, while the smallest differences were seen among the S 5B/Pl rats.

In the rats fed the high-fat ration, body weights and percentage of body fat were larger than in those fed the grain ration. On this basis, the Osborne-Mendel and S 5B/Pl rats represented the two extremes. Although the Wistar-Lewis rats were intermediate between these extremes, they had the second highest percentage of body fat; actually it was not significantly different from that in the Osborne-Mendel rats. These data confirm the fact that body weight per se is not always a good indicator of body fat content even among rats of the same sex fed the same ration.

For the grain-fed rats, the percentages of body fat of the seven strains of rats showed a range which, on an absolute basis, was no more than 7.3 percentage points, while their body weights varied as much as twofold. None of the grain-fed rats differed significantly in percentage of body fat from the Osborne-Mendel rats.

Analysis of variance of these data indicated that body weight but not percentage of body fat was closely associated with strain differences. For body weight, strain differences accounted for 40.2% of the variability, while the ration accounted for 39.7%. For percentage of body fat, strain differences accounted for only 13.5% of the variability while the ration accounted for 74.2%. Except for the Wistar-Lewis rats, a high-fat dietary regimen increased the percentage of body fat to a greater extent in those strains already genetically predisposed to attain a larger body weight. These findings suggest that body weight per se may be influenced to a fairly large extent by genetic factors but that the percentage of fat in the body, and therefore the relative degree of obesity, is likely to be determined to a large extent by the nature of the diet.

CHANGES IN THE BODY DURING THE DEVELOPMENT OF NUTRITIONAL OBESITY

Lean body mass

One of the primary questions we wished to answer by our studies with the obese rat was: is obesity, especially that which is initiated early in life, associated with an increased muscle formation? An answer to that question appeared to be essential since many overweight individuals claimed they were not obese but were just 'large'. That statement implied they had a larger than normal complement of body protein attached to a correspondingly enlarged skeleton. From a theoretical standpoint, such a suggestion appears rational since the obese individual carries a very heavy weight. Such a 'load' constantly carried around, theoretically, should result in the formation of extra muscle tissue. Not only the obese subject, but some investigators have accepted this apparent truism. For instance Passmore et al. (1958) stated that 'in most obese people there may be hypertrophy of skeletal and cardiac muscle, and perhaps of other organs which have to support and move the increased mass of fat'. To evaluate the significance of that possibility requires an experimental model with which it is possible to carry out body composition studies and one in which obesity can be produced by purely dietary means. The Osborne-Mendel rat fulfills both requirements.

When Osborne-Mendel rats fed either the high-fat or grain ration were killed at various time intervals, it became obvious that their body nitrogen content was related to age and not to body weight (Schemmel et al., 1969b). There was an obvious difference in the protein content of male and female rats, with females having less body protein than males (Fig. 1). This difference

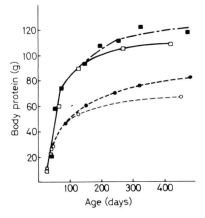

Fig. 1. Body protein content of obese and normal weight Osborne-Mendel rats. □ = Grain, males; ■ = High fat, males; o = Grain, females; ● = High fat, females.

appeared within the first month after weaning, but for both the normal and obese rats of the same sex there was no difference in absolute body protein content until the animals were more than 100 days of age. Thereafter, the obese rats of both sexes had slightly more body protein than their normal weight controls (Fig. 1). Normal weight male rats at 420 days of age had 108 g of protein in their bodies while the obese males had 118 g; for the females of comparable age, the protein contents were 65 and 86 g respectively.

The slightly greater amount of protein in the obese rat was probably associated with the marked expansion of the adipose tissue compartment, skin, blood and internal organs. Although fat comprises the major component of adipose tissue, the cells therein contain a certain amount of protein. The latter contributes its proportionate share of nitrogen to the total protein nitrogen as measured by carcass analysis. The mass of the integument increases as obesity develops. A significant enlargement also occurs in the internal organs and blood volume. For the heart, kidney and adrenals, this enlargement in the obese rats is 1.4 to 1.5 times the weight of the same organs in their normal weight controls; the liver and spleen in the obese rats are 1.75 times as heavy as in the normal weight controls. Since the increase in the weights of the skin and internal organs is not due primarily to an increase in fat, their enlarged size accounts for a certain fraction of the increased nitrogen in the carcasses of the obese rats. How large that fraction is has not been determined.

The increased size of the obese animal is associated with an increased blood volume. This has been demonstrated by a number of techniques. These suggest that for the moderately obese rat the blood volume is 1.5 times that in the normal weight control (Grommet et al., 1973). Again that increase in blood volume is reflected in an increase in blood nitrogenous constituents. As such, the blood nitrogen-containing compounds account for some of the increased carcass nitrogen in the obese animals.

Age influenced the composition of the lean body mass in both the normal and obese rats. This became evident when the percentages of water and protein in the lean body mass were calculated. Lean body mass by definition represented the body weight of the rat when it was killed, minus the weight of body fat and the contents of the gastrointestinal tract.

The water content of the lean body mass changed with the age of the animals and to a lesser extent with the nature of the diet. The water content of the lean body mass of both the grain- and high fat-fed male rats decreased rapidly from a value of 77.5% in the 21-day-old rats to about 72% in the 60-day-old rats (Fig. 2). From then on, the water in the lean body mass of the normal

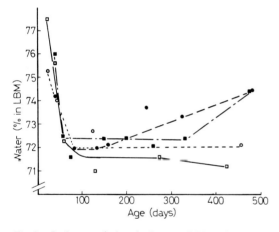

Fig. 2. Body water in lean body mass (LBM) of obese and normal weight Osborne-Mendel rats. Symbols as in Figure 1.

weight male rats slowly and progressively decreased to a value of 71% at age 430 days. During the latter period, the moisture content of the lean body mass in the rats fed the high-fat ration increased to a value of 74.5% when the animals were almost 500 days of age. The moisture content of the lean body mass of the females started at a lower value, 75.5%, than in the males and decreased until the rats were about 100 days of age. There was probably no difference in the moisture content of the grain- and high fat-fed female rats prior to about 300 days of age. Thereafter, the moisture in the lean body mass of the obese rats increased to a value of 74.5%, whereas the water in the lean body mass of the normal weight females remained constant at about 72%.

The unexpected finding was the increase in moisture of the lean body mass as the rats became obese. The marked reduction in this parameter during the early stages of the rat's growth was expected; the plateau reached in the grain-fed rats is what other investigators have observed. However, the increase in percentage of water seen in the lean body mass of the rats fed the high-fat ration is contrary to what initially might be expected. On the basis of work with human subjects, it would be anticipated that if any change occurred, it would be in the opposite direction. The diuretic effect of a high-fat diet has been associated with a certain amount of

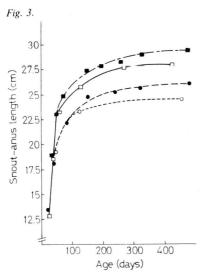

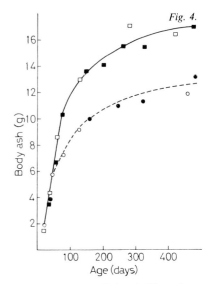

Fig. 3. Snout to anus lengths of obese and normal weight Osborne-Mendel rats. Symbols as in Figure 1.

Fig. 4. Body ash in obese and normal weight Osborne-Mendel rats. Symbols as in Figure 1.

ketosis. In that respect, the rat is unusual since it is resistant to ketosis. Rations which provided as much as 83% of calories from fat when fed to rats never produced ketonuria (Schemmel and Mickelsen, unpublished data).

A plausible explanation for the increased water content of the lean body mass of the obese rats is that obesity is associated with an enlargement of one or more body components that is especially rich in water. Such a compartment could be the blood volume. Obesity in the rat is accompanied by a 50% expansion of the blood volume over that in normal weight rats of the same sex and age (Grommet et al., 1973). The water content of the increased plasma volume would account for much of the 'extra' water present in the lean body mass of the obese rats.

The skeleton

When the size of the obese rat is compared with that of a normal weight littermate (Fig. 3), it appears that the obese rat has the larger skeleton. Actual measurements of nose to anus lengths (Schemmel et al., 1969a) appeared to confirm the impression that obesity might be associated with an increased skeletal size. This was an erroneous impression since total body ash (Fig. 4) in both lean and obese rats was similar when animals of the same age and sex were compared (Schemmel et al., 1969b). Body ash serves as a rough index of skeletal size since Shohl (1939) reported that 83% of the minerals in the body are associated with the skeleton. Further evidence for this came from measurements of the lengths of various bones as indicated by whole body X-rays of lean and obese rats. These measurements indicated no difference between obese and normal weight rats in the lengths of the longer body bones (Schemmel et al., 1969a).

Conclusive evidence for the similarity in skeletal size of obese and normal weight rats came from various types of measurements made on the bones secured from these animals. A special study was performed for this. This required a number of adjustments in the ration so as to provide both the normal weight and obese rats equal daily intakes of calcium, phosphorus and magnesium. These adjustments in mineral contents of the ration were made to ensure that if any skeletal differences were observed they would not be due to variations in intake of any of the major skeletal minerals.

Weaning littermate Osborne-Mendel rats were alternately assigned to the grain or modified high-fat ration (modified only as far as the minerals were concerned). Ten rats from each dietary group were killed at varying time intervals after being on experiment for up to 45 weeks. The skeleton was freed from as much adhering material as could be removed in a short time. The bones were cleaned by exposing them to a colony of dermestid beetles. The latter was carried out for us through the courtesy of Dr. Rollin H. Baker, Director of the Michigan State University Museum. Measurements made on these bones included the length, the widest width, narrowest width and two widths at right angles to each other at the midpoint of the shaft for each of the five major long bones of the right limbs: the humerus, radius, ulna, femur and tibia-fibula.

The results of that study indicated that as the rats became older, the outer dimensions of the long bones increased; however, there was no difference in any of these means when the bones from normal weight and obese rats of the same age were compared. Since the pelvic and pectoral girdles, the scapula and os coxae bear a greater weight in the obese than in the normal weight rat, these bones were also measured. Again there was no difference in the size of the bones except for the os coxae which, in the obese rat, was consistently slightly longer and wider than in the normal weight control. It is impossible to assess the significance of this difference since only these two measurements differed between the obese and normal weight rats; the other 30 outer dimensions of the 8 other bones were the same (Smith et al., 1969). Although the outer dimensions of the bones were the same, the volumes of the femurs of the obese rat were significantly greater than the volumes of these bones from normal weight rats. The larger volume resulted in a lower density.

The breaking strength of the right femur was tested with an apparatus specifically designed for this purpose by Dr. Elwyn R. Miller of the Michigan State University staff. We are deeply grateful to Dr. Miller for his assistance in this phase of the work. Again, there was an improvement in breaking strength of the bones as the animals became older, but in all cases, there was no treatment effect when the bones of the grain-fed rats were compared with those fed the high-fat ration.

These observations suggest that obesity has only a minor effect on muscle and skeletal mass in the obese rat. It has been proposed that the increased nitrogen in the carcasses of the obese rats probably is associated with their enlarged adipose tissue mass. Additional support for that suggestion is provided by our preliminary studies of weight reduction in the obese rats. This involved switching rats weighing 1 kg to the grain ration. When that was done, the obese rats lost weight until they reached 650 g. At the latter weight, the rats' carcasses had essentially the same amount of nitrogen as their normal weight littermates that had never been obese. The extra nitrogen in the obese rats disappeared when the adipose tissue, the skin and internal organs were brought down to a 'normal' weight.

Results on skeletal size in obese rats comparable to ours have been reported by Sokoloff et al. (1960). When an error in their paper is corrected, exactly the same conclusion is reached as that currently being presented. In that paper, the femur lengths of older obese rats were inadvertently compared with those of younger controls. When that error is corrected, no difference is seen in the skeletal size of obese and normal weight rats.

Our studies and those of others indicate that the growth of the skeleton is, to a certain extent, independent of concomitant changes in other compartments of the body. This holds true whether there is an excessive or restricted development of the other body compartments. Continuous growth of the skeleton at a normal or nearly normal rate occurs even when the alterations in growth of the other body components are initiated very early in life during the period when cell numbers are rapidly increasing. Any changes in growth patterns occurring at that time, presumably, influence future trends in growth and development.

Body fat accretion

The deposition of fat in the human body has not been documented partly because of the difficulty in following a group of the same individuals over long periods of time. There are suggestions that, for some people, a rapid accumulation of body fat occurs in early childhood and progressively increases throughout life. For other individuals, especially females, exaggerated fat deposition may be initiated during puberty when girls normally undergo a considerable increase in body fat. The accumulation of body fat may be accentuated by pregnancy, and with each succeeding gestation body weight, and with that body fat, increases. Among males, there is a tendency to 'put on weight' when a sudden and marked reduction in physical activity occurs. Although energy expenditure has been reduced, caloric intake may be maintained near its

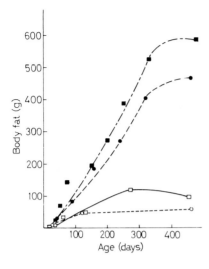

Fig. 5.   Body fat in obese and normal weight Osborne-Mendel rats. Symbols as in Figure 1.

former level with the result that the surplus calories are deposited as fat. Many men experience this difficulty shortly after graduating from college when they assume a relatively sedentary form of living.

Nutritional obesity in the Osborne-Mendel rats is associated with an almost linear increase in body fat as the animals become older (Fig. 5). This is also true, at a much lower level, for the normal weight controls. In other words, feeding the high-fat ration which ultimately resulted in gross obesity in the Osborne-Mendel rats, appeared to accentuate, to a very great extent and at a markedly increased rate, a reaction which normally occurred even in the normal weight controls. For both the obese and normal weight rats, the linearity in body fat accretion with increase in body weight appeared to become accentuated as the animals reached the upper limits of the body weights characteristic of these groups. These curves suggest that the percentage of fat in the additional increments of body weights will be composed of an even greater percentage of fat.

Although there was an almost linear increase in total body fat content as the rats became older, this increase did not occur at a uniform rate for all fat pads. Some indication of the differences at which the fat pads in the normal and obese rats 'matured' can be gleaned from the changes in weight with age of the individual fat pads. In all cases, especially after the body weights of the obese rats began to diverge markedly from those of the normal weight animals of the same age, the weights of the fat pads in the former were 2 to 3 times the weights of the same pads in the latter.

The increase in weight with time of the inguinal, axillary and genital fat pads illustrates the differences in fat deposition within the animals. These fat pads attained their maximum relative weights prior to an age of 100 days in the rats fed the high-fat ration. On the other hand, when rats were fed the low-fat grain ration, the weights of these fat pads continued to increase relatively until the animals were more than 400 days old. (See Figs. 6 and 7 for growth patterns of the inguinal and genital fat depots.) Other fat pads, such as those in the interscapular and perirenal regions, increased in weight throughout life in the high-fat-fed rats, whereas they attained their maximum relative weight (fat pad weight/body weight) early in the life of the grain-fed rats (Schemmel et al., 1970b).

These differences in the relative growth of the fat pads in the normal weight and obese rats have implications for the study of adipose tissue. Our studies point out that no one fat pad is representative of the adipose tissue throughout the body, especially when the animals are fed rations that differ markedly in composition. Not only is this true of the gross weights of these fat pads, but anatomical variations have been reported to exist between abdominal and subcutaneous fat pads (for references see Schemmel et al., 1970b).

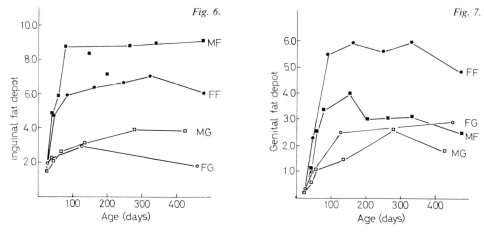

Fig. 6. Inguinal fat depot weight in obese and normal weight Osborne-Mendel rats expressed as g/100 g body weight. Symbols as in Figure 1.

Fig. 7. Genital fat depot weight in obese and normal weight Osborne-Mendel rats expressed as g/100 g body weight. Symbols as in Figure 1.

OBESITY INDUCED EARLY IN LIFE

The large number of recent papers on the long-term effects of nutritional experiences early in life prompted us to explore that approach as a means of initiating obesity in our animals within a short period of time. The 6 to 12 months currently required for the rats to attain weights in excess of 1 kg might be shortened if the development of obesity were started prior to weaning. To that end, pregnant Osborne-Mendel rats were closely observed toward the end of their gestation. As soon as parturition occurred, the litter size was reduced to 8 with an approximately equal distribution of the sexes in each litter. At that time, every alternate dam was switched to the high-fat ration while the others were maintained on the low-fat grain ration. The latter ration had been fed to the rats prior to and during pregnancy since previous work in our laboratory indicated that feeding the high-fat ration interfered with the reproductive process to a certain extent.

By this technique, the dams fed the high-fat ration produced milk that contained almost twice as much fat as that produced by the rats fed the grain ration (Emery et al., 1971; Schemmel et al., 1973). Furthermore, when the young were about 15 days old and had started to eat solid foods, they received the ration available to their dams. The combination of the increased energy content of the milk received and the initiation of the high-fat ration early in life produced weanling rats that weighed 86 g at 3 weeks of age, while the young weaned by the grain-fed rats weighed 58 g. When weaned, the male rats fed the high-fat ration had almost 4 times as much fat in their bodies as the grain-fed controls (Schemmel et al., 1973). This increased fat content was also associated with an increase of almost 30% in body protein.

To explore the effects of these differences in pre-weaning body fat on the subsequent development of obesity, half of the rats in each of the 2 dietary groups were maintained on the ration they received prior to weaning; the remaining rats in each litter were switched to the other ration. These 4 dietary groups were fed their respective rations for as long as 168 days. At intervals during that time, 6 rats from each group were killed for body composition studies. The results indicated that the weights attained after 168 days were the same for both groups fed the grain ration post-weaning; this was also true of both groups fed the high-fat ration post-weaning. In other words, the rats which had been exposed to the high-fat ration prior to weaning and, at that time, had 4 times as much body fat as those fed the grain ration, weighed no more at 168 days than the rats that were first exposed to the high-fat ration at weaning.

These observations also suggest that in the Osborne-Mendel rats the development of an increased body fat content early in life does not destine them to a lifetime of obesity. This was evident by the fact that, even though the rats exposed to the high-fat ration prior to weaning had 4 times as much body fat at weaning as those fed the grain ration, they developed into 'normal weight' animals when fed the grain ration post-weaning.

That the pre-weaning diet did affect the final condition of the animals became evident when the weight gains of individual rats were examined. At the end of the study by far the largest percentage of heavy rats were found among the animals exposed to the high-fat ration prior to weaning and then continued on that ration post weaning. The only other rats that appeared in that category were those that had been fed the grain ration prior to weaning and then transferred to the high-fat ration. In a comparable manner, the lightest weight animals occurred in the group that had been exposed to the grain ration throughout life. At the end of the study, a few of the lightest animals were in the group transferred, at weaning, from the high-fat to the grain ration.

Contrary to expectation, the rats fed the grain ration prior to weaning actually accumulated body fat in the post-weaning period more rapidly per 100 g of body weight than the rats that had been fed the high-fat ration prior to weaning. These observations suggest that in animals as susceptible to nutritional obesity as the Osborne-Mendel rats, the nature of the ration fed post-weaning determines, to a large extent, whether or not adult obesity develops.

The apparent contradiction between our observations and those reported by other investigators (for references see Schemmel et al., 1973) may be attributable to the fact that in most of the previous studies one group of rats was markedly undernourished. This resulted from the fact that many investigators utilized the technique of limiting to a small number the young nursed by one dam while increasing to a very large number those nursed by another female (e.g. 4 vs. 18 young). The large number of young being nursed by one dam were probably underdeveloped at weaning to such an extent that they could not catch up despite the fact that a liberal ration was available to them after weaning in unlimited amounts.

TABLE 2   Energy consumed in the diet and retained in the carcasses of Osborne-Mendel (OM) or S 5B/Pl rats fed low- or high-fat rations

Strain, ration and experimental period (weeks)*	No.	Dietary Energy (kcal/ experimental period)	Retention of dietary energy in carcass (kcal/100 dietary kcal)		
			Total	Fat	Protein
Males					
OM, G, 10	5	5571 ± 273	12.9 ± 0.7	6.7 ± 1.1	6.1 ± 0.1
OM, HF, 10	5	6945 ± 695	28.5 ± 2.0	22.9 ± 2.0	5.5 ± 0.4
S 5B/Pl, G, 10	5	4085 ± 535	10.0 ± 1.5	4.1 ± 1.1	5.9 ± 0.3
S 5B/Pl, HF, 10	5	3908 ± 538	12.8 ± 1.4	6.4 ± 1.8	6.4 ± 0.1
OM, G, 20	5	11226 ± 878	8.3 ± 1.0	4.2 ± 0.9	4.1 ± 0.1
OM, HF, 20	5	13536 ± 700	22.3 ± 3.0	18.7 ± 3.4	3.5 ± 0.4
S 5B/Pl. G, 20	5	8363 ± 230	6.9 ± 0.5	3.1 ± 6.4	3.7 ± 0.1
S 5B/Pl, HF, 20	5	8869 ± 711	8.8 ± 0.6	4.9 ± 0.6	3.9 ± 0.1
Females					
OM, G, 10	5	4360 ± 253	7.9 ± 0.7	4.0 ± 0.7	4.0 ± 0.2
OM, HF, 10	5	4994 ± 400	21.8 ± 1.9	17.8 ± 1.9	4.1 ± 0.3
S 5B/Pl, G, 10	5	2856 ± 133	7.8 ± 0.6	3.7 ± 0.5	4.0 ± 0.2
S 5B/Pl, HF, 10	5	3190 ± 232	8.2 ± 1.1	4.6 ± 1.0	3.6 ± 0.1
OM, G, 20	5	9234 ± 637	6.4 ± 0.6	3.7 ± 0.5	2.7 ± 0.2
OM, HF, 20	5	9976 ± 976	19.8 ± 4.1	17.1 ± 3.7	2.6 ± 0.1
S 5B/Pl, G, 20	4	5388 ± 260	5.2 ± 0.3	2.6 ± 0.2	2.7 ± 0.2
S 5B/Pl, HF, 20	4	6028 ± 63	5.6 ± 0.5	3.2 ± 0.4	2.4 ± 0.1

* Rats were fed the experimental diets from weaning; G = Grain ration; HF = high-fat ration.

CONVERSION OF DIETARY TO CARCASS ENERGY

Strain differences

To secure an explanation for the marked differences in various strains of rats as regards obesity after feeding on a high-fat ration, the efficiency of their food utilization was investigated. This was done for male and female rats of the Osborne-Mendel and S 5B/Pl strains. These strains were studied since they had the greatest and the smallest tendency to increase in weight and accumulate body fat when fed a high-fat ration.

Regardless of whether food intake was measured for 10 or 20 weeks, the Osborne-Mendel rats fed the high-fat ration consumed more calories than their littermates fed the grain ration (Table 2). In the male rats, this difference amounted to more than 20% of the caloric intake of the grain-fed rats. This difference in caloric intake was evident in both the first as well as the second 10-week period. It may have been partially responsible for the obesity which developed in the Osborne-Mendel rats. In the S 5B/Pl male rats, the differences in caloric intake between the grain- and high-fat-fed rats were much smaller than in the Osborne-Mendel rats. In the S 5B/Pl male rats, there was an increased difference in caloric intake in the second 10-week period when compared to the first 10 weeks.

The female Osborne-Mendel rats fed the high-fat ration showed a smaller percentage difference than the males when the intakes of the grain-fed rats were compared with those of the high-fat-fed animals (Table 2). These differences were essentially the same during both the experimental periods. Furthermore, similar differences were seen in both the Osborne-Mendel and S 5B/Pl females as regards caloric intake. Despite the higher caloric intake of the S 5B/PL female rats, they did not deposit any more fat in their bodies than their littermates fed the grain ration. With a similar differential in caloric intake, the Osborne-Mendel female rats ultimately became very obese.

By elimination, the explanation for the differential effect of the two rations on body fat in the two sexes and two strains of rats must involve, primarily, the efficiency with which dietary energy was transposed to depot lipids. Presumably, the rats fed the high-fat ration required little energy to transfer the large percentage of dietary fat to depot lipids. On the other hand, the rats fed the grain ration, according to the hypothesis, had to utilize considerably more energy to convert the dietary starch to depot lipids. Suggestions similar to this have been made by a number of investigators (Baldwin, 1968, 1970; Ball, 1965).

A study was made of the efficiency with which the rats transferred their dietary energy into body energy. For this purpose, the energy deposited in the carcass during the experimental period was determined. The carcass energy in the bodies of the rats (minus that in the gastrointestinal contents) was calculated from body composition analyses of a group of weanling rats. They were comparable in weight and sex to those started on the experiment. The values thus secured for body energy were subtracted from the values for the rats sacrificed at the end of the study.

For estimating the energy content of the carcass, the values 5.4 and 9.3 kcal per g of protein and fat, respectively, were used. These values are based on bomb calorimetry analyses (Schemmel et al., 1972) as well as data from Blaxter and Rook (1953) and Paladines et al. (1964).

Calculations confirmed the existence of a marked difference in the carcass energy of the Osborne-Mendel rats fed the two different rations, but this was not true for the S 5B/Pl rats. When male or female Osborne-Mendel rats were fed the high-fat ration, they retained 2.6 to 3.0 times more dietary energy in their carcasses than the rats of the same strain fed the low-fat grain ration (Table 2). The fraction of dietary energy retained in the carcass was smaller during the second 10 weeks of the study than in the first 10 weeks. This reduction in efficiency occurred in both the high-fat- and grain-fed rats; it was not associated with any change in weight of food consumed. This is likely to be associated with less rapid growth during the second 10-week period.

The energy available from each ration retained in the carcasses of the rats was influenced by the strain, age and sex of the animals. The Osborne-Mendel rats retained a higher percentage of dietary energy in their bodies than the S 5B/Pl rats (Table 2). This difference occurred with both rations and held true for both sexes. The Osborne-Mendel rats fed a high-fat ration retained more than twice as much energy in their bodies than the S 5B/P1 rats fed the same ration. For the grain-fed rats, the differences were much smaller and for the females practically nonexistent. Contrary to many popular statements, male rats were more efficient than females in converting dietary calories to carcass energy. Male Osborne-Mendel rats converted 28.5% of dietary calories ingested during the first 10 weeks of the study to carcass energy, whereas the females converted only 21.8%.

Another measure of bioenergetics, proposed by Forbes et al. (1946a, b), is 'energy output'. This was calculated by subtracting the calories retained in the carcasses of the animals during the experimental period from the caloric intake. These values provide a rough estimate of the energy expended by the animal for maintaining body functions and physical activity. Such calculations indicated that energy output was closely related to lean body mass and was relatively the same for both dietary and strain groups (Table 3). This constancy existed for each half of the experiment.

The outstanding characteristic of energy output was the very great utilization of energy by the females. This difference between males and females for the first 10 weeks ranged from 400 to 700 kcal per 100 g lean body mass and increased to 500 to 1100 kcal for the entire 20-week period. In

TABLE 3   Energy output of Osborne-Mendel (OM) and S 5B/P1 rats*

Strain	Diet	No.	Energy output (kcal)	Lean body mass (g)**	Energy expended/100 g lean body mass (kcal)
Males, 10 weeks					
OM	Grain	5	4855	292	1663
OM	High fat	5	4967	350	1419
S 5B/P1	Grain	5	3675	228	1612
S 5B/P1	High fat	5	3409	227	1502
Males, 20 weeks					
OM	Grain	5	10289	387	2655
OM	High fat	5	10524	422	2494
S 5B/P1	Grain	5	7790	274	2843
S 5B/P1	High fat	5	8087	297	2723
Females, 10 weeks					
OM	Grain	5	4013	195	2058
OM	High fat	5	3902	201	1941
S 5B/P1	Grain	5	2634	130	2026
S 5B/P1	High fat	5	2927	136	2152
Females, 20 weeks					
OM	Grain	5	8639	256	3375
OM	High fat	5	8003	269	2974
S 5B/P1	Grain	4	5104	145	3520
S 5B/P1	High fat	4	5688	147	3869

* For each appropriate experimental group, energy output was calculated by subtracting the energy accumulated in the carcasses from the total energy intake over 10 or 20 weeks. The latter values were calculated by multiplying the amount (g) of food eaten throughout the entire experimental period by 3.62 and 7.18 kcal/g for the grain and high-fat rations, respectively. For definition of energy output, see text (Forbes et al., 1946a, b).
** Carcass weight, exclusive of gastrointestinal contents, minus body fat.

the first 10-week period, the differences between the sexes were greater for the animals fed the high-fat ration than for those fed the grain ration (700 vs. 400 kcal). However, for the 20-week period as a whole, these ration differences disappeared.

SEX DIFFERENCES

An increasing amount of data confirms the observation that, for each age category, women who are of 'normal' body weight have a higher percentage of body fat than men of the same age and also of 'normal' body weight. For Caucasians, this sex difference has been reported to be approximately 2-fold (Brozek et al., 1953). Surprisingly, the difference appears to be even greater for the Chinese. A survey of Chinese medical school personnel on Taiwan indicated that women had 3 times as much body fat as males (Chen et al., 1963). This difference was based both on the values secured from 13 skinfold measurements and on total body fat as estimated from specific gravity and body volume measurements. In 30 males with an average age of 23 years, body fat represented 9.9% of their body weight, whereas in the 29 women with the same average age body fat represented 27.5% of body weight.

The differences in body fat content of males and females become apparent primarily at the time of puberty. Prior to that time, there are only slight differences in the percentage of fat in the bodies of boys and girls (Ljunggren, 1965). There are suggestions that the greater fat content of human females is apparent at birth. Support for this comes from the observations of Owen et al. (1963). The latter is based on the report that body water is a reliable index of lean body mass. Owen et al. found 59.9% water in male newborns, and 56.3% in newborn females. Since body fat is inversely related to body water and lean body mass, these values suggest a slightly higher body fat content in females at birth than in males.

Not only are the amounts and percentages of body fat different in men and women, but so is its distribution. This difference in body build becomes manifest primarily at puberty. It is largely a reflection of the effect of sex hormones in regulating the deposition of fat in various parts of the body. However, once this differential distribution of subcutaneous fat occurs, it does not require the continued presence of the sex hormones to maintain it. This has been emphasized by Hausberger (1965), who pointed out that removal of the ovaries in adult women does not influence the distribution of fat in their bodies. However, if adults are treated with sex hormones, the relative distribution of body fat may be altered. This was reported to have occurred when some men received estrogenic hormones to reduce their serum cholesterol level to minimize the possibility of their developing a cardiovascular attack.

The influence of sex on the susceptibility of an organism to obesity, the relative amount of adipose tissue, its distribution in the body and various physiological parameters associated with its development can be studied in rats which have been made obese by purely dietary means. This procedure obviates the complications associated with many other forms of experimental obesity.

Our studies with the laboratory rat indicate that the development of some fat pads shows indications of sexual dimorphism. This is especially true of the relative weight (i.e. weight of depot as percentage of body weight) of the interscapular depot (Schemmel et al., 1970b). The sexual difference becomes apparent in rats weighing more than 450 g. Up to that weight, the relative weight of this depot increased at essentially the same rate in both males and females. Although there was no statistical difference between the sexes, for each body weight up to 450 g the weight of this depot was slightly greater in the females. However, as the females exceeded their 'normal' adult weight, the relative weight of the interscapular fat pad increased at an accelerating rate to the point where it represented 10% of the body weight in 900 g rats. In contrast, males showed only a slight increase in the relative weight of this fat pad regardless of body weight. This is indicated by the fact that in 450 g male rats, the interscapular fat pad represented 2% of body weight, whereas in 1200 g rats it represented slightly more than 3%.

Three other fat depots show differences between males and females. However, these differences are much less dramatic than for the interscapular fat pad. For instance, the genital fat depot increased in relative weight more rapidly in female than in male rats (Fig. 7). This depot in 200 g female rats represented 2% of body weight, while in 300 g rats it exceeded 5% of body weight. Male rats showed an increase in weight of that fat depot similar to the females up to a body weight of 300 g, when it represented 3% of the rat's weight. Thereafter, the size of the genital fat depot in the male rat increased at the same rate as the body; even at 1200 g, the weight of this depot was 3% of body weight.

A reversal of sexual dimorphism occurred in the inguinal and axillary fat depots. Both of these ultimately became heavier in the male than in the female rats (Schemmel et al., 1970b). For both depots, there was little difference in size until the rats exceeded 300 g. Thereafter, the depot in the males continued to increase in size more rapidly than body weight, attaining a weight equal to 9% of body weight. The inguinal depot in females represented a maximum of 6% of body weight. The axillary fat depot plateaued for the males at about 1.9% of body weight and for the females at 1.3%.

The high-fat diet offered to the rats on an ad lib basis (Schemmel et al., 1969b) produced a variable degree of obesity in the 7 strains of rats studied (Schemmel et al., 1970a). Among the animals of these strains, there was no consistent difference between the males and females in the percentage of body fat, as determined by lipid extraction of an aliquot of the carcass (Mickelsen and Anderson, 1959). Furthermore, there was no consistent tendency for the high-fat ration to favor the deposition of body fat in one sex over the other when compared to the body fat in comparable animals fed the grain ration. This was true of all strains, except the S 5B/Pl. For the S 5B/Pl rats, the ratio of body fat in the high fat-fed/grain-fed animals was slightly greater for the males at both the 10th and 20th weeks of the study; for the other strains, there was no such relationship.

The similarity in the percentage of body fat deposited both in the normal weight and obese male and female rats of the same age is surprising. The 2-fold difference in body fat percentage in men and women has conditioned us to expect the same in animals. This expectation is based on the lipogenic effect associated with the estrogens. Any attempt to explain the situation in rats is further complicated when the overall bioenergetics of male and female rats are considered. One way to calculate energy utilization is by estimating energy output. This has been defined by Forbes et al. (1946a, b) as the difference between the energy intake and that retained in the carcass of the animal. When such calculations are made for the Osborne-Mendel and S 5B/Pl rats (Table 3), it becomes apparent that, although the energy output of male rats was greater than that of the females, the situation was reversed when energy output was calculated on the basis of lean body mass. The latter was used since it is probably the component most intimately involved in energy production (Schemmel et al., 1972).

Although there are only minor differences between the energy outputs per 100 g of lean body mass between the grain- and high-fat-fed rats of the same sex, there is a marked difference between males and females (Table 3). The differences in energy output between the grain and high-fat-fed rats of the same sex range from 110 to 244 kcal for all groups except the 20-week-old˙females; those differences are 349 and 401 kcal. However, the differences between the averages of the males and females of the same age and strain range from 395 to 1146 kcal. In all cases, the energy output of the females is greater than that of the males. These results are unexpected since a similarity in energy output would be anticipated when the calculation was made on the basis of lean body mass. The fact that the females had an almost 25% larger energy output than the males suggests that they had a higher body temperature or were more active than the males, or that they had one or more physiological reactions which were not as efficient as the comparable ones in the male animals. One of these might well have been the conversion of dietary calories to body fat since the males were more efficient in that respect than the females. These findings do suggest that a great deal more work needs to be done on the metabolic differences existing between males and females.

REFERENCES

Angel, J.L. (1949): Constitution in female obesity. Amer. J. phys. Anthropol., 7, 433.

Baldwin, R.L. (1968): Estimation of theoretical calorie relationships as a teaching technique. A review. J. dairy Sci., 51, 104.

Baldwin, R.L. (1970): Metabolic functions affecting the contribution of adipose tissue to total energy expenditure. Fed. Proc., 29, 1277.

Ball, E.G. (1965): Some energy relationships in adipose tissue. Ann. N.Y. Acad. Sci., 131, 225.

Barboriak, J.J., Krehl, W.A., Cowgill, G.R. and Whedon, A.D. (1958): Influence of high-fat diets on growth and development of obesity in the albino rat. J. Nutr., 64, 241.

Blaxter, K.L. and Rook, J.A.F. (1953): The heat of combustion of the tissues of cattle in relation to their chemical composition. Brit. J. Nutr., 7, 83.

Brozek, J., Chen, K.P., Carlson, W. and Bronczyk, F. (1953): Age and sex differences in man's fat content during maturity. Fed. Proc., 12, 21.

Campbell, M.E., Mickelsen, O., Yang, M.G., Laqueur, G.L. and Keresztesy, J.C. (1966): Effects of strain, age and diet on the response of rats to the ingestion of *cycas circinalis*. J. Nutr., 88, 115.

Chen, K.P., Damon, A. and Elliott, O. (1963): Body form, composition, and some physiological functions of Chinese on Taiwan. Ann. N.Y. Acad. Sci., 110, 760.

Ellis, R.W.B. and Tallerman, K.H. (1934): Obesity in childhood: study of 50 cases. Lancet, 2, 561.

Emery, R.S., Benson, J.D. and Tucker, A.A. (1971): Dietary and hormonal effects on extended lactation and lipid metabolism in rats. J. Nutr., 101, 831.

Fino, J.H., Schemmel, R. and Mickelsen, O. (1973): Effect of dietary triglyceride chain length on energy utilization and obesity in rats fed high-fat diets. Fed. Proc., 32, 939.

Forbes, E.B., Swift, R.W., Elliott, R.F. and James, W.H. (1946a): Relation of fat to economy of food utilization. I. By the growing albino rat. J. Nutr., 31, 203.

Forbes, E.B., Swift, R.W., James, W.H., Bratzler, J.W. and Black, A. (1946b): Further experiments on the relation of fat to economy of food utilization. I. By the growing albino rat. J. Nutr., 32, 387.

Frohman, L.A., Bernardes, L.L., Schuatz, J.D. and Burek, L. (1969): Plasma insulin and triglyceride levels after hypothalamic lesions in weanling rats. Amer. J. Physiol., 216, 1496.

Grommet, J., Schemmel, R., Mickelsen, O. and Bull, R.W. (1973): Erythrocyte, plasma and blood volumes of 'normal' weight and obese Osborne-Mendel male rats. Fed. Proc., 32, 939.

Gurney, R. (1936): Hereditary factor in obesity. Arch. intern. Med., 57, 557.

Han, P.W. and Liu, A.G. (1966): Obesity and impaired growth of rats force-fed 40 days after hypothalamic lesions. Amer. J. Physiol., 211, 229.

Hausberger, F.X. (1965): Effect of dietary and endocrine factors on adipose tissue growth. In: Handbook of Physiology, Section 5: Adipose Tissue, p. 527. Editors: A.E. Renold and G.F. Cahill. American Physiology Society, Washington, D.C.

Ingle, D.J. (1949): A simple means of producing obesity in the rat. Proc. Soc. exp. Biol. (N.Y.), 72, 604.

Johnson, S.R., Hogan, A.G. and Ashworth, U.S. (1936): The utilization of energy at different levels of protein intake. Univ. Minn. Coll. Agric. Res. Bull., 246.

Kennedy, G.C. and Parker, R.A. (1963): The islets of Langerhans in rats with hypothalamic obesity. Lancet, 2, 981.

Keys, A., Brozek, J., Henschel, A., Mickelsen, O. and Taylor, H.L. (1950): The Biology of Human Starvation, Vol. I. University of Minnesota Press, Minneapolis.

Ljunggren, H. (1965): Sex differences in body composition. In: Human Body Composition, p. 129. Editor: J. Brozek. Permagon Press, Oxford.

Mayer, J. (1965): Genetic factors in human obesity. Ann. N.Y. Acad. Sci., 131, 412.

Mickelsen, O. (1956): Intestinal synthesis of vitamins in the nonruminant. Vitam. and Horm., 14, 1.

Mickelsen, O. and Anderson, A.A. (1959): A method for preparing intact animals for carcass analyses. J. Lab. clin. Med., 53, 282.

Mickelsen, O., Takahashi, S. and Craig, C. (1955): Experimental obesity. I. Production of obesity in rats by feeding high-fat diets. J. Nutr., 57, 541.

Newman, H.H., Freeman, F.N. and Holzinger (1937): Twins, a Study of Heredity and Environment. University of Chicago Press, Chicago.

Owen, G.M., Jensen, R.L., Thomas, L.N. and Foman, S.J. (1963): Influence of age, sex and diet on total body water of four-to-seven-month-old infants. Ann. N.Y. Acad. Sci., 110, 861.

Paladines, O.L., Reid, J.T., Bensadoun, A. and Van Niekerk, B.D.H. (1964): Heat of combustion values of the protein and fat in the body and wool of sheep. J. Nutr., 82, 145.

Passmore, R., Strong, J.A. and Ritchie, F.J. (1958): The chemical composition of the tissue lost by obese patients on a reducing regimen. Brit. J. Nutr., 12, 113.

Schemmel, R., Mickelsen, O. and Fisher, L. (1973): Body composition and fat depot weights of rats as influenced by ration fed dams during lactation and that fed rats after weaning. J. Nutr., 103, 477.

Schemmel, R., Mickelsen, O. and Gill, J.L. (1970a): Dietary obesity in rats: body weight and body fat accretion in seven strains of rats. J. Nutr., 100, 1041.

Schemmel, R., Mickelsen, O. and Mostosky, U. (1969a): Skeletal size in obese and normal-weight littermate rats. Clin. Orthop. relat. Res., 65, 89.

Schemmel, R., Mickelsen, O. and Mostosky, U. (1970b): Influence of body weight, age, diet and sex on fat depots in rats. Anat. Rec., 166, 437.

Schemmel, R., Mickelsen, O. and Motawi, K. (1972): Conversion of dietary to body energy in rats as affected by strain, sex and ration. J. Nutr., 102, 1187.

Schemmel, R., Mickelsen, O. and Tolgay, Z. (1969b): Dietary obesity in rats: influence of diet, weight, age and sex on body composition. Amer. J. Physiol., 216, 373.

Shohl, A.T. (1939): Mineral Metabolism. Reinhold Publishing Co., N.Y.

Smith, E.H., Schemmel, R., Yang, M.G. and Mickelsen, O. (1969): The effect of dietary obesity on bone growth in the laboratory rat. Fed. Proc., 28, 493.

Sokoloff, L., Mickelsen, O., Silverstein, E., Jay Jr., G.E. and Yamamoto, R.S. (1960): Experimental obesity and osteoarthritis. Amer. J. Physiol., 198, 765.

Tepperman, J. (1957): Etiologic factors in obesity and leanness. Perspect. Biol. Med., 1, 293.

The effects of endurance exercise on body composition*

J. O. Holloszy

Department of Preventive Medicine and Public Health, Washington University School of Medicine, St. Louis, Mo., U.S.A.

During a 6-month long study of the effects of a program of endurance exercise, consisting of running and calisthenics, on a group of previously sedentary, middle-aged men, it was observed that their average body weight did not change (Holloszy et al., 1964). This finding came as a surprise, since our analysis of detailed dietary records kept by the subjects, together with an evaluation of dietary histories, indicated that average caloric intake had remained stable during the study. A rough estimate of the average increase in caloric expenditure resulting from participation in the exercise was 1000 kcal per week per man, or 26,000 kcal for the duration of the study (Holloszy et al., 1964). Assuming that the adipose tissue lost contained 8000 kcal per kg, an average weight loss of 3.25 kg might have been expected. Thus it appeared that body weight had been maintained despite a chronic caloric deficit. A possible explanation for this finding is that, since lean tissue contains only about 1000 kcal per kg, body weight can be maintained constant for a period of time in the face of a caloric deficit if a loss of adipose tissue is compensated for by an increase in lean tissue.

Thus, if the men in the study had lost an average of 3.8 kg of adipose tissue while gaining 3.8 kg of lean tissue on a constant caloric intake, a decrease in body energy stores of approximately 26,600 kcal could have occurred without weight loss. Favoring this possibility is the finding that the men had a significant increase in body specific gravity and a decrease in skinfold thickness measurements (Skinner et al., 1964). These findings, which suggested that individuals are leaner when physically trained than when sedentary, even if body weight is the same, aroused my interest in the effects of endurance exercise on body composition. This interest led to studies on laboratory animals in which food intake and body composition could be measured accurately.

EFFECTS OF A PROGRAM OF RUNNING ON THE BODY COMPOSITION OF YOUNG, GROWING MALE RATS

Young male rats subjected to a program of treadmill running gained significantly less weight than sedentary freely eating animals (Crews et al., 1969). This effect was due not only to the increase in caloric expenditure, but also to a decrease in appetite with a significantly reduced caloric intake. The percentage of the carcass composed of fat was markedly lower in the exercised than in the sedentary animals. Fat constituted only 11.7% of the body weight of the exercised animals compared to 26.4% for the freely eating sedentary and 20% for the paired-weight sedentary controls. In the latter group food intake was restricted so as to maintain body weight the same as that of the exercised animals. Thus, the physically trained group had less fat and more lean tissue than sedentary controls of the same body weight.

* This work was supported by USPHS Research Grant HD01613 and USPHS Training Grant AM05341. The author is the recipient of Research Career Development Award K4-HD-19,573.

EFFECTS OF WEIGHT CHANGES PRODUCED BY EXERCISE OR BY FOOD
RESTRICTION ON BODY COMPOSITION

When the weight of an obese individual is reduced by means of caloric restriction, lean tissue
may account for roughly one-third of the weight loss (Brozek et al., 1963). Our findings
described above (Holloszy et al., 1964; Skinner et al., 1964; Crews et al., 1969), and those of
other investigators (Pă rizková, 1963; Jones et al., 1964), indicating that physically trained
humans and animals are leaner than sedentary controls of the same weight, suggested that
exercise might protect against the loss of lean tissue during weight reduction. This possibility
was investigated in obese rats by comparing the effects on body composition of weight
reduction brought about either by exercise alone or by caloric restriction alone (Oscai and
Holloszy, 1969). Starting at the age of 6 weeks, male rats were fed a diet of high caloric density
for 10 months until they attained an average weight of 706 ± 14 g. They were then divided into 4
groups. An exercising group was subjected to a program of swimming over an 18-week period.
The exercise load was progressively increased until the animals were swimming for 2 hours a
day, 5 days a week, with weights equal to 1% of their body weights attached to their tails. The
exercise resulted in a progressive weight loss. A sedentary, paired-weight group of animals had
their food intake restricted so as to result in a rate of weight loss comparable to that of the
exercisers.

The exercise program resulted in a negative caloric balance secondary to both the increase in
caloric expenditure and to a decrease in appetite with a voluntary reduction in food intake. This
resulted in an average weight loss of 182 ± 19 g. The weight loss of the sedentary, food-restricted
animals closely paralleled that of the exercisers, averaging 182 ± 18 g.

Although the total weight loss and the final body weights of the exercisers and their
sedentary, paired-weight controls were the same, their carcasses were significantly different in
composition. Fat constituted 20.6% and lean tissue 79.4% of the carcasses of the swimmers,
compared to 26.7% fat and 73.3% lean tissue for the paired-weight animals, 35.6% fat and 64.4%
lean tissue for the baseline group, and 43.5% fat and 56.5% lean tissue for the freely eating
controls. In the food-restricted (paired-weight) animals 38% of the body substance lost was lean
tissue while 62% was fat. In contrast, only 22% of the weight loss in the exercised animals was
due to loss of lean tissue. The total amount of protein lost during the period of weight reduction
was more than twice as great for the sedentary, food-restricted rats as for the exercised rats.

These results demonstrate that in obese rats, as in obese humans, one-third or more of the
weight lost as a result of caloric restriction can be accounted for by loss of lean tissue. They also
clearly show that a comparable reduction of weight induced by means of exercise is associated
with loss of significantly less lean tissue and significantly more fat.

In addition to the effect on whole body composition, the response of heart weight to weight
reduction was compared in food-restricted and exercised obese, male rats (Oscai and Holloszy,
1970). The food-restricted animals' heart weights (1220 ± 26 mg) were significantly reduced
compared to those of the baseline group (1555 ± 20 mg). In contrast, the weights of the
swimmers' hearts (1504 ± 71 mg) did not decrease significantly, demonstrating that exercise
protects against the reduction in heart weight associated with loss of body weight.

TISSUE GROUPS CONTRIBUTING TO THE DIFFERENCES IN LEAN BODY MASS
BETWEEN PHYSICALLY TRAINED AND SEDENTARY ANIMALS

Somewhat surprisingly, in view of the above, we have consistently observed that the leg muscles
directly involved in running or swimming do not show hypertrophy in rats subjected to
programs of endurance exercise, but are of the same weight in physically trained and sedentary
rats of the same body weight (Holloszy, 1967; Pattengale and Holloszy, 1967). A study was,
therefore, undertaken to determine which components of the lean body mass account for the
greater percentage of lean tissue in carcasses of animals subjected to endurance exercise.
Previous studies had shown that male rats subjected to prolonged exercise did not increase
their food intake (Oscai et al., 1971) and might even show significant appetite suppression

(Crews et al., 1969; Oscai and Holloszy, 1969). As a result, male rodents subjected to programs of prolonged exercise have lower body weights than freely eating sedentary controls (Crews et al., 1969; Jones et al., 1964; Oscai and Holloszy, 1969; Holloszy, 1967; Oscai et al., 1971). In contrast, female rats subjected to prolonged exercise increase their food intake and gain weight at approximately the same rate as sedentary controls (Oscai et al., 1971). The basis of this interesting sex difference in appetite response to exercise has not yet been elucidated, nor, to our knowledge, is there sufficient information to say whether this is a species-specific peculiarity limited to rats or a more general phenomenon.

Thus two approaches were available to us. One was to restrict the food intake of sedentary male rats so that they would gain weight at the same rate as the exercised animals with which they were paired. The other was to work with female rats that increased their food intake sufficiently in response to exercise to compensate for the increased caloric expenditure. The latter approach seemed preferable since caloric restriction introduces an additional experimental variable that could affect lean body mass. Freely eating female rats were, therefore, used (Oscai et al., 1973). The exercising group was subjected to a swimming program of progressively increasing duration until they were swimming for 6 hours a day, 5 days a week after 5 weeks of training. They were maintained at this level of exercise for an additional 16 weeks. The swimmers significantly increased their food consumption and during the last 16 weeks of the study ate approximately 46% more than the sedentary controls (Oscai et al., 1973).

Whole carcass analyses on a subgroup of the swimmers revealed that fat constituted 9% and lean tissue 91% of the carcass weight, compared to 17.5% fat and 82.5% lean tissue in sedentary controls of the same weight. The weights and the composition of individual hind limb and pelvic girdle muscles were unaffected by the exercise. Also, the weight and composition of the whole 'fore limb unit', which included the scapula and the muscles of the shoulder and pectoral girdles, and of the whole 'hind limb unit', which included those muscles of the pelvic girdle that cross the hip joint, were not significantly different in the two groups. Table 1 summarizes the data on the lean tissue content of the various tissue groups that contributed to the greater lean body mass of the swimmers. It can be seen that, in contrast to the muscles directly involved in the exercise, which did not show hypertrophy, the swimming program apparently induced considerable hypertrophy of the muscles responsible for stabilizing the head, shoulders, spine, pelvis and tail. This is reflected in the approximately 10% greater lean tissue mass of the eviscerated carcass and tail in the exercised group (Table 1).

Available evidence suggests that the effect of exercise on muscle size is determined more by the forcefulness than by the frequency of muscle contractions; numerous muscle contractions do not appear to be any more effective than a few forceful contractions in bringing about or maintaining muscle hypertrophy (De Lorme, 1945; MacQueen, 1954). The muscles directly involved in the swimming, which did not show hypertrophy, were also the weight-bearing or antigravity muscles normally involved in supporting the animal's body weight. Perhaps the rat's everyday activities in its cage provide sufficient anabolic stimulus to the weight-bearing muscles for the repetitive, weak muscle contractions during prolonged endurance exercise not to have an additional effect. On the other hand, contractions sufficiently forceful to provide a stimulus to protein synthesis may occur only rarely in the non-weight-bearing muscles of sedentary rats confined in individual cages. The above finding suggests that in these muscles the isometric contractions involved in stabilizing the head, spine, scapula and tail during swimming provide a sufficient stimulus above usual cage activity to result in hypertrophy.

Of considerable interest is the finding (Table 1) that the weights of a number of organs, including the liver, lungs, kidneys, heart and blood, were 20% to 35% heavier in the exercised than in the control animals. The composition of these organs, which all have a low fat content, was unaffected by the exercise. Two factors must be considered in evaluating the etiology of the increased liver and kidney weights in the exercised animals. One is the 46% greater intake of protein and carbohydrate by the exercised animals. The other is the metabolic stress of the exercise. The latter includes the gluconeogenic stimulus provided by the outpouring of lactate and alanine from the working muscles, as well as the work performed by the kidneys in maintaining acid base balance.

TABLE 1 Contribution of the various tissue groups to the difference in lean body mass between swimmers and sedentary animals

Group	Lean tissue (g)							
	Skin and subcutaneous tissue	Front and hind limb units	Abdominal muscles	Organs*	Abdominal contents**	Tail	Eviscerated carcass***	Total
Swimmers	26.66	48.99	10.14	20.95	14.84	7.16	62.65	191.39
Sedentary	22.99	48.78	9.49	16.27	10.47	6.49	55.90	170.39
Difference	3.67	0.21	0.65	4.68	4.37	0.67	6.75	21.00

* Values are the sum of the fat-free components of the heart, liver, lungs, kidneys and blood.
** The pancreas, intestinal tract, adrenals, uterus, ovaries, mesentery, omentum and retroperitoneal fat were pooled and analyzed together.
*** The remaining carcass consisted of the head and neck, spine (minus tail), rib cage, clavicles, pelvis and remaining attached musculature after removal of front and hind limb units.

DISCUSSION

The finding that physically trained animals are leaner than sedentary controls of the same weight indicates that exercise has a lipid-mobilizing effect. During prolonged exercise fatty acid release from adipose tissue and oxidation by muscle increase markedly (Havel, 1971). This lipolytic action of exercise appears to be mediated, in large part, by increased activity of the sympathetic nervous system, with increased secretion of catecholamines (Havel, 1971; Vendasalu, 1960) and growth hormone (Havel, 1971; Glick et al., 1965). Increased mobilization of fat persists for a considerable period after cessation of exercise (Rodahl et al., 1964). It seems reasonable that the difference in body fat content between trained and sedentary individuals could be a cumulative effect of the daily, exercise-induced episodes of lipolysis.

Our finding that young growing animals subjected to programs of endurance exercise have a greater lean body mass than sedentary controls of the same weight points to a difference in the manner in which the calories available for growth are utilized. Channeling of a greater proportion of available calories into synthesis of lean tissue, with less storage of calories as fat in adipose tissue, as a result of the fat-mobilizing effect of exercise, may be one factor responsible for this difference. Other factors may include a direct effect of exercise on some of the musculature (for example, the heart and the stabilizing muscles of the back in our swimming rats), and the anabolic effect of increased growth hormone secretion associated with exercise.

ACKNOWLEDGMENT

I wish to express my sincere appreciation to L.B. Oscai and P.A. Molé for their collaboration.

REFERENCES

Brozek, J., Grande, F., Anderson, J.T. and Keys, A. (1963): Densitometric analysis of body composition: revision of some quantitative assumptions. Ann. N.Y. Acad. Sci., 110, 113.
Crews III, E.L., Fuge, K.W., Oscai, L.B., Holloszy, J.O. and Shank, R.E. (1969): Weight food intake, and body composition: effects of exercise and of protein deficiency. Amer. J. Physiol., 216, 359.
De Lorme, T.L. (1945): Restoration of muscle power by heavy resistance exercise. J. Bone Jt. Surg., 27, 645.
Glick, S.M., Roth, J., Yalow, R.S. and Berson, S.A. (1965): The regulation of growth hormone secretion. Recent Progr. Hormone Res., 21, 241.
Havel, R.J. (1971): Influence of intensity and duration of exercise on supply and use of fuels. In: Muscle Metabolism During Exercise, p. 315. Editors: B. Pernow and B. Saltin. Plenum Press, New York.

Holloszy, J.O. (1967): Biochemical adaptations in muscle. J. biol. Chem., 242, 2278.

Holloszy, J.O., Skinner, J.S., Toro, G. and Cureton, T.K. (1964): Effects of a six month program of endurance exercise on the serum lipids of middle-aged men. Amer. J. Cardiol., 14, 753.

Jones, E.M., Montoye, H.J., Johnson, P.B., Martin, S.M.J.M., Van Huss, W.D. and Cederquist, D. (1964): Effects of exercise and food restriction on serum cholesterol and liver lipids. Amer J. Physiol., 207, 460.

Macqueen, I.J. (1954): Recent advances in the technique of progressive resistance exercise. Brit. med. J., 2, 1193.

Oscai, L.B. and Holloszy, J.O. (1969): Effects of weight changes produced by exercise, food restriction, or overeating on body composition. J. clin. Invest., 48, 2124.

Oscai, L.B. and Holloszy, J.O. (1970): Weight reduction in obese rats by exercise or food restriction: effect on the heart. Amer. J. Physiol., 219, 327.

Oscai, L.B., Molé, P.A. and Holloszy, J.O. (1971): Effects of exercise on cardiac weight and mitochondria in male and female rats. Amer. J. Physiol., 220, 1944.

Oscai, L.B., Molé, P.A., Krusack, L.M. and Holloszy, J.O. (1973): Detailed body composition analysis on female rats subjected to a program of swimming. J. Nutr., 103, 412.

Pařizková, J. (1963): Impact of age, diet, and exercise on man's body composition. Ann. N.Y. Acad. Sci., 110, 661.

Pattengale, P.K. and Holloszy, J.O. (1967): Augmentation of skeletal muscle myoglobin by a program of treadmill running. Amer. J. Physiol., 213, 783.

Rodahl, K., Miller, H.I. and Issekutz Jr., B. (1964): Plasma free fatty acids in exercise. J. appl. Physiol., 19, 489.

Skinner, J.S., Holloszy, J.O. and Cureton, T.K. (1964): Effects of a program of endurance exercises on physical work capacity and anthropometric measurements of fifteen middle-aged men. Amer J. Cardiol., 14, 747.

Vendsalu, A. (1960): Plasma concentrations of adrenaline and noradrenaline during muscular work. Acta physiol. scand., Suppl., 173, 57.

DISCUSSION

Björntorp: I would like to take the opportunity to ask an expert about the differences in the physical training models in the rat. If one wants to study, for example, blood lipids and glucose tolerance in physically trained rats, which is the preferable model, swimming with loading or running in the treadmill? Does swimming for example introduce unwanted stress effects?

Holloszy: I have the impression that swimming represents a severe 'psychic' stress only in rats that are not experienced swimmers. After a few sessions of swimming, rats appear quite relaxed in the water. They float intermittently, sink to the bottom and push off, and hold on to each other. It is, therefore, much more difficult to standardize the exercise program with swimming than with running. The acute and chronic adaptive responses to prolonged swimming or running are qualitatively similar. However, in our studies, the magnitude of the enzymatic adaptations in the leg muscles are only 1/4 to 1/3 as great in rats subjected to swimming as in rats subjected to our program of running.

Somatotype, growth and physical performance

J.E. Lindsay Carter

Department of Physical Education, San Diego State University, San Diego, Calif., U.S.A.

The technique of somatotyping (Sheldon et al., 1954) has been found useful in describing variation in human populations (Dupertuis, 1963), in examining the relationships between somatotype and growth (Clarke, 1971; Heath and Carter, 1971; Petersen, 1967), occupation (Damon and McFarland, 1955), disease (Parnell, 1958) and physical performance of outstanding athletes (Carter, 1970; Curton, 1951; Štěpnička, 1972; Tanner, 1964). The purpose of this paper was to review the relationships between somatotype, growth and physical performance. From present knowledge, it was hoped to reveal some indication of the regulation of body composition as seen through somatotype.

The concept of somatotyping is appealing because it is a classification of total body form which can be expressed as a simple three-digit number. It provides a Gestalt impression of human physique but it is not limited by placing individuals into discrete categories. As reviewed by Carter and Heath (1971), somatotyping is a generic term embracing a number of different methods which produce different results. Therefore, to be consistent, only studies which have utilized the Heath-Carter somatotype method (Carter, 1972; Heath and Carter, 1967) are cited to illustrate the material in this paper. In the Heath-Carter method the somatotype is defined as a description of present morphological conformation. It is expressed in a three-digit rating representing evaluation of each of three components of the physique (relative fatness, called endomorphy; relative musculoskeletal development, called mesomorphy; relative linearity, called ectomorphy), which describe individual variations in morphology and composition. The present somatotype is rated and no assumptions are made regarding permanence. Because somatotype measures shape and composition and not size, it is useful to accompany somatotype ratings with height and weight values as indicators of body size.

GENETICS AND SOMATOTYPE

Before growth and physical performance can be discussed, the extent to which adult somatotypes are genetically determined needs examination. If all genetic variations should ultimately be traceable to changes at the molecular level, then we may conclude at this time that the appropriate studies in somatotyping have not yet been done. Also needed are family studies over several generations. However, some indication of genetic influence on somatotype may be gathered from examination of different populations.

The values for each somatotype component in various populations range from 1 to 14 for endomorphy, from 1 to 9 for mesomorphy, and from $\frac{1}{2}$ to 9 for ectomorphy. Considerable diversity of mean somatotypes becomes apparent when adult populations of both sexes are analyzed. The mean somatotypes for selected adult samples, as well as mean height and weight, are shown in Figure 1. Several points arise from examining the somatochart. First, sexual dimorphism is apparent within the same ethnic groups – Manus, Eskimo, Hawaiian Japanese, English and U.S. samples (although the latter contain ethnic minorities). Second, except for the Manus females and the Nilote males, females are more endomorphic and less mesomorphic than males. Third, the contrast between the Manus and Nilotes males is striking. The Nilotes are 22 cm taller than the Manus, but are 2 kg lighter. Both have the same endomorphy, so the differ-

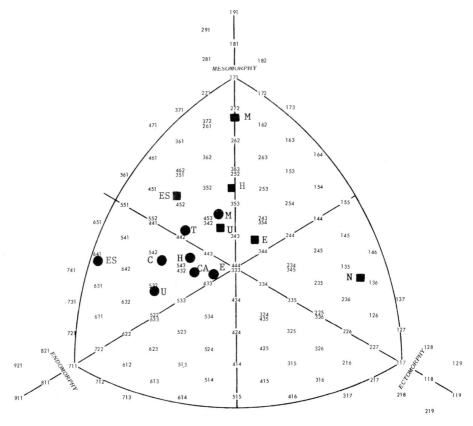

Fig. 1.   Somatochart showing mean somatotypes, height and weight for selected male and female samples. References: [1]Heath et al. (1968); [2]Štěpnička (1972); [3]Clauser et al. (1972); [4]Carter, unpublished data; [5]Heath, personal communication.

Sample	Males = ■ Height (cm)	Weight (kg)	Sample	Females = ● Height (cm)	Weight (kg)
Manus[1] (M)	159.0	60.4	Manus[1] (M)	151.0	45.2
Hawaiian			Hawaiian		
Japanese[1] (H)	166.5	59.0	Japanese[1] (H)	155.5	49.9
Eskimo[1] (ES)	167.0	68.0	Tokyo		
English[1] (E)	180.0	69.0	Japanese[1] (T)	155.0	51.4
U.S.A.[5] (U)	179.0	73.5	Eskimo[1] (ES)	154.0	64.0
Nilote[5] (N)	181.9	58.0	English[1] (E)	164.0	56.3
			Czechoslovakian[2] (C)	164.6	60.3
			U.S.A.F.[3] (U)	162.8	57.7
			Californian[4] (CA)	165.0	58.1

ence is in the reversed dominance of mesomorphy and ectomorphy (the rounded means are 1½-6½-1½ for the Manus, and 1½-3½-6 for the Nilotes). Fourth, the English, U.S. and Nilote males are equally tall but, while the first two are similar in somatotype (3-4½-3½ and 3½-4½-3) and size, the Nilotes are 11-14 kg lighter and differ in somatotype. Fifth, the Manus females are the smallest subjects and show a clear dominance of mesomorphy over the other two components (3-4½-2½). Apart from the Tokyo Japanese (4-4½-2) the remaining female samples are mesoendomorphs, with the Eskimos being the highest in endomorphy (6½-4-1½) and separate from the other samples. Sixth, the English, Czechoslovakian, U.S. Air Force and Californian females are similar in size and somatotype (mean somatotype $4\frac{1}{2}$-$3\frac{1}{2}$-$2\frac{1}{2}$). The Hawaiian Japanese females are close to the Tokyo Japanese in size but are between them and the previous four samples in somatotype ($4\frac{1}{2}$-4-$2\frac{1}{2}$). And seven, although not shown in the Figure, the somatotype distributions of the male Eskimos, Manus and Nilotes do not overlap with each other. The female Eskimos and Manus have some overlap with the other samples but little with each other. The English and U.S. samples (male and female) have the most variable distributions and overlap with all other samples of their own sex. The subject's somatotypes in the above samples reflect both the effects of genetics and of environment, but the quantification of each effect remains difficult. In addition, the question of stability of individual somatotypes is not considered in the above data.

GROWTH AND SOMATOTYPE

Several studies have analyzed the stability of somatotypes during growth. In a study by Zuk (1958), Heath somatotype ratings were made on males and females at ages 12, 17 and 33. The mean somatotype for the males showed relatively stable endomorphy, an increase of $1\frac{1}{2}$ units in mesomorphy, and an increase of $\frac{1}{2}$ and then a decrease of $\frac{1}{2}$ unit in ectomorphy. For females, endomorphy increased by 1 unit, mesomorphy by $\frac{1}{2}$ unit, and ectomorphy decreased by little over $\frac{1}{2}$ unit. Intercorrelations among ages for both sexes ranged from 0.02 to 0.67 for endomorphy, 0.73 to 0.83 for mesomorphy, and 0.71 to 0.84 for ectomorphy. Heath and Carter (1971) studied the somatotype patterns of 438 Manus children aged 15 months to maturity. The subjects were somatotyped in 1954, 1966 and 1968 in a mixed longitudinal cross-sectional study. The somatotype distributions showed little difference between boys and girls from 1 to 4 years; between 5 and 9 years boys and girls shifted toward higher mesomorphy and ectomorphy ratings; between 10 and 18 boys shifted toward higher mesomorphy and ectomorphy ratings, while girls shifted toward higher endomorphy and ectomorphy ratings.

In the Medford Boys' Growth Study, Clarke (1971) annually studied the somatotypes of more than 600 boys aged 7 to 18 years. In a longitudinal study of 100 boys from 9 through 17 years, there were no differences between the mean component ratings for ages 9 to 12, whereas from 12 to 17 about half of the differences on each of the components were significant. Between 9 and 12 almost half of the changes from year to year on the first and third components were greater than 1 unit, while 7% showed a change of 1 unit or more on mesomorphy. In contrast, at ages 12 to 17 almost half of the changes in endomorphy were greater than 1 unit, while approximately one-third of the changes were greater than 1 unit on mesomorphy and ectomorphy. The general trend of the component means (during the adolescent years) was for endomorphy to decrease by $\frac{1}{2}$ unit, mesomorphy to increase by $\frac{1}{2}$ unit, and ectomorphy to show a slight increase, then a slight decrease. Inter-age correlations for single somatotype components showed that the correlations between adjacent ages were between 0.79 and 0.90, but they were reduced to 0.50 and 0.67 at 5 years apart. On an individual basis some somatotypes showed considerable stability (no component rating varied more than $\frac{1}{2}$ unit between years), while others were very unstable (component ratings changed up to 3 units and changes in dominance of components occurred).

PHYSICAL PERFORMANCE AND SOMATOTYPE

Recent studies have concentrated on the relationship of somatotype to physical performance as exhibited in athletic events (Carter, 1970, 1971; Štěpnička, 1972). These events reflect various elements of physical performance such as speed, strength, stamina, suppleness and skill. Furthermore, they provide us with an ideal set of functional tasks against which we can theoretically specify optimal body structure. In terms of somatotype, which ones are most successful and at which sports?

In studies of over 1000 male and female athletes, from club to Olympic level in more than a dozen sports, it has been shown that athletes from some sports have limited and distinctive somatotype means and distribution, while those from other sports are similar (Carter 1970, 1971). In general, athletes are more mesomorphic and less endomorphic than their parent populations, but vary on ectomorphy. For males, the average mesomorphy is approximately $5\frac{1}{2}$, but some groups have means close to 4 and some as high as $7\frac{1}{2}$ on mesomorphy. Individual ratings of 7 to 9 on mesomorphy are found in some wrestlers, weightlifters, gymnasts, footballers (U.S.A.) and field event athletes. Some athletes, such as those in distance running and basketball, have mesomorphy ratings lower than 4, but these are usually accompanied by low endomorphy and high ectomorphy ratings. For most athletes in vigorous sports the average endomorphy is 2 to $2\frac{1}{2}$, and the lowest values possible (1 and $1\frac{1}{2}$) are found with considerable frequency in some events. Ratings of 4 to 5 on endomorphy are found in combination with higher mesomorphy and very low ectomorphy in some wrestlers, weightlifters, shot-putters, and U.S. football players. The mean somatotype and mean height and weight for selected samples of athletes are shown in Figure 2. The U.S. footballers ($4\text{-}6\frac{1}{2}\text{-}1\frac{1}{2}$) and field events throwers ($3\text{-}6\frac{1}{2}\text{-}1\frac{1}{2}$) are identical on mesomorphy and ectomorphy but the former are more endomorphic. Except for the heavyweight class in boxing, weightlifting and wrestling (including sumo), these are the largest of athletes. The means on endomorphy and ectomorphy for the wrestlers ($2\text{-}6\text{-}1\frac{1}{2}$) and weightlifters ($2\text{-}7\frac{1}{2}\text{-}1$) are affected somewhat by the great variability in the amount of endomorphy of some heavyweights, but the level of mesomorphy is typical for most weight classes. The rowers ($2\text{-}5\frac{1}{2}\text{-}2\frac{1}{2}$) and the sprinters ($2\text{-}5\frac{1}{2}\text{-}3$) have almost identical somatotype means and distributions but the sprinters are smaller by 12 cm and 16 kg. Golf emphasizes skill more than other qualities of physical performance, and their somatotype ($4\text{-}5\text{-}2\frac{1}{2}$) is close to the U.S. university student somatotype ($3\frac{1}{2}\text{-}4\frac{1}{2}\text{-}3$).

Although less is known about the somatotypes of female athletes, present indications are that they show similar trends, but differ quantitatively, from males. The mean for most female athletes on mesomorphy is $4\frac{1}{2}$, while the least mesomorphic are around $2\frac{1}{2}$ and the most mesomorphic about 6. In contrast to men, successful female athletes have higher endomorphy than mesomorphy, however, most are endomesomorphs or ectomesomorphs. Figure 2 illustrates somatotype and size data for selected female samples of athletes. Long and high jumpers ($3\text{-}3\text{-}4\frac{1}{2}$) and ballet dancers ($2\frac{1}{2}\text{-}3\text{-}4$) are more ectomorphic than other athletes and reference samples. The mean somatotypes lie on or close to the ectomorphic axis, indicating that for each group the ratings on endomorphy and mesomorphy are relatively close, while the values for ectomorphy may be low or high. The throwers are the tallest, heaviest, most mesomorphic and endomorphic and least ectomorphic of the athletes. The Czech sprinters and modern gymnasts have similar somatotypes ($3\frac{1}{2}\text{-}4\frac{1}{2}\text{-}2\frac{1}{2}$), but the sprinters are larger by 3 cm and 3 kg.

We have seen that some sports require certain groups of somatotypes for successful performance. Is this somatotype pattern prerequisite to success at all levels of performance? Studies on groups of U.S. footballers, distance runners, gymnasts and rowers at different levels reveal that these sports have definite and limited distribution patterns of somatotype at lower levels but that the patterns become narrower as the level increases (Carter, 1971; Carter et al., 1971). In the Medford Growth Study, outstanding athletes showed greater mesomorphy and lower endomorphy than non-athletes (Clarke, 1971). Ross and Day (1972) showed that in competitive skiers (boys and girls aged 6-14) ectomesomorphy was related to success. Přízková (1963) reported that intensity of training changes the percentage of fat and lean body mass, therefore, these changes could be reflected in somatotype changes. Although appropriate longitudinal studies have not been made on athletes, recent studies of both young and old men undergoing physical training showed that somatotypes changed significantly (Carter and Phillips, 1969; Carter and Rahe, 1973). The values of the first component for the athletes in this paper are consistent with the fat percentage in athletes. We would expect that the lowest values on endomorphy might be found in endurance type athletes whose training and competition involves high energy cost over long periods. The fact that, in addition to distance runners, we find gymnasts and wrestlers and weightlifters below middle weight class, all sharing the lowest levels of endomorphy, indicates the importance of other factors. In gymnastics, wrestling and weightlifting it is desirable to be as strong as possible, yet as light as possible, while distance runners should have great cardiovascular endurance and be as light as possible. In practice, both training and diet are used to control excess fat.

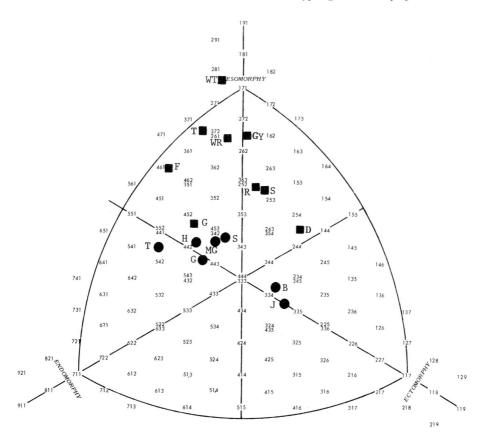

	Males = ■			Females = ●	
Sample	Height (cm)	Weight (kg)	Sample	Height (cm)	Weight (kg)
Weight-lifters (WT)[1]	167.9	73.1	Field throwers (T)[1]	168.0	69.0
U.S. footballers (F)[1]	184.4	94.4	Professional golfers (G)[1]	167.6	62.4
Field throwers (T)[1]	189.2	100.3	Jumpers (J)[1]	168.0	52.1
Rowers (R)[1]	191.2	88.6	Ballet dancers (B)[1]	162.8	50.1
Distance runners (D)[1]	176.5	63.2	Handballers (H)[4]	165.6	62.0
Golfers (G)[1]	181.4	87.0	Sprinters (S)[4]	166.7	60.1
Wrestlers (WR)[1]	173.2	77.2	Modern gymnasts (MG)[4]	163.5	57.1
Gymnasts (GY)[2]	165.1	68.1			
Sprinters (S)[3]	179.5	72.5			

Fig. 2. Somatochart showing mean somatotypes, height and weight for selected male and female samples. References: [1]Carter (1971); [2]Carter et al. (1971); [3]Carter, unpublished data; [4]Štěpnická (1972).

CONCLUSIONS

On the basis of available evidence the following conclusions seem warranted. (1) There are considerable somatotype differences between some populations, and sexual dimorphism is apparent within populations. (2) Both growth and training can change somatotype. (3) Success in different sports favors certain somatotypes. (4) Some somatotype distributions of athletes become narrower with higher levels of competition. (5) Female somatotypes show similar patterns to males but the somatotype ratings differ in magnitude. (6) Physical activities that place a premium on strength, power, speed, or endurance tend to limit successful participation to the somatotypes that are best suited, or best developed, to fulfill the physical requirements of the activity. (7) The indications are that athletes are both born and made, but that genetics prescribes the limits.

REFERENCES

Carter, J.E.L. (1970): The somatotypes of athletes – A review. Hum. Biol., 42, 535.

Carter, J.E.L. (1971): Somatotype characteristics of champion athletes. In: Anthropological Congress Dedicated to Ales Hrdlicka. Editor: V.V. Novotny. Academia, Czechoslovak Academy of Sciences, Prague.

Carter, J.E.L. (1972): The Heath-Carter Somatotype Method. San Diego, California, published by the author.

Carter, J.E.L. and Heath, B.H. (1971): Somatotype methodology and kinesiology research. Kinesiology Rev., 2, 10.

Carter, J.E.L. and Phillips, W.H.(1969): Structural changes in exercising middle-aged males during a 2-year period. J. appl. Physiol., 27, 787.

Carter, J.E.L., Sleet, D.A. and Martin, G.N. (1971): Somatotypes of male gymnasts. J. Sport Med. (Torino), 11, 2.

Carter, J.E.L. and Rahe, R.H. (1973): Effects of stressful underwater demolition training on body structure. Paper presented at Californian Association for Health, Physical Education and Recreation, San Diego.

Clarke, H.E. (1971): Physical and Motor Tests in the Medford Boys' Growth Study. Prentice-Hall, Englewood Cliffs.

Clauser, C.E., Tucker, P.E., McConville, J.T., Churchill, E., Larbach, L. and Reardon, J. (1972): Anthropometry of Air Force Women (AMRL-TR-70-5). Aerospace Medical Research Laboratory, Wright-Patterson Air Force Base, Ohio.

Cureton, T.K. (1951): Physical Fitness of Champion Athletes. University of Illinois Press, Urbana, Ill.

Damon, A. and McFarland, R.A. (1955): The physique of bus and truck drivers: with a review of occupational anthropology. Amer. J. phys. Anthropol., 13, 711.

Dupertuis, C.W. (1963): A preliminary somatotype description of Turkish, Greek and Italian military personnel. In: Anthropometric Survey of Turkey, Greece and Italy, Chapter 4. Editors: H.T.E. Hertzberg, E. Churchill, C.W. Dupertuis, R.M. White and A. Damon. MacMillan, New York.

Heath, B.H. and Carter, J.E.L. (1967): A modified somatotype method. Amer. J. phys. Anthropol., 27, 57.

Heath, B.H. and Carter, J.E.L. (1971): Growth and somatotype patterns of Manus children, Territory of Papua and New Guinea: Application of a modified somatotype method to the study of growth patterns. Amer. J. phys. Anthropol., 35, 49.

Heath, B.H., Mead, M. and Schwartz, T. (1968): A somatotype study of a Melanesian population. In: Proceedings, VIII Congress of Anthropological and Ethnological Sciences, Tokyo. Vol. I, p. 9. Science Council of Japan, Tokyo.

Pǎrizková, J. (1965): Physical activity and body composition. In: Human Body Composition. Editor: J. Brozek. Pergamon Press, London.

Parnell, R.W. (1958): Behaviour and Physique. Edward Arnold, London.

Petersen, G. (1967): Atlas for Somatotyping Children. C.C. Thomas, Springfield, Ill. and Van Gorcum, Assen, Netherlands.

Ross, W.D. and Day, J.A.P. (1972): Physique and performance of young skiers. J. Sport Med. (Torino), 12, 30.

Sheldon, W.H., Dupertuis, C.W. and McDermott, E. (1954): Atlas of Men. Harper Bros., New York.

Štěpnička, J. (1972): Typological and motor characteristics of athletes and university students. (In Czech). Charles University, Prague.

Tanner, J.M. (1964): The Physique of the Olympic Athlete. George Allen and Unwin, London.

Zuk, G.H. (1958): The plasticity of the physique from early adolescence through adulthood. J. genet. Psychol., 92, 205.

Physical exercise and body composition

J. H. Wilmore

Human Performance Laboratory, Department of Physical Education, University of California at Davis, Davis, Calif., U.S.A.

Over the past 10 years, it has been demonstrated unequivocally that exercise is a primary factor in both the control and alteration of body composition. The actual mechanisms which initiate these basic alterations are unknown at the present time, but it is becoming increasingly apparent that they are considerably more complex than a simple imbalance between caloric intake and caloric expenditure.

At the request of the organizing body of this meeting, this paper is limited solely to the presentation of data from studies in which the present author was involved. It must be recognized that this represents only a small part of the total body of knowledge presently available in this area of study.

INVESTIGATIONS

In 1970, Wilmore et al. reported the results of a 10-week program of jogging as it influenced various parameters of body composition. Fifty-five men, 17-59 years of age, participated in a walk-jog-run program 3 days per week for 12-24 minutes per day. The average jogging speed was 200 m / minute (7.5 m.p.h.), or a total distance of 51.8 miles in 413 minutes. The results from this study are presented in the upper part of Table 1. Body density was determined by the hydrostatic weighing technique and relative fat by the equation of Siri (1956). The specific technique used in this and the subsequent studies to be reported in this paper are identical and are reviewed in detail in a forthcoming publication (Behnke and Wilmore, 1974).

Small but significant reductions were noted in body weight and in relative and absolute body fat, with lean weight remaining essentially the same. The loss of 1.15 kg of fat, while not spectacular, is of a substantial magnitude when the intensity and duration of the exercise program are considered. This is equivalent to a 0.25 lb fat loss per week or a 13.0 lb fat loss per year.

In a similar study (Wilmore, 1971; Wilmore and Haskell, 1971), 44 subjects, with a mean age of 48.1 years, participated in individual (non-group) exercise programs for a period of 12-15 weeks. The specific effects of exercise alone cannot be evaluated in this study since many subjects also experienced dietary intervention during the same interval.

The exercise program was individually prescribed on the basis of the individual's endurance capacity ($\dot{V}O_{2\,max}$). The program consisted of walking, jogging, running, bicycling or swimming for 15-25 minutes per day, 3 days per week at 75% of his $\dot{V}O_{2\,max}$, as assessed by the pulse rate equivalent.

The results from this study are presented in the middle section of Table 1. Fat reduction amounted to 3.54 kg or 17.5%, while weight decreased only 2.9 kg, the net difference being the result of a slight gain in lean body weight. While this loss was mediated by both diet and exercise, the dietary counseling was oriented more towards proper food selection than toward a major caloric reduction in the daily food intake; thus these changes are probably largely the result of the exercise program.

In a later study, Wilmore et al. (unpublished data) observed changes in body composition in middle-aged males following a 6-month program of walking, jogging and running for 3 days per week 25-30 minutes per day, at an intensity of 75% of the measured $\dot{V}O_{2\,max}$. The results from

TABLE 1   Body composition changes with physical exercise in previously sedentary males

Variable	Pre-Training		Post-Training		Δ	$\% \Delta$
	Mean	σ	Mean	σ		
Berkeley Group (Wilmore et al., 1970) (n = 55)						
Weight (kg)	79.6	10.1	78.6	9.1	—1.0	—1.3
Lean body weight (kg)	64.2	6.3	64.4	6.2	0.2	0.3
Fat weight (kg)	15.4	n.a.	14.2	n.a.	—1.2	—7.8
Density (g/ml)	1.056	0.013	1.059	0.013	0.003	0.3
Relative fat (%)	18.8	5.9	—7.8	5.7	—1.0	—5.9
Palo Alto Group (Wilmore and Haskell, 1971; Wilmore, 1971) (n = 44)						
Weight (kg)	81.4	10.6	78.5	10.2	—2.9	—3.5
Lean body weight (kg)	61.1	6.1	61.8	6.4	0.7	1.1
Fat weight (kg)	20.3	n.a.	16.8	n.a.	—3.5	—17.5
Density (g/ml)	1.043	n.a.	1.051	n.a.	0.008	0.8
Relative fat (%)	24.5	5.2	20.9	4.9	—3.6	—14.7
Davis Group (Wilmore et al., unpublished data) (n = 32)						
Weight (kg)	80.5	10.1	77.9	9.6	—2.6	—3.2
Lean body weight (kg)	61.9	6.7	62.9	6.5	1.0	1.6
Fat weight (kg)	18.6	6.7	15.0	6.1	—3.6	—19.4
Density (g/ml)	1.047	n.a.	1.056	n.a.	0.008	0.8
Relative fat (%)	22.7	6.2	18.8	6.0	—3.9	—17.2

n.a. = not available

this study are presented in the lower section of Table 1. They tend to confirm the results of the previous two studies.

Moody et al. (1972) observed alterations in body composition in obese and normal high school girls who participated in a daily program of walking and jogging over a period of 15 or 29 weeks. The girls started by walking and jogging 1 mile in equal proportions, and this was gradually increased until the daily distance required was 3-3.5 miles, jogging at least 75% of this distance. The results of this study are presented in Table 2.

The loss in body fat amounted to 2.31 kg for the 15-week obese group and 2.66 kg for the 29-week obese group. The normal group experienced no significant changes in their body composition as a result of the exercise program. In fact, within this group, 4 lean girls actually gained fat weight during the 15-week program.

This latter finding led to a study (Wilmore, 1973a) regarding the influence of exercise on the ability to gain weight in chronically underweight women. Six women with a history of frequent episodes of weight loss and an inability to gain weight were placed on a 15-week program of controlled treadmill exercise for 30 minutes per day, 5 days per week in an attempt to assist them to gain weight.

A slow, but steady trend of weight gain was observed over the course of the study. Four of the girls had achieved an increase of more than 1 kg in weight by the end of the seventh week. Of the 2 girls who showed no weight gain, 1 was probably suffering from anorexia nervosa, in which weight changes are closely related to the emotional state. The second girl missed approximately 50% of the training sessions and was very uncooperative. All weight gains were a result of increases in lean body weight, as the fat weight remained essentially unchanged.

Brown and Wilmore (1970) studied the influence of weight training on the body composition and strength of 5 female athletes who were national women's or age group champions in the shot put, discus or javelin. These girls were lifting for an average of 3 days per week over a period of 7 months. Three of these girls increased their body weight during the 7-month program, but only 2 demonstrated an increase in their fat weight (~ 2 kg), and both of these girls were more than 2 standard deviations below the mean for girls of this age (Wilmore and Behnke, 1970). Only 1 of the girls had a substantial increase in her lean body weight (3.9 kg), while the other 4 remained essentially unchanged.

TABLE 2 Body composition changes in normal and obese girls resulting from a program of jogging (Moody et al., 1972)

	No.	Density (g/ml)		Weight (kg)		Lean body weight (kg)		Relative fat (%)	
		Mean	σ	Mean	σ	Mean	σ	Mean	σ
Normal group (< 30% relative fat)									
Initial	12	1.044	0.014	59.43	9.06	45.00	4.68	24.3	6.0
Final	12	1.046	0.011	58.84	10.35	45.11	5.65	23.3	5.0
△		0.002		−0.59		0.11		−1.0	
t		1.14		0.89		0.21		1.10	
Obese group (> 30% relative fat)									
15 weeks									
Initial	28	1.013	0.009	71.60	10.90	43.76	5.10	38.9	4.7
Final	28	1.018	0.010	70.52	11.08	44.89	5.15	36.4	4.9
△		0.005		−1.08		1.13		−2.5	
t		4.74*		2.13*		3.77*		4.64*	
29 weeks									
Initial	19	1.012	0.010	71.48	11.68	43.51	4.94	39.1	5.0
Midpoint	19	1.017	0.010	71.03	12.03	44.90	4.97	36.8	4.8
Final	19	1.019	0.009	70.33	10.07	45.02	4.55	36.0	4.2
△ Initial-final		0.007		−1.15		1.51		−3.1	
t Initial-final		4.57*		1.19		3.74*		4.47*	

* Significant at the 0.05 level.

In a recent study, Wilmore (1973b) evaluated alterations in the body composition of a group of 47 women and 26 men as a result of a 10-week program of weight training. The subjects trained for an average of 30 minutes per day, 2 days per week. The results from this program are presented in Table 3. The groups made similar relative gains in strength and absolute gains in body composition. Weight remained constant, while lean body weight and fat weight were significantly increased and decreased respectively. Muscular hypertrophy was evident in both groups, was confined basically to the upper extremity, and was of substantially greater magnitude in the males.

TABLE 3   Body composition changes with a 10-week weight training program (Wilmore, 1973b)

	Women (n = 47)		Men (n = 26)	
	Mean	σ	Mean	σ
Weight (kg)				
Initial	57.95	6.56	72.93	10.76
Final	57.90	6.58	73.20	10.70
$\triangle$	0.05		0.27	
% $\triangle$	—0.08		0.37	
t	—0.20		0.87	
Lean body weight (kg)				
Initial	43.57	3.95	62.98	7.73
Final	44.63	4.08	64.17	7.68
$\triangle$	1.06		1.19	
% $\triangle$	2.42		1.90	
t	7.34*		4.67*	
Absolute body fat (kg)				
Initial	14.38	4.05	9.95	4.56
Final	13.30	3.90	9.02	4.34
$\triangle$	—1.08		—0.93	
% $\triangle$	—7.54		—9.30	
t	—5.35*		—3.05*	
Relative body fat (%)				
Initial	24.51	4.79	13.24	4.83
Final	22.65	4.62	11.92	4.56
$\triangle$	—1.86		—1.32	
% $\triangle$	—7.59		—9.99	
t	—6.41*		—3.23*	

* Significant at the 0.05 level.

Lastly, some indication of the role of exercise in the regulation of body composition can be found by observing the body composition of an atypical group of individuals. Table 4 provides unpublished data on 19 outstanding female and 7 outstanding older male distance runners. With only one exception (NJ), the values for relative body fat are all substantially below the mean values of 14.6 and 25.7 found for large groups of normal young men and women respectively (Wilmore and Behnke, 1969, 1970). Undoubtedly, a certain degree of this leanness is attributable to genetic factors, i.e. you must have a certain body type to be a successful distance runner. However, this excessive leanness is probably largely the result of the tremendous mileage averaged by these runners in their weekly workouts, i.e. 50-100 miles or more per week. Most of these individuals also mentioned that they gained weight very rapidly during periods of inactivity or low intensity training.

TABLE 4 Body composition assessment of outstanding female and male distance runners

Subjects	Age (years)	Height (cm)	Weight (kg)	Density (g/ml)	Fat (%)	Lean body weight (kg)	Fat (kg)
Females							
LM	21	167.6	49.93	1.081	7.8	46.04	3.89
LF	13	172.7	57.30	1.060	16.9	47.57	9.73
DH	20	170.2	52.53	1.058	17.9	43.12	9.41
SS	18	167.6	58.90	1.077	9.6	53.24	5.66
DB	28	162.8	50.72	1.085	6.4	47.46	3.26
NH	15	163.8	44.25	1.065	14.9	37.66	6.59
KS	13	167.6	44.10	1.059	17.4	36.43	7.67
FL	18	162.8	46.28	1.068	13.5	40.03	6.25
FC	31	171.5	51.97	1.078	9.0	47.28	4.69
CP	21	—	55.50	1.058	18.0	45.54	9.96
CT	21	173.4	54.59	1.065	14.6	46.60	7.99
NG	18	160.0	47.88	1.078	9.3	43.43	4.45
KF	37	154.9	53.64	1.058	18.0	43.99	9.65
DG	35	168.4	55.04	1.072	11.9	48.51	6.53
BW	34	166.4	52.92	1.064	15.2	44.88	8.04
CB	24	172.7	52.61	1.085	6.2	49.34	3.27
RQ	11	160.0	37.65	1.061	16.4	31.49	6.16
KM	36	182.9	61.54	1.060	17.2	50.98	10.56
EC	36	182.9	65.44	1.065	15.0	55.65	9.79
Males							
JK	41	184.2	71.63	1.087	5.3	67.83	3.80
PW	41	172.7	53.22	1.083	7.1	49.46	3.76
PM	43	181.7	61.86	1.108	−3.2	61.86	—
BF	46	175.5	67.20	1.084	6.6	62.77	4.43
WF	63	176.2	73.46	1.077	9.8	66.25	7.21
NJ	72	169.2	62.78	1.054	19.6	50.46	12.32
DF	13	179.0	65.22	1.083	7.1	60.59	4.63

INTERPRETATIONS

It appears from the limited data presented in this paper that exercise does mediate a basic change in body composition. There is generally an increase in the lean body weight which is probably a result of muscular hypertrophy. The muscular hypertrophy is possibly associated with an increase in the level of serum human growth hormone (HGH), which has been shown to rise during exercise and remain elevated for several hours during recovery (Hunter and Greenwood, 1964; Roth et al., 1963; Sutton et al., 1968). HGH is considered a protein anabolic hormone, and, in animals receiving growth hormone, the increased deposition of protein is accompanied by a loss of carcass fat (Turner, 1960). This latter finding could be related to the loss of fat which has been a consistent finding in the studies reviewed above.

The loss of body fat is also associated with an increased expenditure of calories. However, the caloric equivalent of the amount of exercise actually performed is under that necessary to account for the fat loss in some studies. It is possible that there is a concomitant decrease in caloric intake with exercise, which results in a caloric imbalance of the magnitude necessary to explain the fat loss. The work of Mayer and his colleagues at Harvard (Mayer, 1968) would support this latter concept. They have shown that exercise up to 1 hour in duration per day tends to suppress the appetite. There is also the possibility that exercise stimulates lipolytic action (Crews et al., 1969; Parízková and Stankova, 1964).

The increase in weight noted in the lean subjects as a result of their exercise programs appears to be somewhat paradoxical. However, the gains are primarily the result of gains in lean body weight with little change in fat weight. There is also some evidence which indicates that a conservation of energy accompanies training in these subjects, i.e. they are able to perform standardized submaximal bouts of exercise at a reduced oxygen cost (Wilmore, 1973a).

In summary, no conclusions can be drawn at this time relative to the mechanisms responsible for these fat and lean weight changes. It is probable, however, that fat loss and lean body weight gain result from a complex series of biochemical reactions which will be totally understood only after many additional years of research.

REFERENCES

Behnke Jr., A.R. and Wilmore, J.H. (1974): Evaluation and Regulation of Body Build and Composition. Prentice-Hall, Englewood Cliffs, N.J.

Brown, C.H. and Wilmore, J.H. (1970): Weight training in female athletes. Paper presented to the American College of Sports Medicine, Albuquerque, New Mexico, May, 1970.

Crews, E.L., Fuge, K.W., Oscai, L.B., Holloszy J.O. and Shank, R.E. (1969): Weight, food intake and body composition: effects of exercise and protein deficiency. Amer. J. Physiol., 216, 359.

Hunter, W.M. and Greenwood, F.C. (1964): Studies on the secretion of human pituitary growth hormone. Brit. med. J., 1, 804.

Mayer, J. (1968): Overweight. Prentice-Hall, Englewood Cliffs, N.J.

Moody, D.L., Wilmore, J.H., Girandola, R.N. and Royce, J.P. (1972): The effects of a jogging program on the body composition of normal and obese high school girls. Med. Sci. Sports, 4, 210.

Parízková, J. and Stankova, L. (1964): Influence of physical activity on a treadmill on the metabolism of adipose tissue in rats. Brit. J. Nutr., 18, 325.

Roth, J., Glick, S.M., Yalow, R.S. and Berson, S.A. (1963): Secretion of human growth hormone: physiologic and experimental modification. Metabolism, 12, 577.

Siri, W.E. (1956): Body Composition from Fluid Spaces and Density. Berkeley, California. Report, Donnor Laboratory of Medical Physics, University of California, Berkeley, California.

Sutton, J., Young, J.D., Lazarus, L., Hickie, J.B. and Maksvytis, J. (1968): Hormonal changes during exercise. Lancet, 2, 1304.

Turner, C.D. (1960): General Endocrinology, 3rd ed. W.B. Saunders, Philadelphia.

Wilmore, J.H. (1971): Individual fitness program prescription. Paper presented at the Symposium on Organization and Conduct of Fitness Programs, American College of Sports Medicine, Toronto, Canada, May, 1971.

Wilmore, J.H. (1973a): Exercise-induced alterations in weight of underweight women. Arch. phys. Med., 54, 115.

Wilmore, J.H. (1973b): Alterations in strength, body composition and anthropometric measurements consequent to a 10-week weight training program. Med. Sci. Sports, in press.

Wilmore, J.H. and Behnke, A.R. (1969): An anthropometric estimation of body density and lean body weight in young men. J. appl. Physiol., 27, 25.

Wilmore, J.H. and Behnke, A.R. (1970): An anthropometric estimation of body density and lean body weight in young women. Amer. J. clin. Nutr., 23, 267.

Wilmore, J.H. and Haskell, W.L. (1971): Use of the heart rate-energy expenditure relationship in the individualized prescription of exercise. Amer. J. clin. Nutr., 24, 1186.

Wilmore, J.H., Royce, J., Girandola, R.N., Katch, F.I. and Katch, V.L. (1970): Body composition changes with a 10-week program of jogging. Med. Sci. Sports, 2, 113.

Hormonal disturbances in obesity*

R. K. Kalkhoff

Metabolism Division, Department of Medicine, Medical College of Wisconsin, Clinical Research Center of Milwaukee County General Hospital and Deaconess Hospital, Milwaukee, Wis., U.S.A.

Human obesity is associated with a variety of hormonal disturbances. Since the early work of Yalow and Berson (1960) and Karam et al. (1963), increased basal and postprandial insulin concentrations in obese subjects have been documented by several laboratories (Rabinowitz, 1970). Coupled with this abnormality is the commonly observed finding of suppressed plasma growth hormone responses to induced hypoglycemia (Roth et al., 1963). In addition, our laboratory has recently reported alterations of plasma alpha cell glucagon in this condition (Kalkhoff et al., 1973).

The present review of our experience in this research area contrasts the hormonal profiles of obese and muscular overweight subjects and emphasizes the relationship between successful weight reduction and correction of hormonal disturbances in our obese population.

STUDIES OF OBESE OVERWEIGHT AND MUSCULAR OVERWEIGHT MEN

Insulin and growth hormone hemodynamics were assessed in lean men who were matched in age to overweight obese and muscular male subjects (Kalkhoff et al., 1971). Data recorded in Figures 1-3 reveal that no major deviations from normal occurred in the muscular men despite body weight that exceeded ideal values by approximately 40%. This contrasted to findings in equally overweight obese men, who demonstrated basal and post-challenge hyperinsulinemia during glucose and tolbutamide tolerance tests and subnormal increments of growth hormone after insulin-induced hypoglycemia.

Triceps skin-fold thickness, a satisfactory estimate of total body fatness (Seltzer et al., 1965), was 4.5 to 8-fold greater in the obese men relative to control and muscular individuals, respectively. Thus, insulin and growth hormone disturbances in obesity relate to factors contributing to excessive deposition of adipose tissue per se, since people who are comparably overweight only because of increased lean tissue mass do not differ significantly from normal control subjects.

It was also of interest that fasting plasma or serum concentrations of lipids were generally normal in each of the three study groups. However, levels of free fatty acids, cholesterol and triglyceride were consistently in a higher range in the obese men (Table 1). This finding may have relevance to lipid disturbances described in the obese state generally and may be implicated in the increased frequency of atherosclerosis and vascular complications reported in this condition.

EFFECTS OF WEIGHT REDUCTION ON PLASMA GROWTH HORMONE DISTURBANCES IN THE OBESE SUBJECT

Plasma growth hormone responses to hypoglycemia produced with intravenous insulin were

* Studies from the author's laboratory that are cited in this publication were supported by a research grant (AM10305) and clinical research center grant (RR00058) from the United States Public Health Service and by Tops Club, Inc., Obesity and Metabolic Research Program of Deaconess Hospital.

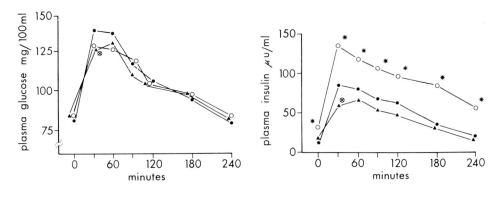

Fig. 1.   Plasma glucose and insulin responses during 100 g oral glucose tolerance tests. Each group represents 10 healthy, young adult men. The control group weighed 74 ± 2 kg and were 5 ± 2% above ideal body weight (IBW, Metropolitan Life Insurance Tables, 1959). The obese overweight men weighed 104 ± 4 kg or 42 ± 4% above IBW. The muscular overweight men weighed 95 ± 5 kg or 39 ± 3% above IBW. Triceps skinfold thicknesses in mm for control, obese and muscular subjects were 7 ± 0.5, 31 ± 2 and 4 ± 0.6, respectively. Asterisks denote significance of the difference between means of obese men and corresponding values of control or muscular men (P < 0.05). Encircled x indicates a significant difference between means of control and muscular men (P < 0.05). (From R.K. Kalkhoff and C.A. Ferrou, 1971, *New Engl. J. Med.;* courtesy of the editors.)

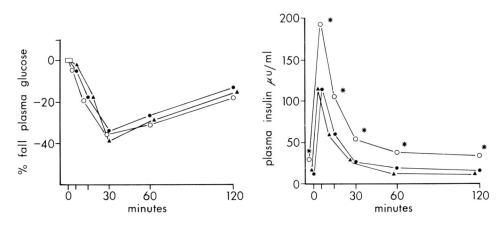

Fig. 2.   Plasma glucose and insulin responses during intravenous tolbutamide (1 g) tolerance tests. See legend to Figure 1. (From R.K. Kalkhoff and C.A. Ferrou, 1971, *New Engl. J. Med.;* courtesy of the editors.)

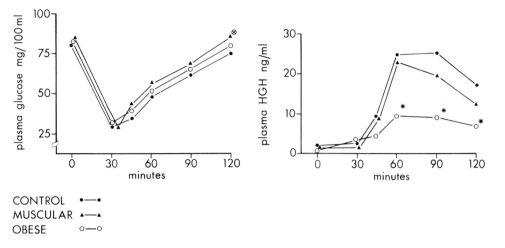

CONTROL •—•
MUSCULAR ▲—▲
OBESE   o—o

Fig. 3. Plasma glucose and growth hormone (HGH) responses to intravenous insulin. See legend to Figure 1. (From R.K. Kalkhoff and C.A. Ferrou, 1971, *New Engl. J. Med.;* courtesy of the editors.)

TABLE I   Fasting substrate and insulin concentrations (mean ± S.E.M.) in control, muscular overweight and obese overweight subjects

	Control	Muscular	Obese
Plasma glucose (mg/100 ml)	81 ± 1	83 ± 1	83 ± 2
Plasma insulin (µU/ml)	10 ± 1	14 ± 2	32 ± 6*
Plasma free fatty acids (µmoles/l)	328 ± 28	362 ± 40	508 ± 42*
Serum cholesterol (mg/100 ml)	195 ± 9	209 ± 9	234 ± 14*
Serum triglyceride (mg/100 ml)	61 ± 10	57 ± 14	104 ± 6*

Asterisks indicate significance of the difference between means of control and obese groups (P < 0.05).

determined in 6 obese, middle-aged subjects (125 ± 10 kg) before and after an average weight reduction of 39 kg (Kalkhoff et al., 1971). The characteristic diminished growth hormone increment was observed in this group before treatment when compared to 10 non-obese control subjects (Fig. 4). The growth hormone profile was restored to a normal response curve in all thinned obese individuals, which is consistent with a previous experience reported by Lessof et al. (1966). However, only 2 of the 6 obese patients achieved ideal body weight, whereas the remaining 4 remained substantially overweight (30 to 54% above ideal body weight) at the time of the second test. Sims et al. (1968) have also demonstrated marked suppression of growth hormone responses in normal volunteers after weight gains of only 15% during a phase of forced feeding. The results of their study and our own suggest that the growth hormone defect relates less to the magnitude of obesity than to a state of prolonged, excessive caloric intake. Similarly, correction of the abnormality may follow a suitable period of caloric deficit even though obesity still persists.

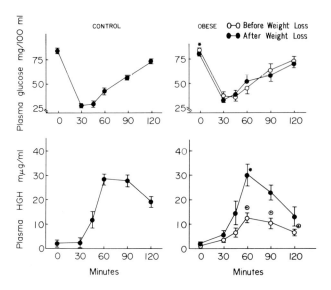

Fig. 4. Plasma glucose and growth hormone (HGH) responses to intravenous insulin in 10 non-obese control subjects (*left*) and 6 obese subjects before and after an average weight reduction of 39 kg (*right*). Values are mean ± S.E.M. Asterisks indicate significance of the difference between means before and after treatment of obesity, ($P < 0.05$). Encircled + denotes significant differences between means of control and obese groups ($P < 0.05$). (From R.K. Kalkhoff et al., 1971, *Diabetes;* courtesy of the editors.)

EFFECTS OF OBESITY AND WEIGHT REDUCTION ON PANCREATIC ALPHA AND BETA CELL FUNCTION

Detailed studies of 6 obese women who lost an average of 38 kg, and who reduced from $65 \pm 2\%$ to $2 \pm 1\%$ above ideal body weight are depicted in Figures 5-7, along with data pertaining to randomly selected obese and lean control women (Kalkhoff et al., 1973). As anticipated, excessive insulin responses to oral glucose were decreased to a control range after weight loss (Fig. 5), which is similar to findings reported earlier (Kalkhoff et al., 1971; Farrant et al., 1969).

Standard infusions of arginine, 30 g over a 30 minute period, were employed to stimulate pancreatic beta and alpha cell secretion concomitantly (Kalkhoff et al., 1973). An exaggerated plasma insulin response to arginine provocation was associated with a greatly augmented increment of plasma glucagon in the obese subjects. Subsequent weight reduction lowered the plasma insulin and glucagon curves to the range of non-obese control women (Fig. 6).

These results might implicate hyperglucagonemia in the genesis of endogenous insulin resistance and hyperinsulinemia in the obese state, since glucagon is a potent diabetogenic agent (Unger, 1972). However, additional data derived from these investigations do not support this conclusion.

Basal levels of plasma glucagon in all non-diabetic obese patients were well within a normal range of our immunoassay (40 to 120 pg/ml) despite significantly elevated fasting concentrations of insulin. Moreover, intravenous infusions of glucose suppressed plasma glucagon to the same extent in obese and lean subjects (Fig. 7), unlike what is observed in uncontrolled diabetes, which is believed to represent a state of inappropriate glucagon secretion (Unger, 1972). Ingestion of a 400 g lean beef meal, a more physiologic stimulus for pancreatic insulin and glucagon secretion than a pharmacologic infusion of arginine, failed to increase plasma glucagon to a greater extent in the obese subjects (Fig. 8).

It appears that most indices of pancreatic alpha cell function are normal in non-diabetic obesity. The physiologic meaning of glucagon hypersecretion during infusions of arginine is uncertain. It may represent a compensatory adjustment that protects the obese individual from

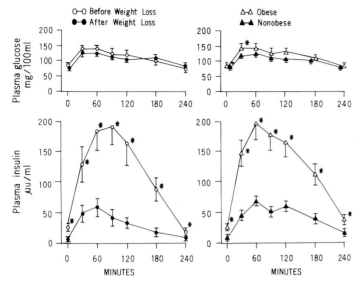

Fig. 5. Plasma glucose and insulin concentrations during 100 g oral glucose tolerance tests in 6 obese women before and after weight loss (*left*) and in 9 obese and 8 non-obese women (*right*).Values are mean ± S.E.M. and asterisks indicate significance of the difference between corresponding means within a graph (P < 0.05). Weight of obese subjects on the left before treatment was 99 ± 9 kg (65 ± 2% above IBW) and fell to 62 ± 2 kg (2 ± 1% above IBW). Triceps skin-fold thicknesses were 43 ± 2 and 14 ± 1 mm, respectively. The obese and non-obese women (*right*) weighed 100 ± 2 kg (67 ± 1% above IBW) and 53 ± 2 kg (−6 ± 2% below IBW), and skin-fold thicknesses were 41 ± 3 mm and 13 ± 1 mm, respectively. (From R.K. Kalkhoff et al., 1973, *New Engl. J. Med.;* courtesy of the editors.)

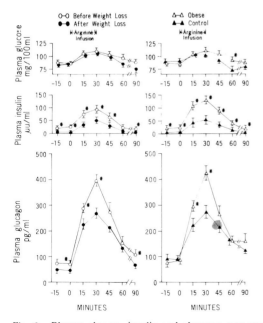

Fig. 6. Plasma glucose, insulin and glucagon responses to 30 minute infusions of arginine (30 g). See legend to Figure 5. (From R.K. Kalkhoff et al., 1973, *New Engl. J. Med.;* courtesy of the editors.)

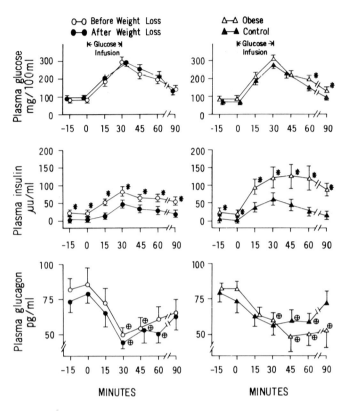

Fig. 7.   Plasma glucose, insulin and glucagon concentrations during 30-minute infusions of glucose (45 g). Encircled + indicates a significant decrease of glucagon below basal levels ($P < 0.05$). See legend to Figure 5. (From R.K. Kalkhoff et al., 1973, *New Engl. J. Med.;* courtesy of the editors.)

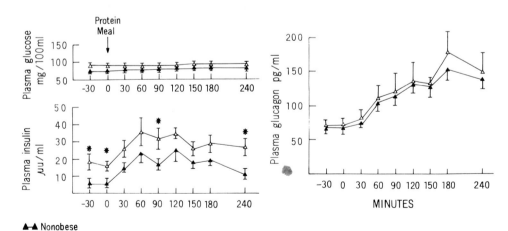

Fig. 8.   Plasma glucose, insulin and glucagon responses to a 400 g lean beef meal in 9 obese and 8 non-obese subjects. Data relate to the same subjects shown on the right in Figures 5-7. See legend to Figure 5. (From R.K. Kalkhoff et al., 1973, *New Engl. J. Med.;* courtesy of the editors.)

developing delayed hypoglycemia after relatively greater insulin responses to arginine as opposed to a protein meal (Unger, 1972). Similar exaggerated insulin-glucagon increments during arginine infusion have been reported in other states associated with some form of insulin antagonism, including acromegaly and that induced by glucocorticoid or growth hormone administration in normal subjects (Goldfine et al., 1972; Marco et al., 1973). In other studies of non-diabetic obese patients, intravenous alanine, unlike arginine, induced a subnormal rise of plasma glucagon even though insulin secretion was greatly augmented (Wise et al., 1973). Thus, interpretation of amino acid infusion data in their present form is difficult in a physiological context and must be deferred pending additional studies.

CONCLUDING REMARKS

Plasma insulin and growth hormone disturbances in human obesity are acquired metabolic changes, since they are reversible following caloric restriction and weight reduction and inducible after excessive caloric intake and fat tissue weight gain.

In the past it has been popular to regard hyperinsulinemia of obesity as a metabolic adaptation to peripheral insulin resistance, because there is evidence for the latter in both skeletal muscle and the adipocytes of obese individuals (Salans et al., 1968; Rabinowitz and Zierler, 1962). Other studies suggest that chronic over-eating, particularly of carbohydrate, promotes changes within the endocrine pancreas that lead to islet hypertrophy (Ogilvie, 1933) and insulin hypersecretion (Grey et al., 1971). Increased quantities of circulating insulin, in turn, may generate insulin antagonism in tissues of the obese host rather than vice versa (Mahler et al., 1971). From this standpoint, it is also likely that the nutritional status of the obese patient is a more critical determinant of the growth hormone defect than is the actual degree of adipose tissue mass, for reasons stated earlier in this discussion. Although it is not certain what factors directly mediate growth hormone and insulin changes in the obese individual, the role of alpha cell glucagon in this process, if any, has not been established by this investigation.

REFERENCES

Farrant, P.C., Neville, R.W.J. and Stewart, G.A. (1969): Insulin release in response to oral glucose in obesity: the effect of reduction of body weight. Diabetologia, 5, 198.

Goldfine, I.D., Kirsteins, L. and Lawrence, A.M. (1972): Excessive glucagon responses in active acromegaly. Hormone metab. Res., 4, 97.

Grey, N. and Kipnis, D.M. (1971): Effect of diet composition on the hyperinsulinemia of obesity. New Engl. J. Med., 285, 827.

Kalkhoff, R.K. and Ferrou, C.A. (1971): Metabolic differences between obese overweight and muscular overweight men. New Engl. J. Med., 284, 1236.

Kalkhoff, R.K., Gossain, V. and Matute, M.L. (1973): Plasma glucagon in obesity. Response to arginine, glucose and protein administration. New Engl. J. Med., 289, 465.

Kalkhoff, R.K., Kim, H.J., Cerletty, J. and Ferrou, C.A. (1971): Metabolic effects of weight loss in obese subjects. Changes in plasma substrate levels, insulin and growth hormone responses. Diabetes, 20, 83.

Karam, J.H., Grodsky, G.M. and Forsham, P.H. (1963): Excessive insulin response to glucose as measured by immunochemical assay. Diabetes, 12, 197.

Lessof, M.H., Young, M.S. and Greenwood, F.C. (1966): Growth hormone secretion in obese subjects. Guy's Hosp. Rep., 115, 65.

Mahler, R.J. and Szabo, O. (1971): Restoration of insulin sensitivity in the obese mouse following suppression of pancreatic islet cell hyperplasia. Diabetes, 20, Suppl. 1, 336.

Marco, J., Calle, C., Roman, D., Diaz-Fierros, M., Villanueva, M.L. and Valverde. I.(1973): Hyperglucagonism induced by glucocorticoid treatment in man. New Engl. J. Med., 288, 128.

Ogilvie, R.F. (1933): The islands of Langerhans in 19 cases of obesity. J. Path. Bact., 37, 473.

Rabinowitz, D. (1970): Some endocrine and metabolic aspects of obesity. Ann. Rev. Med., 21, 241.

Rabinowitz, D. and Zierler, K.L. (1962): Forearm metabolism in obesity and its response to intra-arterial insulin. Characterization of insulin resistance and evidence for adaptive hyperinsulinism. J. clin. Invest., 41, 2173.

Roth, J., Glick, S.M., Yalow, R.S. and Berson, S.A. (1963): Secretion of human growth hormone: physiologic and experimental modification. Metabolism, 12, 577.

Salans, L.B., Knittle, J.L. and Hirsch, J. (1968): The role of adipose cell size and adipose tissue insulin sensitivity in the carbohydrate tolerance of obesity. J. clin. Invest., 47, 153.

Seltzer, C.C. and Mayer, J. (1965): A simplified criterion of obesity. Postgrad. Med., 38, A101.

Sims, E.A.H., Goldman, R.F., Gluck, C.M., Horton, E.S., Kelleher, P.C. and Rowe, D.C. (1968): Experimental obesity in man. Trans. Ass. Amer. Phycns, 81, 153.

Unger, R.H. (1972): Circulating pancreatic glucagon and extra pancreatic glucagon-like materials. In: Endocrine Pancreas, Handbook of Physiology: Endocrinology, Vol. I, p. 529. Editors: D.A. Steiner and N. Freinkel. Williams and Wilkins, Baltimore.

Wise, J.K., Handler, R. and Felig, P. (1973): Evaluation of alpha-cell function by infusion of alanine in normal, diabetic and obese subjects. New Engl. J. Med., 288, 487.

Yalow, R.S. and Berson, S.A. (1960): Plasma insulin concentrations in non-diabetic and early diabetic subjects. Diabetes, 9, 254.

Adrenocortical response to lysinevasopressin in obesity: further evidence for abnormal ACTH regulation in obese subjects with striae*

F. Ceresa, A. Angeli, G. Boccuzzi, R. Frairia and **D. Bisbocci**

Istituto di Patologia Speciale Medica and Istituto di Semeiotica Medica, University of Turin, Turin, Italy

Plasma cortisol response to synthetic 8-lysinevasopressin (LVP) was studied in 20 normal subjects (11 men and 9 women, aged 17–52 years), in 10 cases of simple obesity (6 men and 4 women, aged 23–38 years) and in 15 cases of obesity with striae (6 men and 9 women, aged 14–34 years). In obese subjects with striae at least 5 of the following signs were present: moon face, acne, striae, hypertension, truncal distribution of fat, decreased glucose tolerance, urinary 17-

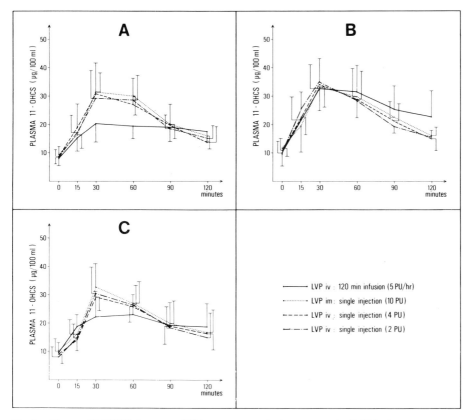

Fig. 1. Plasma 11-OHCS (mean ± S.D.) response to lysinevasopressin (LVP). A = normal subjects. B = patients with obesity with striae. C = patients with simple obesity.

* This work has been supported by the Consiglio Nazionale delle Ricerche: Contract No. 71.00795.04 115.562.

OHCS output over 0.12 mg/kg body weight/24 hours. Cushing's syndrome was ruled out by the presence of adrenal circadian rhythm and by the results of ACTH stimulation and dexamethasone suppression tests.

Various experimental procedures were used. LVP was administered as a single intravenous injection of 2 and 4 pressor units (pU), as a single intramuscular injection of 10 pU and as a continuous intravenous infusion over 2 hours in doses of 2.5 and 5 pU/hour. In all cases LVP was injected or infused at 5 p.m. Blood samples were drawn at zero time and then at 15, 30, 60, 90, and 120 minutes after injection or beginning the infusion. Plasma cortisol was measured fluorometrically as 11-hydroxycorticosteroids (Angeli et al., 1971). Student's t test for non-paired comparisons was used to establish the significance of differences between groups and $P < 0.05$ was regarded as significant.

In normal subjects plasma cortisol rose rapidly following a single i.v. or i.m. injection of LVP and usually reached a maximum at 30 minutes, after which it progressively declined. Considerable individual variability was present. The infusion of 5 pU/hour over 2 hours resulted in a different pattern of response, with maximum increment over values at zero time less marked than after a single injection and in most cases with a steady state in the circulating level of cortisol up to the end of the infusion (Fig. 1). When the smaller amount of 2.5 pU/hour was infused the response in some subjects was similar to that recorded with the infusion of 5 pU/hour, while in the remainder no significant variation of the circulating corticosteroid was noted.

In simple obese subjects the plasma cortisol response to LVP was similar to that of the normal group, both after injection and after constant i.v. infusion (Fig. 1). In obese subjects with striae the response was different. As shown in Figure 1, the plasma cortisol curves obtained with single injection and with 2-hour infusion appeared the same; plasma 11-OHCS rose rapidly to a maximum and then declined, even though LVP administration was continued for up to 120

TABLE 1   Mean ± 1 S.D. plasma 11-OHCS increments over values at zero time during 2-hour intravenous infusion of 5 pU/hour of lysinevasopressin

Group	Plasma 11-OCHS (μg/100 ml)	Increase (μg/100 ml)				
	0	15 minutes	30 minutes	60 minutes	90 minutes	120 minutes
A. Normal subjects						
Mean (n = 20)	7.46	7.04	12.68	12.52	12.06	10.22
± S.D.	2.50	2.80	5.5	4.0	3.7	3.27
B. Obese subjects with striae						
Mean (n = 10)	10.22	10.53	22.78	22.12	13.81	12.13
± S.D.	3.46	8.3	9.1	7.3	4.8	6.42
C. Simple obese subjects						
Mean (n = 7)	9.54	7.82	12.97	13.28	9.94	9.4
± S.D.	2.96	3.88	7.5	5.6	5.6	6.1
A vs. B	n.s.*	n.s.	$P < 0.01$	$P < 0.01$	n.s.	n.s.
A vs. C	n.s.	n.s.	n.s.	n.s.	n.s.	n.s.
B vs. C	n.s.	n.s.	$P < 0.01$	$P < 0.01$	n.s.	n.s.

* n.s. = not significant (P > 0.05).

minutes. In some subjects infused with 2.5 pU/hour no increase in plasma cortisol was noted, while in the remainder the infusion of this dose resulted in a marked response, superimposable on that observed with the other experimental procedures. When various procedures were utilized in the same subject, the same response was obtained. Statistical comparison of mean increments over values at zero time gave significant differences in the case of the 2-hour infusion of 5 pU/hour at 30 and 60 minutes both from normal and simple obese subjects (Table 1).

DISCUSSION

The specific mechanism by which LVP stimulates ACTH secretion has not yet been identified and hypothalamic as well as pituitary sites of action of LVP have been suggested (Hedge et al., 1966; Yates et al., 1971). Studies on the inactivation of LVP following i.v. administration in man have demonstrated a particularly short half-life of the peptide (Fabian and Forsling, 1968). Therefore, it seems likely that after i.v. infusion the concentration of LVP acting on the hypothalamus or pituitary is reduced as compared with that obtainable after a single injection, because a steady state in the circulating level of the infused drug is attained very rapidly.

In our normal subjects the release of ACTH following LVP was largely controlled by the amount of the drug operating at tissue level, so that the plasma cortisol response following i.v. infusion of 2.5-5 pU/hour was different from that following single i.v. or i.m. injection. No significant changes in this control were noted in simple obesity, suggesting that abnormally high caloric intake and obesity per se do not affect the response. In the obese with striae, while the minimum effective dose of infused LVP did not appear to be different from normal, the flexibility of the response was lost and the response was always characterized by a prompt and massive discharge of ACTH, which in turn caused a peak followed by a progressive fall in the circulating level of cortisol.

It may be suggested that in specific obese subjects with striae the inappropriate secretion of ACTH following LVP infusion depends on some abnormal regulation of ACTH. Earlier findings have indicated that these cases have hyperresponsiveness to metyrapone, with wide fluctuations at different hours of the day (Ceresa et al., 1965), an abnormal pattern of the adrenal circadian rhythm (Linquette et al., 1968) and increased 'basal' ACTH activity with reduced sensitivity to corticoid inhibition (Ceresa et al., 1970). On the other hand, sensitivity of the adrenal cortex to exogenous ACTH was found to be normal (Gogate and Prunty, 1963; Summers et al., 1964).

In conclusion, our data on plasma cortisol response to LVP support the thesis that a true hyperfunction of the pituitary-adrenal axis, even if transient and spontaneously reversible (Summers et al., 1964), is operating in obesity with striae and may explain the occurrence of a clinical picture somewhat resembling Cushing's syndrome.

REFERENCES

Angeli, A., Boccuzzi, G., Bisbocci, D. and Frairia, R. (1971): Caratteristiche della risposta cortisolemica iniziale dopo stimolazione impulsiva con corticotropina sintetica beta-1-24. Boll. Soc. ital. Biol. sper., 47, 122.

Ceresa, F., Angeli, A., Boccuzzi, G. and Perotti, L. (1970): Impulsive and basal ACTH secretion phases in normal subjects, in obese subjects with signs of adrenocortical hyperfunction and in hyperthyroid patients. J. clin. Endocr., 31, 491.

Ceresa, F., Strumia, E., Angeli, A. and Dellepiane, M. (1965): Behaviour of the feedback mechanism controlling ACTH secretion in normal subjects and under different endocrine conditions, and relative clinical implications. In: Proceedings, II International Congress of Endocrinology, p. 1027. Editor: S. Taylor. ICS 83, Excerpta Medica, Amsterdam.

Fabian, M. and Forsling, M.L. (1968): The half-times (t$\frac{1}{2}$), clearance, binding and stability of neurohypophysial hormones in human plasma. J. Physiol. (Lond.), 198, 19.

Gogate, A.A. and Prunty, F.T.G. (1963): Adrenal cortical function in obesity with pink striae in the young adult. J. clin. Endocr., 23, 747.

Hedge, G.H., Yates, M.B., Marcus, R. and Yates, F.E. (1966): Site of action of vasopressin in causing corticotrophin release. Endocrinology, 79, 328.

Linquette, M., Fossati, P., Decoulx, M., Racadot, A., Fourlime, J.C. and Lefebrvre, J. (1968): Etude des 17-OHCS urinaires, du cortisol plasmatique et de son rythme circadien au cours du jeune chez l'obèse. Lille méd., 13, 33.

Summers, V.K., Sheehan, H.L., Hipkin, L.J. and Davis, J.C. (1964): Differential diagnosis of Cushing's syndrome and obesity associated with striae. Lancet, 2, 1079.

Yates, F.E., Russel, S.M., Dallman, M.F., Hedge, G.A., McCann, S.M. and Dhariwal, A.P.S. (1971): Potentiation by vasopressin of corticotropin release induced by corticotropin-releasing factor. Endocrinology, 28, 3.

Corticotropic function in human obesity

P. Vague[1], G. Lombardi[2], M. Minozzi[2], A. Valette[1] and C. Oliver[1]

[1]*Endocrinology Clinic, Hôpital de la Conception, Marseilles, France and* [2]*Riparto di Endocrinologia, Istituto di Clinica Medica, Università degli Studi, Naples, Italy*

An exaggerated secretion of cortisol demonstrated by a high production rate is observed in some but not all obese subjects, without a clear relation to the percentage of excess weight (Boyer et al., 1968, 1970; Vague et al., 1971a). The mechanism may theoretically be related to an increased sensitivity of the adrenals to endogenous ACTH, or to hyperactivity of the pituitary or both. While the second hypothesis is favoured, no direct evidence has been offered.

In this study the plasma ACTH levels assessed by radioimmunoassay (Vague et al., 1971b) in basal and post-stimulation conditions have been compared in normal weight and non-diabetic obese subjects.

BASAL VALUES

Between 8 and 8.30 a.m. in fasting and recumbent normal weight healthy subjects 90% of the values were in the range 10-60 pg/ml. In the same conditions elevated values higher than 60 pg/ml were demonstrated in approximately one-third of the obese patients (Fig. 1).

PLASMA   ACTH   8-8.30 a.m.

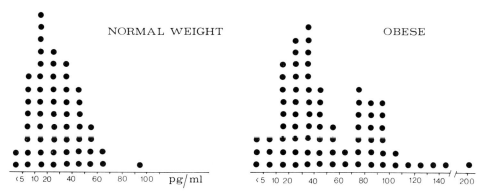

Fig. 1. Individual basal plasma ACTH levels in pg/ml (10^{-12} g/ml) in normal weight and non-diabetic obese subjects.

When the obese were divided into those with normal and those with elevated basal ACTH levels (males being excluded for sake of homogeneity) (Table 1), it appeared that, while age and relative body weight were similar in the two groups, the urinary excretion of 17-OH Porter and Silber chromogens expressed in mg/day or in mg/g of creatinine was significantly higher in the group with elevated basal plasma ACTH levels.

TABLE 1   Plasma ACTH in obese females at 8-8.30 a.m.

	$\leqslant$ 60 pg/ml	> 60 pg/ml	Significant difference
No.	37	20	
Age (years)	27.8 ± 9.8 S.D.	26.5 ± 8.6 S.D.	—
% ideal body weight	155.6 ± 24.2 S.D.	161 ± 36 S.D.	—
Urinary 17-OH			
mg/day	5.05 ± 0.42 S.E.M.	6.87 ± 0.94 S.E.M.	P < 0.025
mg/g creatinine	3.84 ± 0.26 S.E.M.	5.24 ± 0.49 S.E.M.	P < 0.005

Furthermore, in various normal subjects, basal plasma ACTH and cortisol were positively correlated (Fig. 2). The same was observed in obese subjects.

It appears from these results that in obese subjects the level of adrenocortical function is directly related statistically to the basal level of plasma ACTH and that the exaggerated cortisol production observed is related to elevated plasma ACTH levels.

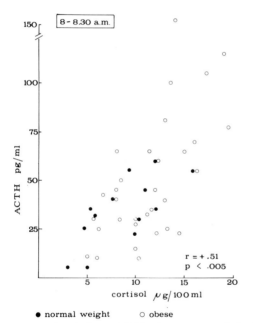

● normal weight   ○ obese

Fig. 2.   Basal plasma ACTH values versus cortisol values (Boyer et al., 1970) in the same sample in various normal weight healthy subjects and non-diabetic obese subjects.

ACTH RESPONSE TO STIMULATION

Plasma ACTH levels depend on a circadian rhythm (Vague et al., 1973) and present intermittent, apparently spontaneous bursts of secretion, but two main mechanisms are able to superimpose their effects on the basal secretion. These are the feedback mechanism in response to modification in cortisol secretion and the mechanism of response to stress in the liberal sense of the word.

We have compared the pituitary reactivity of obese and normal weight subjects to the feedback mechanism and to two types of stress stimuli, one entirely physiological – a mild muscular exercise, the other more aggressive – insulin-induced hypoglycemia.

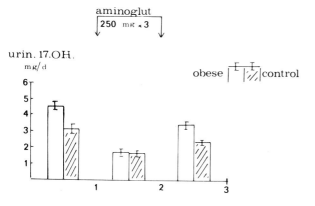

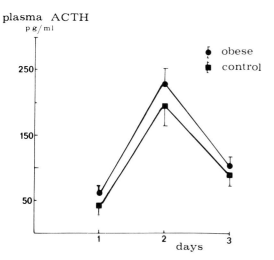

Fig. 3. Urinary Porter and Silber chromogens before, during and after administration of aminoglutethimide. Plasma ACTH levels at 8.00 a.m. before and after administration of aminoglutethimide.

The administration of low doses of aminoglutethimide (250 mg every 8 hours per 24 hours) induced a diminution of cortisol secretion, as shown by the similar fall in urinary 17-OH excretion in obese and normal weight subjects (Fig. 3). The resulting rise in plasma ACTH was not significantly different in the two groups.

A mild exercise (80 Watts for 10 minutes on a bicycle ergometer) during the afternoon to minimize the possible interference by spontaneous bursts of secretion, which are known to be more frequent in the morning, among other metabolic or hormonal modifications (Heim, 1973), a rise in plasma ACTH levels identical in normal weight and obese patients (Fig. 4).

Insulin-induced hypoglycemia (below 45 mg/100 ml) has been shown to cause a rise in plasma ACTH levels similar in both normal weight and obese subjects (Bell et al., 1970), and our results were in agreement with this finding. However, when the obese subjects were analysed not as a group but individually, it was observed that the maximal value of plasma ACTH after insulin administration was found in the obese who demonstrated a high urinary 17-OH excretion the day before the test (Fig. 5).

As a group the obese did not appear to have an exaggerated pituitary reactivity to the stimuli used, but on an individual basis the higher their basal 17-OH excretion, the higher their ACTH response to hypoglycemia.

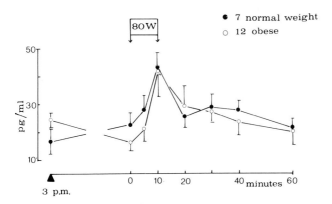

Fig. 4.   Effect of 10 minutes exercise on a bicycle ergometer on plasma ACTH levels.

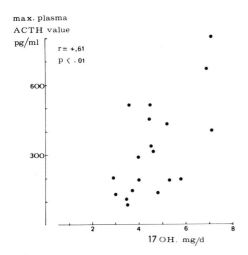

Fig. 5.   Maximal plasma ACTH level during the course of insulin-induced hypoglycemia (< 45 mg/100 ml) versus urinary 17-OH excretion the day before the test in obese subjects.

CONCLUSION

Among obese non-diabetic subjects one-third have elevated basal plasma ACTH levels and demonstrate an exaggerated adrenocortical function which may be directly related to the ACTH levels. However, the stimulation of ACTH secretion by the feedback mechanism, mild muscular exercise or insulin-induced hypoglycemia does not promote an exaggerated ACTH response among obese patients as a group versus normal weight individuals. The ACTH reponse to insulin-induced hypoglycemia appears to be correlated to the basal excretion of urinary 17-OH.

REFERENCES

Bell, J.P., Donald, R.A. and Espiner, E.A. (1970): Pituitary response to insulin-induced hypoglycemia in obese subjects before and after fasting. J. clin. Endocr., 31, 546.

Boyer, J., Clement, M., Vague, P. and Vague, J. (1968): Le taux de production de cortisol dans les obésités féminines. In: Atti, XII Congresso Nazionale della Società Italiana di Endocrinologia: Componenti Endocrine nelle Obesità, Catania, 1968, p. 197. La-Pigraf, Milan.

Boyer, J., Girodengo, M., Dzieniszewski, J. and Vague, J. (1970): La réponse sécrétoire cortico-surrénale des sujets obèses à l'injection de 1-24 corticotrophine. Ann. Endocr. (Paris), 31, 869.

Heim, M. (1973): Réponses métaboliques et hormonales à l'exercice chez les sujets normaux, obèses et diabètiques insulino-dépendants. Thesis, Marseilles.

Vague, J.,Vague, P., Boyer, J. and Cloix, M.C. (1971a): Anthropometry of obesity. Diabetes, adrenal and beta cell functions. In: Diabetes, p. 517. Editor: R.R. Rodriguez. I.C.S. 231. Excerpta Medica, Amsterdam.

Vague, P., Oliver, C. and Bourgoin, J.Y. (1973): Circadian rhythm in plasma ACTH in healthy adults. Chronobiology, 21, 112.

Vague, P., Oliver, C., Jaquet, P. and Vague, J. (1971b): Le dosage radio-immunologique de l'ACTH plasmatique. Rev. europ. Etud. clin. biol., 16, 485.

Prolactin release after insulin-induced hypoglycaemia in obese subjects*

G. Copinschi, M. L'Hermite, R. Leclercq, E. Virasoro and **C. Robyn**

School of Medicine, Université Libre, Brussels, Belgium

It has been shown recently that in normal men insulinic hypoglycaemia is systematically followed by a significant increase in plasma levels of immunoreactive prolactin (Copinschi et al., 1972) and of growth hormone (HGH) and corticotrophin. In obese patients, the corticotrophin response to hypoglycaemia is normal (Bell et al., 1970), but the HGH release is blunted when compared to normal subjects (Beck et al., 1964). The purpose of the present study was to investigate the prolactin response to hypoglycaemia in obese male subjects.

MATERIAL AND METHODS

Six normal men, aged 21-24, and 4 obese males, aged 22-39, were investigated. The body weight of the obese subjects ranged from 151 to 185% (mean 165%) and that of the normal subjects from 84 to 103% (mean 93%) of their ideal weight (Lorentz, 1929). All tests were performed after an overnight fast. An indwelling plastic catheter was inserted into a forearm vein at 8 a.m. Insulin (Actrapid, Novo) was injected through this catheter at 9 a.m. Normal subjects were given 0.12 U/kg and obese patients 0.30-0.33 U/kg in order to attain a similar degree of hypoglycaemia. Blood samples were obtained at frequent intervals from 60 minutes before to 180 minutes after insulin injection. Blood sugar was determined with a Technicon autoanalyzer (Hoffman, 1937). Plasma HGH was measured by radioimmunoassay (Virasoro et al., 1971). Plasma prolactin was determined using a radioimmunoassay developed for ovine (Davis et al., 1971) and extended to human prolactin (L'Hermite et al., 1972a); in this assay, there is no crossreaction between prolactin and HGH (L'Hermite et al., 1972b). Plasma levels of prolactin were evaluated by reference to a serum pool, rich in prolactin, which was arbitrarily considered to contain 1 U of immunoreactive prolactin/ml.

RESULTS

Results are summarized in Figure 1. After the injection of insulin, blood sugar fell markedly from normal basal levels in all subjects. Changes were similar in normal and obese men. In both groups, mean minimal levels were reached at 25 minutes. Plasma levels of HGH remained unchanged for 20 minutes after insulin injection, then increased rapidly in normal subjects to reach a mean peak value of 36.7 ± 7.7 ng/ml at 60 minutes. In contrast, obese patients showed little increase, reaching a maximal mean level of 11.3 ± 2.9 ng/ml at 50 minutes. Basal concentrations of plasma prolactin were lower in obese than in normal men, averaging respectively 99 ± 45 mU/ml and 262 ± 60 mU/ml. Levels of prolactin changed little for 20 minutes, then rose rapidly in both groups to reach mean peak values of 633 ± 199 mU/ml at 50 minutes in normal subjects and 495 ± 78 mU/ml at 75 minutes in obese patients; however, the magnitude of the increase over basal values was similar in both groups.

* This work was supported by a grant from the Ford Foundation to Professor P.O. Hubinont, and was partially performed under contract of the Ministère belge de la Politique Scientifique within the framework of the association Euratom-University of Pisa-University of Brussels.

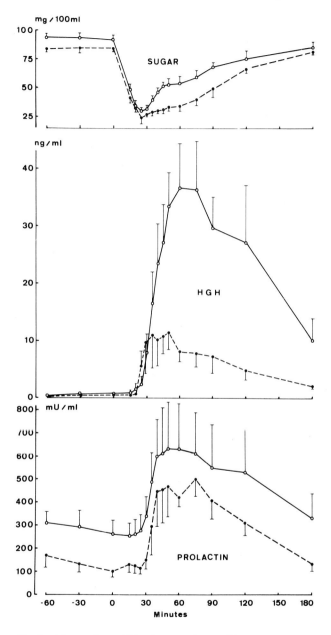

Fig. 1. Mean values of blood sugar, plasma HGH and prolactin after insulin injection in normal (solid lines) and obese (broken lines) men. Vertical lines denote standard error of the mean.

DISCUSSION

The present study suggests that basal values of plasma prolactin are lower in obese than in normal men. However, the prolactin response to hypoglycaemia is not impaired in obesity, in contrast to the HGH response. Therefore, the hypoglycaemic stimulation of prolactin release can be used as a provocative test of hypothalamo-pituitary function in obese patients.

This dissociation of HGH and prolactin secretion in obesity awaits a physiologic explanation. The fact that a stress such as hypoglycaemia raises prolactin more than HGH values in obese patients would favor increasing lipid deposition, since HGH is lipolytic (Raben and Hollenberg, 1959) and prolactin possibly lipogenic in man (Bern and Nicoll, 1968).

REFERENCES

Beck, P., Koumans, J.H.T., Winterling, C.A., Stein, M.F., Daughaday, W.H. and Kipnis, D.M. (1964): Studies of insulin and growth hormone secretion in human obesity. J. Lab. clin. Med., 64, 654.

Bell, J.P., Donald, R.A. and Espiner, E.A. (1970): Pituitary response to insulin-induced hypoglycemia in obese subjects before and after fasting. J. clin. Endocr., 31, 546.

Bern, H.A. and Nicoll, C.S. (1968): The comparative endocrinology of prolactin. Recent Progr. Hormone Res., 24, 681.

Copinschi, G., L'Hermite, M., Vanhaelst, L., Leclercq, R., Bruno, O.D., Golstein, J. and Robyn, C. (1972): Libération de prolactine par hypoglycémie insulinique chez l'homme. C.R. Acad. Sci. (Paris), 275, 1419.

Davis, S.L., Reichert, L.E. and Niswender, G.D. (1971): Serum levels of prolactin in sheep as measured by radioimmunoassay. Biology of Reproduction, 4, 145.

Hoffman, W.S. (1937): A rapid photoelectric method for the determination of glucose in blood and urine. J. biol. Chem., 120, 51.

L'Hermite, M., Delvoye, P., Nokin, J., Vekemans, M. and Robyn, C. (1972a): Human prolactin secretion as studied by radioimmunoassay: some aspects of its regulation. In: Prolactin and Carcinogenesis, p. 81. Editors: A.R. Boyns and K. Griffiths. Alpha Omega Alpha, Cardiff.

L'Hermite, M., Vanhaelst, L., Copinschi, G., Leclercq, R., Golstein, J., Bruno, O.D. and Robyn, C. (1972b): Prolactin release after injection of thyrotrophin-releasing hormone in man. Lancet, 1, 763.

Lorentz, F.H. (1929): Ein neuer Constitutionindex. Klin. Wschr., 8, 348.

Raben, M.S. and Hollenberg, C.H. (1959): Effect of growth hormone on plasma fatty acids. J. clin. Invest., 38, 484.

Virasoro, E., Copinschi, G., Bruno, O.D. and Leclercq, R. (1971): Radioimmunoassay of human growth hormone using a charcoal-dextran separation procedure. Clin. chim. Acta, 31, 294.

Effects of thyrotropin-releasing hormone on serum thyrotropin and thyroxine levels in obesity

E. Martino, F. Franchi, A. Pinchera, J.H. Romaldini, P. Biagioni, A.M. Loi and L. Baschieri

Centro per la Prevenzione del Gozzo, Istituto di Medicina del Lavoro and I Clinica Medica, University of Pisa, Pisa, Italy

Impairment of the pituitary-thyroid axis resulting from hypothalamic disorders has been implicated in the pathogenesis of obesity. This hypothesis has so far received no support from conventional tests of thyroid activity. In the present investigation the function of the pituitary-thyroid axis has been evaluated by the use of synthetic thyrotropin-releasing hormone (TRH).

Twelve obese subjects of both sexes with no evidence of other metabolic disorders were studied. No medication was given for at least 6 months prior to this study. Twenty normal subjects were used as controls. Measurements of serum thyrotropin (TSH) and thyroxine (T_4) levels and of the free T_4 index (FTI) were made before and at various times after intravenous administration of a standard dose (200 μg) of synthetic TRH. Basal values of TSH, T_4 and FTI and levels of TSH after TRH showed no differences between the two groups. An increase in serum T_4 and FTI following the injection of TRH was noted in the control but not in the obese subjects. Five of the latter groups were tested again after the administration of 5 μg of TRH per kg of body weight. Under these conditions the TSH changes showed no difference with respect to the standard TRH dose, but a significant rise of T_4 and FTI was observed. Relevant data are summarized in Table 1.

The present study indicates that obesity may be associated with a diminished thyroxine response to TRH, whereas TSH secretion appears to be normal. Among other explanations, these results may suggest an impaired thyroid response to circulating TSH, as recently reported by York et al. (1972) in genetically obese rats. Further studies are required to elucidate this problem.

TABLE 1   Basal values of circulating TSH, T_4 and FTI and changes occurring after TRH administration in 12 obese and 20 normal subjects (mean ± S.E.)

	TSH		T_4		FTI	
	Basal	Peak after TRH	Basal	2 hr after TRH	Basal	2 hr after TRH
	(μU/ml)	(μU/ml)	(μg/100 ml)	(% increase)		(% increase)
Controls (Body weight: 64 ± 2 kg)	2.57 ± 0.28	12.97 ± 1.94*	7.0 ± 0.3	14.4 ± 2.2*	6.3 ± 0.4	21.7 ± 2.4*
Obese subjects (Body weight: 100 ± 5 kg)	2.5 ± 0.2	12.84 ± 1.45* 14.7 ± 1.6**	7.5 ± 0.5	2.5 ± 4* 12.3 ± 9.8**	6.9 ± 0.3	3.7 ± 5.4* 16.7 ± 5.2**

* Values after 200 μg of TRH.
** Values after 5 μg of TRH per kg body weight (5 subjects).

REFERENCE

York, D.A., Hershman, J.M., Utiger, R.D. and Bray, G.A. (1972): Endocrinology, 90, 67.

The behaviour of somatomedin in obese subjects

G. Giordano, E. Foppiani, F. Minuto, M. Comaschi and **M. Marugo**

Cattedra di Endocrinologia, Istituto Scientifico di Medicina Interna, University of Genoa, Genoa, Italy

With reference to the hypothesis that in all probability somatomedin activity and non-suppressible insulin-like activity (NSILA) are two biological activities of the same molecule (Hall and Uthne, 1971), we studied the behaviour of somatomedin in a group of obese patients on a low calorie diet. It is known that fasting is the only condition in which a significant increase of NSILA occurs (Froesch et al., 1967).

Our study was performed on 9 obese women who were aged between 18 and 45 years, overweight by 48% to more than 80%, and inpatients of our hospital. These subjects were not receiving any form of therapy. Blood samples were collected at 8 a.m. from an indwelling intravenous needle in basal conditions and after an 8-day low calorie diet (600 cal/day). Growth hormone (HGH) and insulin (IRI) were estimated by radioimmunoassay and somatomedin by a 6 point symmetrical radiobiossay according to Daughaday et al. (1959), modified by Alford et al. (1972). Statistical analysis was performed by Student's t test.

The data obtained are shown in Table 1 and Figure 1. A weight loss, ranging from 1.7 to 14.1 kg, with no side effects was observed in all patients. Basal HGH was 1.74 ± 0.84 ng/ml before

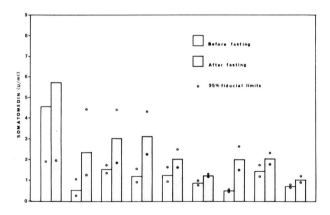

Fig. 1. Somatomedin levels before and after dieting in 9 obese subjects.

dieting and 0.80 ± 0.65 ng/ml afterwards; the difference between these two values was statistically significant (P < 0.05). IRI was 20.79 ± 9.86 μIU/ml before and 31.35 ± 17.05 μIU/ml after dieting (not statistically significant). Somatomedin increased 1.05 U/ml (fiducial limits 0.52-2.11) to 2.07 (1.10-3.92, P < 0.05). The somatomedin percentage increase (calculated on the basis of the mean values of each subject) was 124% ± 119% (P < 0.01).

We think that, in spite of the lack of data on NSILA in these subjects, the somatomedin increase reflects a parallel NSILA increase. This is a further demonstration that somatomedin and NSILA are two activities contained in the same molecule.

TABLE 1 Growth hormone, insulin and somatomedin levels in obese subjects before and after an 8-day low-calorie diet

Subject	% over-weight	Weight (kg)		HGH (ng/ml)		IRI (μIU/ml)		Somatomedin (U/ml)			
		Before	After	Before	After	Before	After	Before	Fiducial limit	After	Fiducial limit
C.A.	48	73.600	71.700	1.25	0.74	11.50	7.60	4.56	1.90 — 10.96	5.72	1.95 — 16.82
L.G.	66	87.700	84.000	1.60	0.55	9.40	15.00	0.53	0.26 — 1.06	2.34	1.25 — 4.42
B.C.	64	95.700	91.600	1.02	0.21	40.33	49.00	1.53	1.35 — 1.73	3.03	1.85 — 4.91
M.C.	>80	150.800	138.900	1.30	0.00	16.25	64.00	1.19	0.91 — 1.55	3.13	2.27 — 4.32
Z.D.	>80	114.800	110.000	0.80	0.00	15.15	21.03	1.25	0.95 — 1.65	2.02	1.62 — 2.51
B.R.	59	83.800	69.700	2.36	1.18	22.00	29.50	0.87	0.76 — 0.99	1.23	1.19 — 1.28
G.R.	57	86.600	84.900	3.40	1.65	17.50	31.00	0.50	0.56 — 0.44	1.99	1.49 — 2.66
L.A.	>80	98.500	93.400	2.50	1.60	31.00	34.00	1.44	1.19 — 1.75	2.04	1.77 — 2.33
P.L.	>80	165.200	156.600	1.40	1.25	24.00	31.00	0.71	0.64 — 0.79	1.04	0.92 — 1.18
Mean				1.74	0.80	20.79	31.35	1.05	0.52 — 2.11	2.07	1.10 — 3.95
S.D.				0.84	0.65	9.86	17.05				
P					< 0.05		not significant			< 0.05	

We also wish to point out that HGH decreased parallel to the somatomedin increase. If confirmed, these results would suggest the possibility that factors other than HGH may induce somatomedin synthesis in man.

REFERENCES

Alford, F.P., Bellair, J.T., Burger, H.G. and Loevett, N. (1972): A simplified assay for somatomedin. J. Endocr., 54, 365.

Daughaday, W.H., Salmon Jr., W.D. and Alexander, F. (1959): Sulfation factor activity of serum from patients with pituitary dwarfism. J. clin. Endocr., 19, 743.

Froesch, E.R., Burgi, H., Muller, W.A., Humbel, R.E., Jacob, A. and Labhart, A. (1967): N.S.I.L.A. of human serum: purification physiochemical and biological properties and its relation to total serum I.L.A. Recent Progr. Horm. Res., 23, 565.

Hall, K. and Uthne, K. (1971): Some biological properties of purified sulfation factor from human plasma. Acta med. Scand., 190, 137.

Massive obesity and thyroid function

M. Faggiano[1], **C. Carella**[1], **T. Criscuolo**[1], **P. Jaquet**[2] and **M. Minozzi**[1]

[1]*Divisions of Endocrinology and of Medical Genetics, I Faculty of Medicine, University of Naples, Naples, Italy and* [2]*Division of Endocrinology, University of Marseilles, Marseilles, France*

In massive obesity abnormalities in plasma insulin and growth hormone variation under different stimuli have been described. Although normal thyroid function is usually reported in obesity, Scriba et al. (1967) found a significant reduction in protein bound iodine, and Bray et al. (1971) have discussed the occurrence of inadequate tissue levels of thyroid hormones during caloric restriction. Ten male obese subjects, ranging in age from 16 to 38 years, with an overweight range of 51-127% (Metropolitan Life Insurance Co. Tables) received a rapid i.v. injection of 400 μg synthetic thyrotropin-releasing hormone (TRH) (Hoffman-La Roche), and thyrotropin (TSH) plasma levels were measured 10, 20, 30, 45, 60, 120, 180 and 300 minutes after injection. Basal levels of TSH were similar in normal (1.9 μIU/ml $\pm$ 0.18 S.E.M. H. TSH 68/38 MRC) and obese subjects (3.0 $\pm$ 0.96).

Mean maximal plasma TSH increment after TRH (Fig. 1a) was 16.1 $\pm$ 2.4 μIU/ml in obesity versus 13.2 $\pm$ 0.8 in normals (P > 0.05). In only 1 case (Fig. 1b) was plasma TSH response to

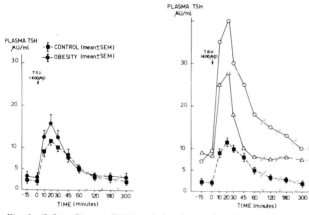

Fig. 1a (left). Plasma TSH variation in obesity after a 400 μg i.v. injection of TRH
Fig. 1b (right). Abnormal TRH response in 1 case of obesity (a) compared to a typical case of mild hypothyroidism (b).

TRH exaggerated and prolonged as in mild hypothyroidism. These results confirm the absence of thyrotropic dysfunction in obesity. The unique excessive response to TRH may be considered a casual association.

REFERENCES

Bray, G.A., Raben, M.S., Londono, J. and Callagher Jr, T.F. (1971): Effects of triiodothyronine, growth hormone and anabolic steroids on nitrogen excretion and oxygen consumption of obese patients. J. clin. Endocr., 33, 293.

Scriba, P.C., Richter, J., Horn, K., Beckebans, J. and Schwarz, K. (1967): Zur Frage der Schilddrüsenfunktion bei Adipositas. Klin. Wschr., 45, 323.

Regulation of the adipose mass: histometric and anthropometric aspects

J. Vague, P. Rubin, J. Jubelin, G. Lam-Van, F. Aubert, A.M. Wassermann and **J. Fondarai**

Endocrinology Clinic, Hôpital de la Timone, Marseilles, France

Adipose mass may be defined as the total quantity of triglycerides (TG) in the body. These substances are water insoluble and the number of their molecules is unimportant. Physiologically only the weight of TG and their surface area in contact with other elements of the medium must be considered. TG are present but invisible in most tissues (but absent in the brain). Their level is about 1% (wet weight) in liver and muscle, around 15 g and 300 g respectively, a reserve of energy which is immediately, but probably only in part, available. During severe work one-third of these TG and two-thirds of glycogen disappear in skeletal muscle (Fröberg et al., 1971).

In poikilothermic species provided with a liver, hepatocytes normally contain droplets of TG, but there is negligible adipose tissue. In all homeothermic species TG are almost entirely contained in adipocytes, in which they constitute a single droplet, the immediately reactive part of which is limited to its surface. The numerous droplets of TG in brown adipose tissue represent an intermediate stage in this distribution pattern. Without forgetting the diffuse TG in most tissues, adipose mass can be grossly estimated as the total mass of TG contained in adipocytes. This parameter will be particularly considered in this paper.

RESULTS OF DETERMINATION OF ADIPOSE MASS BY AN ANTHROPOMETRIC METHOD

Method

From a comparative study (Vague et al., 1969a) of various methods of determining the adipose mass in man, we concluded that the measurement of the skinfold thickness at the proximal parts of the arm and thigh and of the circumference of the limb at this level was among the most accurate methods and certainly the quickest and most practical. This technique (Fig. 1) gives the following data: (1) the brachial adipo-muscular ratio (BAMR); (2) the femoral adipo-muscular

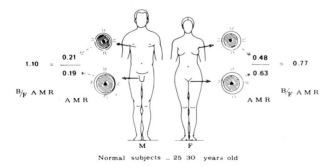

Normal subjects _ 25 30 years old

Fig. 1. Measurement of the adipo-muscular ratio (AMR) in normal subjects 25-30 years old. B/F AMR = ratio of brachial AMR to femoral AMR. See text.

ratio (FAMR); (3) the ratio of BAMR to FAMR (B/F AMR); (4) the mean AMR, i.e. mean of BAMR and FAMR; (5) the percentage of adipose mass in total body weight, derived from mean AMR (Vague et al., 1969a) multiplied by mean percentage of fat in adipose tissue (0.80) and by the density of fat (0.92) (= body fat percentage); (6) the adipose mass (= body weight × fat percentage).

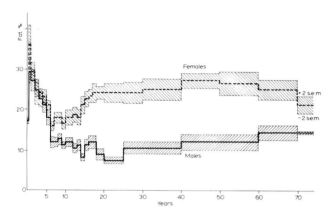

Fig. 2. Development of the body fat percentage in males and females with age.

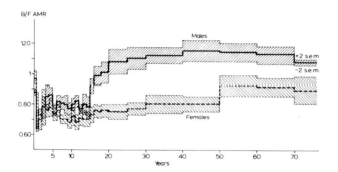

Fig. 3. Development of the ratio of brachial to femoral AMR (B/F AMR) in males and females with age.

Adipose mass as a function of age and sex

Constantly fed by maternal blood, the human fetus is deprived of fat reserves until the end of the fifth month. This reserve is still very small at the end of the eighth month, which partially explains the premature child's fragility. Our measurements in both sexes from birth to old age gave the curves shown in Figure 2. Identical for both sexes at birth (17%), the percentage of fat increases in the relatively motionless baby during the first year and decreases afterwards in relation to the musculature and motility increase, which initiates a competition between fat and muscle which will last for life. At the age of 5 the adipose mass returns to the level at birth (17%). From this age the percentage of fat in boys becomes smaller than in girls, attaining half of this value at 15 years. This relative proportion remains the same between the sexes, though the adipose mass increases with age. The late apparent decrease in females is due to the loss of total body weight, the components of which are described below.

From puberty adipose mass predominates on the upper part of the body in males and on the lower part in females. This difference decreases with age but never entirely disappears, while fat rises in the female between the ages of 50 and 60 to reach the same level found in 15-year-old boys (Fig. 3). These two different types of fat distribution are the rule in the majority of adult males and females. However, a minority of subjects of each sex has the fat distribution of the other sex. So android (or male type) and gynoid (or female-type) obesity and intermediate forms will be observed in both sexes.

Adipose mass and muscular activity

The development of adipose mass with age in both sexes in itself supplies the evidence that its value is inversely related to muscular activity. Measurement of the adipose mass was performed in 67 top-level athletes: 31 males (25 of international and 6 of national standard) and 36 females (4 of international and 32 of national standard). They were aged between 20 and 30, performing various athletic activities and of varying weights.

It was observed that adipose mass in athletes was much less than in the normal population, an obvious phenomenon already reported by many authors (Behnke and Royce, 1966; Bessou, 1970; Fischer et al., 1970; Jokl, 1964; Kireilis and Cureton, 1947; Parizkova, 1959; Pascale et al., 1955; Riendau et al., 1958; Thompson et al., 1956; Tanner, 1952, 1959; Vague et al., 1969a; Wilmore et al., 1970). This adipose mass was inversely correlated to the degree of training and to constitutional athletic qualities. It was at its lowest level in Olympic athletes.

The relationship between the logarithm of the adipose mass or the mean AMR in various athletes and their weight index (ratio of measured weight to ideal weight according to the Lorentz formula modified as a function of the bihumero-bitrochanter width (Vague and Jouve, 1958)) is given by a straight line from the judokas and light runners to weight lifters and heavy-weight boxers (Fig. 4). The slope of this straight line represents the development of the adipose mass of athletes as a function of their weight, all other things being equal, especially their athletic capacity and training. In other words, the heavier an athlete is as a consequence of his

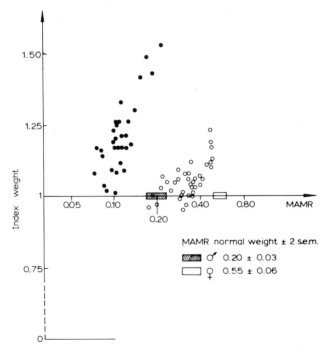

Fig. 4.   Relationship of mean AMR and weight index in athletic males (●), athletic females (o) and normal subjects.

muscular development, the more his percentage of fat seems to approximate that measured in the less heavy part of a normal population.

The definition of normal weight as a function of height according to the Lorentz formula has the disadvantage that it gives too high values in longilineal subjects and too low values in brevilineal subjects. Our correction using the bihumero-bitrochanter width (Vague and Jouve, 1958) avoids this error, but only in the mean occidental population without special muscular training. The weight of the majority of athletes, with a smaller adipose mass than normal, is above this value.

Thus there is a problem of how to define the normal weight of heavy athletes and that of the subjects in a normal population who have increased musculature. The most precise definition would be the body weight for which the fat weight would not be greater than in the normal subjects in the same category. Under these conditions the mutual growth of fat and muscle would forbid sufficient muscular development for athletic or professional performance.

The best definition of normal weight for heavy muscular subjects seems to be the weight for which the mean AMR does not exceed its level in the normal subjects in the same category. In our experience, the adipose mass of top-level heavy athletes usually remains within these limits.

Adipose mass and innervation
Innervation, especially with adrenergic nerves, acts in the same way as physical exercise. Adrenergic activity tends to reduce adipose mass and develops musculature, according to much experimental (Clement, 1950; Fredholm, 1970; Toldt, 1870) and clinical data (Balasse, 1968; Sdrobici et al., 1967; Vague et al., 1969a).

Adipose mass and weight variations
The variation in the adipose and muscular mass during weight gain or loss depends on the circumstances of these weight changes. Various possibilities exist, from gross protein and fat loss during total fasting and protein gain during a low-calorie, high-protein diet with exercise to the exclusive deposition of fat during overfeeding without exercise and fat and muscular gain during the same overfeeding with exercise. Clinical experience demonstrates that in major obesity the weight gain can be almost exclusively ascribed to fat, and that in extreme emaciation fat disappears, at least two-thirds of the body protein being necessary for survival (Cahill et al.,

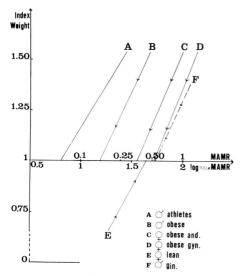

Fig. 5. Relationship of mean AMR and weight index during weight decrease or increase (see text). A = straight line drawn by the coordinates of mean AMR and weight index in athletic males of various weights (see Fig. 4).

1970). It was thought useful, therefore, to determine the proportion of fat and muscle involved in weight gain or loss due almost exclusively to variations in food intake.

In 41 obese subjects (12 android males and 1 gynoid male; 13 android and 15 gynoid females) and in 21 lean females, we measured mean AMR during weight loss or gain by under- or overfeeding 3-4 times (the protein intake remaining normal) and plotted this as a function of weight index. The curve is logarithmic (Fig. 5) and is similar to that obtained from the measurement of athletes of various weights with identical athletic qualities (Fig. 4).

These data allow us to conclude that: (1) the evolution of fat and muscle mass proceeds according to a definite law when only calorie intake varies with a normal protein intake; and (2) this evolution leads to the concept of a normal weight AMR and to the possibility of estimating the AMR value for every weight. In any individual these values appear to be more or less permanent characteristics which are inclined to reappear whatever may be the artificial variations imposed on a subject. In 25-30-year-old males and females of a normal population, normal weight mean AMR was 0.20 and 0.55 respectively.

For the calculation of the normal weight AMR the most convenient mathematical model is the following equation:

$$\frac{MW}{NW} = K \log (100\ MAMR) - K \log (100\ NWMAMR) + 1$$

where MW = measured weight, NW = normal weight (Vague and Jouve, 1958), $\frac{MW}{NW}$ = weight index, MAMR = mean adipo-muscular ratio, NWMAMR = normal weight adipo-muscular ratio. K = slope of the straight line drawn according to the following coordinates: $\log (100\ MAMR)$ on the abscissa and $\frac{MW}{NW}$ on the ordinate,

or $K \log (100\ NWMAMR) = K \log (100\ MAMR) - \frac{MW}{NW} + 1$

or $\log (100\ NWMAMR) = \log (100\ MAMR) - \frac{MW}{KNW} + \frac{1}{K}$

Under these conditions, the degree of error is less than 0.001 and the slope K gives the AMR at normal weight.

ACTION OF HORMONES ON THE ADIPOSE MASS

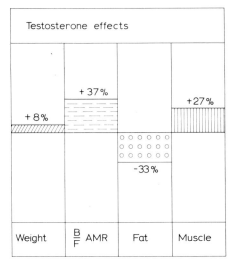
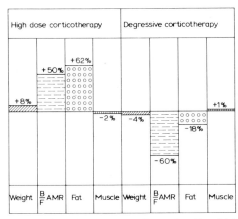

Fig. 6 (left).   Effect of testosterone (1 g per month) in eunuchoid patients. B/F AMR = ratio of brachial to femoral AMR. See text.

Fig. 7 (right).   Effect of corticosteroids. See text and legend to Figure 6.

Three hormones, the secretion of which is often disturbed and which are frequently used therapeutically, have important effects on the AMR. In eunuchoid patients long-acting testosterone (1 g monthly for 1 year or more) develops musculature and reduces adipose mass much more in the lower part than in the upper part of the body (Vague et al., 1968, 1969a) (Fig. 6). Corticosteroids (at doses equivalent to 30 mg prednisone daily for 3 months or more) increase the adipose mass much more in the upper part than in the lower part of the body and reduce musculature (Vague et al., 1969a, 1971b) (Fig. 7). In hyperthyroidism the emaciation is a consequence of muscle loss more than of fat loss. The recovery following surgery, [131]Iodine or antithyroid drugs increases musculature to a much greater extent than adipose mass (Vague et al., 1973) (Fig. 8). This is an additional demonstration of the strongly proteolytic and weakly lipolytic effect of thyroid hormones when used as therapeutic agents in obesity and argues against too frequent use of them.

Some of the factors which determine the development of muscle and fat for a given weight are outlined in Figure 9.

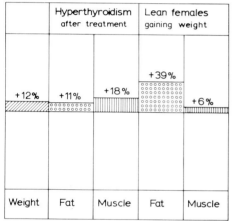

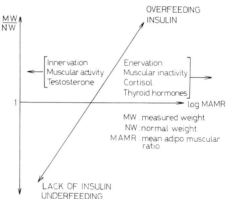

Fig. 8 (left). Effect of treatment for hyperthyroidism.

Fig. 9 (right). Main factors influencing the coordinates of log mean AMR and weight index.

MEASUREMENT OF ADIPOSE MASS BY COMBINED CYTOMETRIC AND ANTHROPOMETRIC METHODS

Cytometric methods of measuring adipocytes involve the second aspect of adipose mass, the surface area of the TG mass in contact with the molecular system able to act on it. For cell diameters usually ranging between 60 and 120 μm, the corresponding volumes will be 116 and 929 pl respectively and the surface areas 11,304 and 45,330 μm^2. The multiplication of adipocytes during growth in man (Bonnet et al., 1970; Brook et al., 1972; Hirsch et al., 1966; Hirsch and Knittle, 1970; Sjöström et al., 1972) and animals (Hirsch and Han, 1969; Knittle and Hirsch, 1968; Lemonnier, 1970b, 1971) and the distinction between hyperplastic and hypertrophic obesity (Björntorp et al., 1971a), the latter with particular metabolic complications, are well known today. We have previously demonstrated the same metabolic complications in android obesity, which predominates on the upper part of the body (Vague, 1949, 1953, 1956, 1960; Vague et al., 1964, 1971a). We thought it useful to study the cell sizes of adipocytes in the subcutaneous areas where the differences between males and females are most clear-cut, i.e. the deltoid and retro-trochanteric areas, which were also measured with respect to the skinfold thickness.

Method

In 58 females and 38 males adipose tissue samples were taken by needle biopsy prepared according to the technique of Sjöström et al. (1971) modified by the use of a Nageotte cell. 100 to 130 cells were measured for each sample. This technique gave:

1. The mean volume of deltoid (DV) and trochanteric (TV) adipocytes.

2. The arbitrary mean adipocyte volume (MV): $MV = \dfrac{DV + TV}{2}$.

3. The number of adipocytes constituting the theoretical column of the adipose tissue thickness, a number which is arbitrary in absolute terms but certainly correct in relative terms, of deltoid area (DN):

$$= \frac{\text{thickness of deltoid adipose tissue}}{\text{mean diameter of deltoid adipocytes}}$$

and trochanteric area (TN):

$$= \frac{\text{thickness of retro-trochanteric adipose tissue}}{\text{mean diameter of retro-trochanteric adipocytes}}$$

4. Mean number (MN), also arbitrary, of adipocytes: $\dfrac{\text{adipose mass}}{MV \times 0.92}$ (0.92 = density of TG).

Correlation studies were performed between DV, TV, MV, DN, TN and MN on the one hand and adipose mass and B/F AMR on the other.

Results

Adipocyte volume and adipose mass   There was a good correlation between deltoid adipocyte volume and adipose mass (Fig. 10), but not between the trochanteric adipocyte volume and this mass. The subjects in whom fat predominated in the upper part of the body, in other words android subjects of both sexes, had larger deltoid adipocytes than gynoid subjects of both sexes in whom fat predominates in the lower part of the body (Fig. 10). In the trochanteric area there was no difference in adipocyte volume between the various types of adipose mass distribution (Fig. 10). Mean adipocyte volume was related to adipose mass, and especially for the lower values of this mass (Fig. 11).

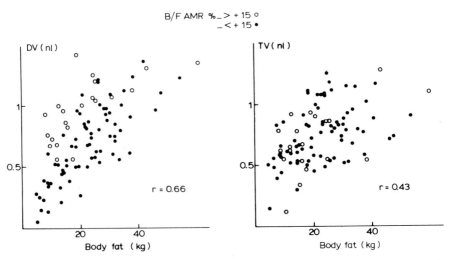

Fig. 10.   Deltoid (DV) and trochanteric (TV) adipocyte volume as a function of body fat in android (o) and gynoid (●) subjects.

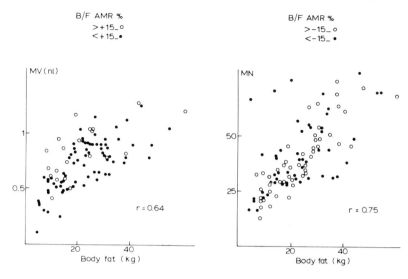

Fig. 11. Mean adipocyte volume (MV) and number (MN) (10⁹) as a function of body fat in android (o) and gynoid (●) subjects.

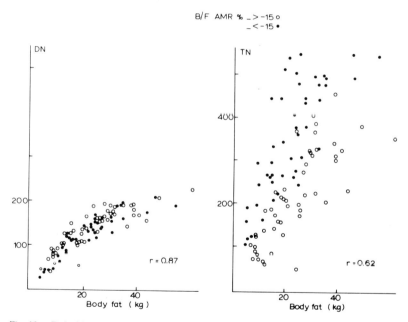

Fig. 12. Deltoid (DN) and trochanteric (TN) adipocyte numbers (10⁹) as a function of body fat in android (o) and gynoid (●) subjects.

Adipocyte number and adipose mass The number of deltoid adipocytes was well correlated and the number of trochanteric adipocytes less well correlated with adipose mass (Fig. 12). The android subjects had less trochanteric adipocytes than gynoid subjects (Fig. 12). The mean adipocyte number was related to the adipose mass when the latter was large (Fig. 11). Every major obesity involves an increased number of adipocytes, probably constituted before the end

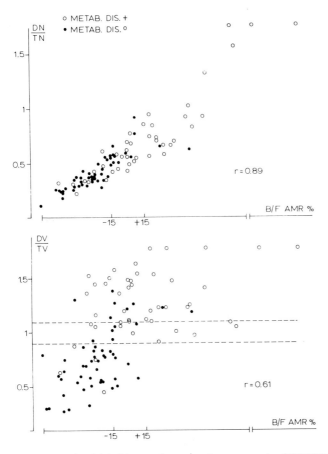

Fig. 13. Ratio of deltoid to trochanteric adipocyte number (DN/TN) and ratio of deltoid to trochanteric adipocyte volume (DV/TV) as a function of percentage brachial to femoral AMR (B/F AMR) in subjects with (o) and without metabolic disorders (●).

of growth. Women have twice the adipose mass of men because they have twice as many adipocytes.

Comparative estimations of volume and number of deltoid and trochanteric adipocytes and topography of adipose mass   Subjects with fat predominating in the upper part of the body had larger deltoid adipocytes (Fig. 10) and fewer trochanteric adipocytes (Fig. 12) than the others. The ratio of deltoid to trochanteric adipocyte number (DN/TN) was well correlated with the predominance of fat in the upper part of the body (Fig. 13). The ratio of deltoid to trochanteric adipocyte volume (DV/TV) was also significantly correlated with this parameter, but to a lesser extent (Fig. 13). The ratios DV/TV and DN/TN increased with the predominance of fat in the upper part of the body, but, whereas high DV/TV were the consequence of the increase in numerator, a decrease in the denominator was the cause of high DN/TN values (Figs. 10, 12 and 13).

Effect of testosterone and cortisol on adipocyte volume and number (Fig. 14)   We can report preliminary findings on a small number of cases. Long-acting testosterone in 3 eunuchoids (1 g monthly for 6 to 12 months) had little effect on adipocyte volume but decreased their number

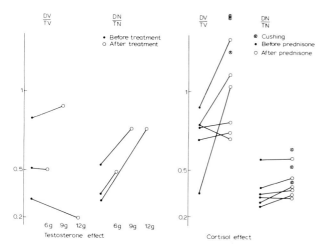

Fig. 14. Effects of testosterone and cortisol on the ratio of deltoid to trochanteric adipocyte volume (DV/TV) and of deltoid to trochanteric adipocyte number (DN/TN) in eunuchoidism, Cushing's syndrome and corticotherapy.

considerably in the trochanteric area. This phenomenon is probably the cause of the difference in adipose mass between males and females, which starts at puberty.

In 3 cases of Cushing's disease adipocyte volume was much larger in the deltoid than in the trochanteric area. The number of adipocytes in the two areas was in accordance with the topographical distribution of fat.

In 6 children receiving 2 mg/kg prednisone daily for 2 months the ratio DV/TV was markedly increased in 3 cases without modification of the number of adipocytes.

ANTHROPOMETRIC AND CYTOMETRIC MEASUREMENTS OF PATHOGENIC ADIPOSE MASS

Anthropometric measurements

In previous papers (Vague, 1947, 1949, 1953, 1956; Vague et al., 1964, 1967a, 1969a, b) we believe we have demonstrated that the effect of fat overload on the onset, evolution and complications of diabetes, hyperuricemia, hyperlipemia and atherosclerosis is not related to the degree of obesity but to its android character, in other words to the predominance of fat in the upper part of the body. It has also been demonstrated that this diabetogenic effect of android obesity is independent of family history of diabetes (Vague et al., 1971a). It was necessary to verify and complete these data with respect to the adipose mass itself and its topography.

163 obese subjects between 40 and 50 years old were studied. They included 80 males (38 with overt diabetes, i.e. fasting hyperglycemia but not requiring insulin, and 42 non-diabetics with a normal oral glucose tolerance test) and 83 females (50 diabetics and 33 with a normal oral glucose tolerance test). The age range of 40 to 50 was chosen to obtain a homogeneous population and because it is the usual time of the early onset of overt diabetes in obese subjects.

We plotted the ratio of the measured adipose mass (r BF) to normal adipose mass (i BF) against the ratio of the measured (r) B/F AMR to the normal (i) B/F AMR for sex and age. When a comparable number of diabetics and non-diabetics of both sexes are found below the 45° line of the first quadrant, we find that in males only diabetics and in females a large majority of diabetics are found above that line (Figs. 15 and 16).

This distribution once more demonstrates that: (1) major obesity may or may not lead to diabetes in the age group 40 to 50 when fat overload does not predominate in the upper part of

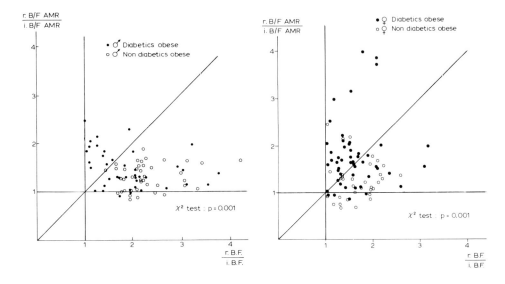

Fig. 15 (left).   Relationship between the ratios measured body fat (r BF) to normal body fat (i BF) and measured brachial-femoral AMR (r B/F AMR) to normal brachial-femoral AMR (i B/F AMR) in obese diabetic and non-diabetic males.

Fig. 16 (right).   Same relationship as in Figure 15 for obese diabetic and non-diabetic females.

the body, and (2) on the other hand, even when it is not very great, obesity leads to diabetes when it predominates in the upper part of the body.

Similar findings resulted from the anthropometric study of hyperlipemia (Types III and IV in Fredrickson's classification). In 16 females (9 with overt diabetes and 4 with chemical diabetes) 12 had hyperandroid fat topography. In 35 males (17 with overt diabetes and 4 with chemical diabetes) 30 had hyperandroid fat topography.

Cytometric measurement

Among the 96 subjects in whom anthropometric and adipocyte measurements were performed, 18 had overt diabetes and 27 had abnormal and 51 normal glucose tolerance tests (Fig. 13). Almost all diabetics had deltoid adipocytes larger than trochanteric ones. Almost all non-diabetics had trochanteric adipocytes larger than deltoid ones. As regards the number of deltoid and trochanteric adipocytes, there was no relationship with diabetes, but all non-diabetics did have low B/F AMR and low DN/TN.

In contrast to the response to glucose, the insulin response to tolbutamide is independent of chemical, subclinical and mild diabetes (Ramahandridona, 1972; Vague, 1974). For this reason the value iT (plasma insulin area for the first 10 minutes after venous tolbutamide injection) was chosen for evaluation of insulin secretion in 32 of our subjects, 9 males and 23 females. Insulin response to tolbutamide increased in relation to the increase in the ratio DV/TV (Fig. 17).

In summary, diabetes in obese subjects is associated with the predominance of fat in the upper part of the body, and with adipocytes larger and relatively greater in number in the deltoid than in the trochanteric area. Hyperinsulinism in obese subjects is correlated with larger adipocytes in the deltoid than in the trochanteric area.

Hyperlipemia (Frederickson Type III and IV) was found in 7 of these cases with increases of DV/TV and DN/TN ratios.

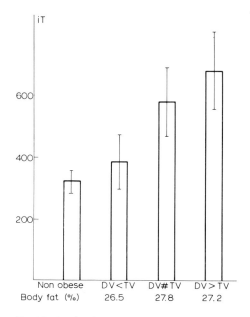

	Non obese	DV<TV	DV#TV	DV>TV
Body fat (%)		26.5	27.8	27.2

Fig. 17. Ratio of deltoid to trochanteric adipocyte volume (DV/TV) and plasma insulin area for the first 10 minutes after i.v. injection of tolbutamide (iT).

MECHANISM OF THE PATHOGENIC EFFECT OF THE PREDOMINANCE OF ADIPOSE MASS IN THE UPPER PART OF THE BODY AND OF ENLARGED DELTOID ADIPOCYTES

The relationship between the predominance of fat in the upper part of the body and enlarged deltoid adipocytes on the one hand and the metabolic complications of obesity on the other is certain. Regarding its mechanism we shall only make the following brief comments.

The diabetogenic effect of the predominance of fat in the upper part of the body is not correlated with familial history of diabetes (Vague et al., 1971a). An equivalent demonstration for hyperlipemia, hyperuricemia and atherosclerosis has still to be obtained.

Insulin resistance of adipocytes seems proportional to their increase of volume and therefore to the relative decrease of their surface (Björntorp et al., 1971a, c; 1972c; Gordon, 1970). One can conceive that this resistance stimulates insulin hypersecretion, the self-increase of obesity and eventually β-cell exhaustion.

A decrease of the number of adipocytes in the lower part of the body is observed concomitantly with an increase of adipocyte volume in the upper part in obesity with metabolic disorders. The above-mentioned effects of testosterone and cortisol lead us to think that possibly these two hormones are responsible for the former and latter phenomena, respectively.

In obese subjects in a homogeneous series (Vague et al., 1971a) the elevated cortisol production rates and the high plasma insulin levels found, especially when fat predominated in the upper part of the body, appeared to be closely related without mediation by the weight. For these reasons the minor but normally present and long-acting hyperadrenocorticism of diabetogenic obesity probably plays a part in the hypertrophy of upper body adipocytes and consequently in the constitution of diabetes and related metabolic disorders.

CONCLUSIONS

Evaluation of the adipose mass as measured by the skinfold thickness at the four proximal points of the arms and thighs and of the circumference of the limb at this level permit us to delineate the topography of adipose tissue, which allows us to predict the metabolic complications of obesity.

The development of adipose mass measured in this way remains a feature from birth to old age. During maturity the adipose mass in females is twice that in males. Fat predominates on the lower part of the body in the former and on the upper part in the latter.

Physical exercise increases musculature and decreases adipose mass. The respective development of muscle and adipose mass in athletes of various weights and in subjects whose weight increases or decreases only as a result of over- or underfeeding (with a correct protein intake) is ruled by a definite law. The evolution of AMR as a function of weight is in both cases exponential. The knowledge of this evolution allows us to calculate for any subject in determined conditions a normal weight AMR.

Cytometric determination of adipocyte size performed in deltoid and retrotrochanteric areas represents a useful approach in addition to anthropometric measurement of adipose mass. The predominance of fat in the upper part of the body in the majority of males and in android females is correlated with an increase of deltoid adipocyte volume and a relative decrease of trochanteric adipocyte number.

Testosterone increases musculature and decreases adipose mass, especially in the lower part of the body. It also decreases the number of adipocytes, especially in the retro-trochanteric area. This hormone is probably responsible for the differences in adipose mass and topography in males and females after puberty.

Cortisol excess decreases musculature and increases adipose mass, especially in the upper part of the body, with a specific increase in deltoid adipocyte volume.

Thyroid hormone excess decreases musculature much more than adipose mass, thus increasing the AMR.

The adipose mass is regulated by a neuro-hormono-enzymatic system, which explains its usual stability. The concept of a necessary adipose mass in determined conditions is a consequence of this regulatory system. Loss of this mass without modification of its determining conditions leads to stress and often to nervous breakdown.

Mechanical complications of obesity are in direct ratio to overload. Metabolic complications (diabetes, hyperlipemia, hyperuricemia, atherosclerosis) are increased with increasing adipose mass, but much more by the predominance of fat in the upper part of the body and the increased volume of deltoid adipocytes, both of which characterize android obesity. The adipose mass excess is more pathogenic when it predominates in the upper part of the body and when genetic predispositions to such metabolic disorders are greater.

Our data suggest that the increased volume of adipocytes in the upper part of the body and the consequent relative decrease of their surface might be responsible for insulin resistance, hyperinsulinism and β-cell exhaustion. Cortisol seems to be involved in this increase of deltoid adipocyte volume.

The treatment of obesity must be the more drastic the more the adipose mass is pathogenic. On the other hand, in order to avoid a nervous breakdown, it must be cautious when the enlarged adipose mass is needed by the obese patient in his situation. The aim must always be not so much to reduce the excess adipose mass itself but rather to correct the internal and external conditions which have increased this mass.

REFERENCES

Balasse, E. (1968): Influence of norepinephrine, growth hormone and fasting on FFA mobilization and glucose metabolism in lean and obese subjects. Diabetologia, 4, 20.

Behnke, A.R. and Royce, J. (1966): Body size, shape and composition of several types of athletes. J. Sport Med. (Torino), 6, 75.

Bessou, M. (1970): Masse grasse corporelle et aptitudes sportives. Thesis, Faculté de Médecine, Toulouse.

Björntorp, P., Bengtsson, C., Blohme, G., Jonsson, A., Sjöström, L., Tibblin, E., Tibblin, G. and Wilhelmsen, L. (1971a): Adipose tissue fat cell size and number in relation to metabolism in randomly selected middle-aged men and women. Metabolism, 20, 927.

Björntorp, P., Berchtold, P. and Tibblin, G. (1971c): Insulin secretion in relation to adipose tissue in men. Diabetes, 20, 65.

Björntorp, P., Fahlen, M., Grimby, G., Gustafson, A., Holm, J., Renstrom, P. and Schersten, T. (1972a): Carbohydrate and lipid metabolism in middle-aged, physically well-trained men. Metabolism, 21, 1037.

Björntorp, P., Grimby, G., Sanne, H., Sjöström, L., Tibblin, G. and Wilhemsen, L. (1972c): Adipose tissue fat cell size in relation to metabolism in weight-stable, physically active men. Hormone Metab. Res., 4, 182.

Björntorp, P., Gustafson, A. and Persson, B. (1971b): Adipose tissue fat cell size and number in relation to metabolism in endogenous hypertriglyceridemia. Acta med. scand., 190, 363.

Björntorp, P., Jonsson, A. and Berchtold, P. (1972b): Adipose tissue cellularity in maturity onset diabetes mellitus. Acta med. scand., 191, 129.

Bonnet, F., Gosselin, L., Chantraine, J. and Senterre, J. (1970): Adipose cell number and size in normal and obese children. Rev. europ. Etud. clin. biol., 15, 1101.

Brook, C.G.D., Lloyd, J.K. and Wolf, O.H. (1972): Relation between age of onset of obesity and size and number of adipose cells. Brit. med. J., 2, 25.

Cahill Jr., G.F., Marliss, E.B. and Aoki, T.T. (1970): Fat and nitrogen metabolism in fasting man. In: Adipose Tissue. Regulation and Metabolism Functions, p. 181. Editors: B. Jeanrenaud and D. Hepp. G. Thieme, Stuttgart and Academic Press, New York.

Clement, G. (1950): La mobilisation des graisses de réserve chez le rat. Influence du système nerveux sympathique. Arch. Sci. physiol., 4, 13.

Fischer, G., Israel, S., Strauzenberg, S.E. and Thierbach, P. (1970): Messungen des Depotfetts bei Sportlern. Theor. Prax. Korperkultur, 19, 1084.

Fredholm, B.B. (1970): Studies on the sympathetic regulation of circulation and metabolism in isolated canine subcutaneous adipose tissue. Acta physiol. scand., Suppl. 354, 47 p.

Fröberg, S.O., Carlson, L.A. and Ekelund, L.G. (1971): Local lipid stores and exercise. Advanc. exp. Med. Biol., 11, 307.

Gordon, E.S. (1970): Metabolic aspects of obesity. Advanc. metab. Dis., 4, 229.

Hirsch, J. and Han, P.W. (1969): Cellularity of rat adipose tissue: effects of growth, starvation and obesity. J. Lipid Res., 10, 77.

Hirsch, J. and Knittle, J.L. (1970): Cellularity of obese and non-obese human adipose tissue. Fed. Proc., 29, 1516.

Hirsch, J., Knittle, J.L. and Salans, L.B. (1966): Cell lipid content and cell number in obese and non-obese human adipose tissue. J. clin. Invest., 45, 1013.

Jokl, E. (1964): Physiology of Exercise. C.C. Thomas, Springfield, Ill.

Kireilis, R.W. and Cureton, T.K. (1947): The relationships of external fat to physical education activities and fitness tests. Res. Quart. Amer. Ass. Hlth phys. Educ., 18, 123.

Knittle, J.L. and Hirsch, J. (1968): Effect of early nutrition on the development of rat epididymal fat pads: cellularity and metabolism. J. clin. Invest., 47, 2091.

Lemonnier, D. (1970a): Cellularité et caractères morphologiques du tissu adipeux du rat rendu obèse par un régime hyperlipidique. Arch. Anat. Morph. exp., 59, 1.

Lemonnier, D. (1970b): Augmentation du nombre et de la taille des cellules adipeuses dans l'obésité nutritionnelle de la souris. Experientia (Basel), 26, 974.

Lemonnier, D. (1971): Sex difference in the number of adipose cells from genetically obese rats. Nature (Lond.), 231, 50.

Parizkova, J. (1959): The development of subcutaneous fat in adolescents and the effect of physical training and sport. Physiol. bohemoslov., 8, 112.

Pascale, L.R., Frankel, T., Grossman, M.I., Freeman, S., Fellner, I.L., Bond, E.E., Ryan, R. and Bernstein, L. (1955): Changes in Body Composition of Soldiers During Paratrooper Training. Report No. 156, Denver Medical Nutrition Laboratory.

Ramahandridona, G. (1972): Réponse insulinique différente au glucose et à la tolbutamide, un marqueur génétique du diabète sucré? Thesis, Marseilles.

Riendau, R.P., Welch, B.E., Crisp, C.E., Crowley, L.V., Griffin, P.E. and Brockett, J.E. (1958): Fat and skill. Res. Quart. Amer. Ass. Hlth phys. Educ., 29, 200.

Sdrobici, D., Bonaparte, H., Pieptea, R. and Sapatino, V. (1967): Rôle des catécholamines dans la mobilisation des graisses du panicule adipeux chez les obèses soumis au jeûne. Nutr. et Dieta (Basel), 9, 271.

Sjöström, L., Björntorp, P. and Vrana, J. (1971): Microscopic fat cell size measurements on frozen-cut adipose tissue in comparison with automatic determinations of osmium-fixed fat cells. J. Lipid Res., 12, 521.

Sjöström, L., Smith, U., Krotiewski, M. and Björntorp, P. (1972): Cellularity in different regions of adipose tissue in young men and women. Metabolism, 21, 1143.

Tanner, J.M. (1952): The effects of weight training on physique. Amer. J. Physiol. Anthropol., 10, 427.

Tanner, J.M. (1959): The measurement of body fat in man. Proc. Nutr. Soc., 18, 148.

Thompson, C.W., Buskirk, E.R. and Goldman, R.F. (1956): Changes in body fat estimated from skinfold measurements of college basket ball and hockey players during a season. Res. Quart. Amer. Ass. Hlth phys. Educ., 27, 418.

Toldt, C. (1870): Beiträge zur Histologie und Physiologie des Fettgewebes. Akad. Wissensch. Wien., 62, 445.

Vague, J. (1947): La différenciation sexuelle, facteur déterminant des formes de l'obésité. Presse méd., 55, 339.

Vague, J. (1949): Le diabète de la femme androide. Presse méd., 57, 835.

Vague, J. (1953): La Différenciation Sexuelle Humaine. Ses Incidences en Pathologie. Masson et Cie, Paris.

Vague, J. (1956): The degree of masculine differentiation of obesities: a factor determining predisposition to diabetes, atherosclerosis, gout and uric calculous diseases. Amer. J. clin. Nutr., 4, 20.

Vague, J. (1960): Los cuatros estados pancreaticos de la diabetes por obesidad androide. Rev. ibér. Endocr., 7, 27.

Vague, J., Boyer, J., Jubelin, J., Nicolino, J. and Pinto, C. (1969a): Adipomuscular ratio in human subjects. In: Physiopathology of Adipose Tissue, p. 360. Editor: J. Vague. Excerpta Medica, Amsterdam.

Vague, J., Boyer, J., Vague, P., Clement, M. and Codaccioni, J.L. (1967a): Les frontières de la maladie de Cushing. In: Compte Rendus, IXè Réunion des Endocrinologistes de Langue Française, p. 61. Editor: A. Soulaire. Masson et Cie, Paris.

Vague, J., Codaccioni, J.L., Boyer, J. and Vague, P. (1967b): Relationship between plasma insulin and cortisol production rate in diabetic and non-diabetic obese subjects. In: Abstracts, VI Congress of the International Diabetes Federation, Stockholm, 1967, p. 165. ICS 140, Excerpta Medica, Amsterdam.

Vague, J., Codaccioni, J.L., Teitelbaum, M., Vague, P., Bernard, P.M. and Boyer, J. (1965): Clinical and biological peculiarities of diabetogenic obesities. In: Proceedings. II International Congress of Endocrinology London, 1964, p. 977. Editor: S. Taylor. ICS 83. Excerpta Medica, Amsterdam.

Vague, J. and Jouve, A. (1958): Les relations de l'obésité et de l'artériosclérose. Docum. sci. Guigoz, (Basel), 41, 1.

Vague, J., Jubelin, J. and Boyer, J. (1971b): Effets de la corticothérapie sur le rapport adipo-musculaire. Ann. Endocr. (Paris), 32, 388.

Vague, J., Jubelin, J. and Boyer, J. (1973); Evolution du rapport adipo-musculaire dans l'hyperthyroïdie. Ann. Endocr. (Paris), 34, 216.

Vague, J., Nicolino, J. and Pouch, J.C. (1968): Effets de la testostérone sur le rapport adipo-musculaire chez les hommes eunuchoides. Ann. Endocr. (Paris), 29, 370.

Vague, J., Vague, P., Boyer, J. and Cloix, M.D. (1971a): Anthropometry of obesity, diabetes, adrenal and beta-cell functions. In: Diabetes, p. 517. Editor: R.R. Rodriguez. ICS 231. Excerpta Medica, Amsterdam.

Vague, P., Boeuf, G., Depieds, R. and Vague, J. (1969b): Plasma insulin levels in human obesity. In: Physiopathology of Adipose Tissue, p. 203. Editor: J. Vague. Excerpta Medica, Amsterdam.

Vague, P., Ramahandridona, G., Aprile, N. and Vague, J. (1974): Differential insulin response to glucose and tolbutamide. An hereditary trait for diabetes mellitus. Diabetologia, in press.

Wilmore, J.H., Royce, J., Girandola, R.N., Katch, F. and Katch, V.L. (1970): Body composition changes with a 10-week program of jogging. Med. Sci. Sport., 2, 113.

Adipose mass and blood lipids

Adipose tissue and liver lipogenesis and triglyceride secretion in obese-hyperglycemic (ob/ob) mice: possible relationship with hyperinsulinemia*

B. Jeanrenaud, E.G. Loten and **F. Assimacopoulos-Jeannet**

Laboratoires de Recherches Médicales, University Medical School, Geneva, Switzerland

In this paper I would like to summarize part of the work carried out in our laboratories by Dr. E.G. Loten and F. Assimacopoulos-Jeannet.

As is well known, the ob/ob Bar Harbor mouse (C57BL6J ob/ob) provides an interesting model of diabetes in that it presents at the same time obesity, hyperglycemia and hyperinsulinemia. It thus has several characteristics in common with human maturity onset diabetes (Cameron et al., 1972). Two main aspects of the metabolic disorders prevailing in these ob/ob mice have been recently studied in our laboratories. In the first series of experiments we attempted to measure the contribution of a variety of tissues to lipogenesis in the whole animal (Loten et al., 1973). In another series we studied the metabolism of lipids in perfused livers of non-obese and obese hyperglycemic mice (Assimacopoulos-Jeannet et al., 1974). The main observations of these two studies are summarized below.

CONTRIBUTION OF VARIOUS TISSUES TO LIPOGENESIS IN THE WHOLE ANIMAL

The method used in these experiments (Loten et al., 1973) has been to administer trace amounts of [¹⁴C]glucose intravenously to the mice and to measure labeled glucose carbon incorporation into lipid fractions in different tissues at various times. Using this approach, it could be seen that the ob/ob mice incorporated more lipids into both liver and adipose tissue than lean mice. Furthermore, the saponifiable fatty acids accounted, both in liver and adipose tissue, for a much higher proportion of total radioactivity in the lipids of the obese animals than in lean animals.

In view of these results it was of further interest to investigate whether the large incorporation of labeled glucose carbon into adipose tissue triglycerides in the ob/ob mice was in situ lipogenesis or merely uptake from circulating lipids synthesized in the liver. To answer this question a functional hepatectomy was carried out by tying ligatures around the portal vein and the hepatic artery. Under these conditions, although the livers of ob/ob mice synthesized virtually no lipid from glucose, the lipid synthesis in adipose tissue was almost as high as in the ob/ob mice in which the liver had *not* been excluded from the circulation. This suggests that the increased recovery of glucose carbon in the lipids of adipose tissue of ob/ob mice represents mainly lipids synthesized in situ rather than lipids synthesized in the liver and subsequently taken up by the adipose tissue.

In a further series of experiments we hypothesized that the anomalies observed in both liver and adipose tissue of ob/ob mice might be related, at least in part, to the high endogenous plasma insulin levels known to prevail in these animals. To check this possibility we treated ob/ob mice with the β-cell toxic agent, streptozotocin, a treatment which lowered serum insulin levels as well as pancreatic insulin content and made the animals glucosuric. Following this treatment lipogenesis from glucose in both liver and adipose tissue of ob/ob mice was drasti-

* This work has been supported by Grant No. 3.552.71 from the Fonds National Suisse de la Recherche Scientifique, Berne, Switzerland and by a grant-in-aid from Nestlé Alimentana, Vevey, Switzerland

cally curtailed, thus indicating that the decrease in plasma insulin levels produced by streptozotocin treatment had resulted in a partial reversal of the abnormally high lipogenesis in these two
tissues.

LIPOGENESIS AND LIPID SECRETION BY PERFUSED LIVERS OF OB/OB MICE

It has been reported previously that lipogenesis is increased in livers of ob/ob mice (Winand,
1970). Indeed, the hepatocytes of these mice are characterized by a fatty infiltration com-
prising numerous fat droplets of various sizes scattered throughout the parenchyma. These
observations clearly indicate that livers of ob/ob mice present anomalies of the regulation of
carbohydrate and lipid metabolism, the cause of which is still unknown. One of the difficulties
encountered in previous investigations pertaining to the liver metabolism of these animals has
been the lack of a satisfactory method for the systematic in vitro study of this organ. Recently,
we have developed a method permitting the in situ perfusion of livers of small rodents such as
the mouse (Assimacopoulos-Jeannet et al., 1973). Using this technique we have studied some
aspects of the regulation of lipogenesis and triglyceride secretion in perfused livers of non-obese
and obese (ob/ob) mice.

Experiments have been carried out by measuring the incorporation of tritium from ^3H_2O into
hepatic lipids as well as into the newly synthesized (i.e. labeled) triglycerides secreted into the
perfusate. It was found that lipogenesis from glucose (11 mM) was 2-3 times greater in livers of
ob/ob mice than in those of normal controls. Basal lipogenesis (i.e. no added substrate) was also
much higher in livers of ob/ob mice than in those of controls, and the addition of acetate,
pyruvate or fructose resulted in increases in lipogenic rates that were always greatest in livers of
ob/ob mice. The secretion of newly synthesized triglycerides was considerably higher in livers of
ob/ob mice than in controls and was proportional to hepatic lipogenesis, the relative proportion of the newly synthesized triglyceride released to that stored within the liver remaining about
the same (20-23%). When total secretion (i.e. unlabeled) of triglycerides was measured it was
observed that perfused livers of non-obese mice secreted triglycerides at a low rate in the
absence of fatty acid in the medium, a secretion that was strongly stimulated by the addition of
albumin-bound fatty acids. In contrast, basal secretion of triglycerides by livers of ob/ob mice
was considerably higher than that of controls, as high in fact as that observed in control livers
perfused without fatty acid. However, the high triglyceride secretion of livers from ob/ob mice
was not further stimulated by the addition of fatty acids in the medium.

To investigate the possibility that the anomalies observed in perfused livers of the ob/ob mice
were possibly related to chronic hyperinsulinemia, ob/ob mice were treated with streptozotocin. Seven days after this treatment, livers were perfused, and lipogenesis as well as triglyceride secretion was again measured. It was found that lipogenesis of livers from streptozotocin-
treated ob/ob mice was considerably lower than that of livers from non-treated obese animals.
Moreover, the high triglyceride release previously observed in livers of ob/ob mice perfused
without added fatty acid was restored to normal by streptozotocin treatment and could be
stimulated by the addition of fatty acids, whereas this stimulation could not be obtained in livers
of non-treated ob/ob mice.

CONCLUSION

These in vivo and in vitro experiments indicate that lipogenesis of adipose tissue and liver are
greatly increased in ob/ob mice. In addition, not only is lipogenesis increased in perfused livers
of ob/ob mice but also triglyceride secretion. All these anomalies can be partly reversed by
making the obese animals relatively insulin-deficient. This suggests that the metabolic
anomalies described might well be related, at least in part, to the fact that ob/ob mice are
permanently exposed to very high levels of circulating insulin. It is difficult to assess the
relevance of the present findings to the development of obesity. It is conceivable, however, that
the genetic lesion of these animals causes, via as yet unknown mechanisms, a primary elevation
of plasma insulin which in turn induces fat deposition in adipose tissue as well as increased
synthesis and release of lipids by the liver.

REFERENCES

Assimacopoulos-Jeannet, F., Exton, J.H. and Jeanrenaud, B. (1973): Control of gluconeogenesis and glycogenolysis in the perfused livers of normal mice. Amer. J. Physiol., 225, 25.

Assimacopoulos-Jeannet, F., Singh, A., Le Marchand, Y., Loten, E.G. and Jeanrenaud, B. (1974): Abnormalities in lipogenesis and triglyceride secretion by perfused livers of obese-hyperglycemic (ob/ob) mice: relationship with hyperinsulinemia. Diabetologia, in press.

Cameron, D., Stauffacher, W. and Renold, A.E. (1972): In: Handbook of Physiology, Section 7: Endocrinology, Vol. I. Endocrine Pancreas, p. 611. Editors: R.O. Greep and E.B. Astwood.

Loten, E.G., Rabinovitch, A. and Jeanrenaud, B. (1974): In vivo lipogenesis of lean and obese-hyperglycemic (ob/ob) mice: possible role of hyperinsulinemia. Diabetologia, in press.

Winand, J. (1970): Aspects qualitatifs et quantitatifs du métabolisme lipidique de la souris normale et de la souris congénitalement obèse. Médico-monographies d'agrégés, Arscia, Brussels.

Morphological and temporal characteristics of obesity as determinants of plasma triglyceride in a free living population

M. Stern, P. Wood, A. King, K. Osann, J. Farquhar and **A. Silvers**

Department of Medicine, Stanford University School of Medicine, Palo Alto, Calif., U.S.A.

Obesity has traditionally been regarded as having adverse health consequences. However, in several recent publications, the role of obesity as an *independent* risk factor in the development of cardiovascular disease has been questioned. It is increasingly being recognized that obesity is a heterogeneous disorder, and it is therefore possible that certain types of obesity are associated with an increased risk of cardiovascular disease, whereas other types are not.

In the studies to be presented, we have examined the way in which this heterogeneity influences the relationship between obesity and plasma triglyceride concentration, an important cardiovascular risk factor. The heterogeneity of obesity was considered under the following headings: centralized vs. generalized distribution of fat; early vs. late onset; and hyperplastic (or increased number of fat cells) vs. hypertrophic (or increased size of fat cells).

A random sample consisting of 622 men and women, age range 35 to 60, residing in Modesto, California were invited to participate in the study. 155 men and 199 women were eventually examined, representing an overall response rate of 54%. Height and weight were determined in all subjects and each subject was then asked to recall how much he or she weighed at age 18. Relative weight – defined as actual divided by ideal weight – was calculated, using standard height-weight tables, both for the current weight and for the weight at age 18. Body mass index was calculated as weight divided by height squared. Skinfold thickness was determined using Lange skinfold thickness calipers over the forearm and the subscapular regions. Blood specimens were obtained after an overnight fast followed by a fat-free breakfast and plasma triglyceride concentration was determined using an automated method.

The means and standard deviations of the experimental variables are shown in Table 1.

TABLE 1   Mean and standard deviation of experimental variables

	Men (n = 155)	Women (n = 199)
Plasma triglyceride (mg/100 ml)	147 ± 83	122 ± 76*
Forearm skinfold thickness (mm)	4.6 ± 1.5	9.8 ± 6.3**
Scapula skinfold thickness (mm)	20.0 ± 8.7	21.7 ± 11.9
Relative weight	1.2 ± 0.18	1.2 ± 0.24
Relative weight at age 18	1.0 ± 0.18	1.0 ± 0.21

* $P < 0.01$,   ** $P < 0.001$.

Plasma triglyceride concentration was significantly higher in men than in women. Forearm skinfolds were more than twice as thick in women than in men and this difference was highly significant. Scapula skinfold, however, was only slightly thicker in women than in men and this difference was not statistically significant. Relative weights, both current and at age 18, were essentially identical in the two sexes. The data demonstrate the tendency toward acquired weight gain after the age of 18 in both sexes in this population.

Since plasma triglyceride concentration is known to have a log normal distribution, the

TABLE 2 Correlation of adiposity indices with log triglyceride

	Men	Women
Forearm skinfold thickness	0.10	0.19*
Scapula skinfold thickness	0.30**	0.45**
Body mass index	0.29**	0.35**
Relative weight	0.29**	0.36**

* $P < 0.01$, ** $P < 0.001$.

logarithms of the triglyceride values (log TG) were used in the analyses to be presented. The relationship between the log TG and various measures of adiposity are shown in Table 2. In general, somewhat higher correlations were observed in women than in men. If one uses scapula skinfold thickness, body mass index or relative weight as measures of adiposity, the data would suggest that differences in adiposity account for roughly 9% of the observed variance in log TG in men, and 16% in women.

In an effort to explain a greater percentage of the total variance in log TG, we carried out several types of multi-variate analysis. The following linear model was eventually selected as the one which best described the data:

$$\log TG = a + \beta_0 X_2 + [\beta_1 + \beta_2 X_2] X_1$$

Log TG was pictured as being related to a primary variable, X_1, indicative of the degree of adiposity. The regression coefficient for this relationship, however, was depicted as not being a constant, but rather a linear function of some other variable, X_2, which we have termed a 'modifying influence' or 'modifying variable'.

In the initial analysis, we set X_1 equal to scapula skinfold thickness, and X_2 equal to forearm skinfold thickness. The results are shown in Table 3. For men, β_1 is significantly positive, indi-

TABLE 3 Standardized regression coefficients for regression of scapula and forearm skinfold thicknesses on log triglyceride

	β_0	β_1	β_2	Multiple R
Men	0.454*	1.056**	−1.143**	0.40
Women	−0.089	0.893**	−0.014	0.45

* $P < 0.05$. ** $P < 0.001$.

cating that scapula skinfold thickness makes a positive contribution to triglyceride variance. The modifying coefficient, on the other hand, β_2, is significantly negative. Thus, increasing degrees of forearm skinfold thickness tend to weaken the positive relationship between log TG and scapula skinfold thickness. The multiple correlation coefficient for the overall relationship was 0.40. In women, a different situation prevailed. Although a positive relationship was noted between scapula skinfold thickness and log TG, as in men, no modifying influence of forearm skinfold thickness could be discerned.

These results are in general agreement with those of Albrink and Meigs (1965) in showing that, for men, plasma triglyceride concentration is more closely related to centralized than to generalized obesity. However, these workers studied only men, and therefore did not observe the interesting sex difference which we have observed.

TABLE 4 Standardized regression coefficients for regression of current relative weight and relative weight at age 18 on log triglyceride

	β_1	β_2	Multiple R
Men	0.621**	—0.391*	0.33
Women	0.538**	—0.220	0.36

* $P < 0.05$, ** $P < 0.001$.

Albrink and co-workers have also suggested that the anatomical distribution of fat is related to the age of onset of obesity, with congenital, or early onset, obesity having a more generalized distribution, and acquired, or adult onset, obesity having a more centralized distribution. We examined this hypothesis using our model by setting the primary variable equal to current relative weight, and the modifying variable equal to relative weight at age 18. The results are shown in Table 4. For men, a significant, positive relationship was noted between log TG and current relative weight, and once again the effect of the modifying variable – in this case, relative weight at age 18 – was to weaken the primary relationship. Stated another way, increasing degrees of relative weight at age 18 appeared to 'protect' against the hypertriglyceridemic effects of current relative weight. A similar trend appeared to be present in women although the 'protective' effect of relative weight at age 18 did not achieve statistical significance.

TABLE 5 Standardized regression coefficients for regression of total body fat and fat cell number on log triglyceride

	β_1	β_2	Multiple R
Men	1.907*	—1.683*	0.62
Women	1.3	—1.7	0.48

*$P [0.01$.

In a separate study involving 35 non-randomly selected subjects (23 men and 12 women), fat cell size and number were determined using the osmium tetroxide technique of Hirsch and Gallian (1968). Total body fat was also determined in these individuals by underwater weighing. The resulting data were analyzed using our model by setting the primary variable equal to total body fat and the modifying variable equal to total body fat cell number. The results, shown in Table 5, are analogous to those already presented. In men, increasing numbers of fat cells appeared to 'protect' against the hypertriglyceridemic effect of increasing total body fat. Similar trends were observed in women, but, again, the results did not achieve statistical significance.

In conclusion, we believe that the relationship between adipose tissue and plasma triglyceride concentration is not a simple one, but rather that it is influenced by various modifying factors. Among these are the anatomical distribution of the fat, the age of onset of obesity and the cellularity of the adipose tissue. The influence of these modifying variables is most easily demonstrated in men. Although some or all of these factors may also be operative in women, their effects appear to be less pronounced.

REFERENCES

Albrink, M.J. and Meigs, J.W. (1965): The relationship between serum triglycerides and skinfold thickness in obese subjects. Ann. N.Y. Acad. Sci., 131, 673.
Hirsch, J. and Gallian, E. (1968): Methods for the determination of adipose cell size in man and animals. J. Lipid Res., 9, 110.

DISCUSSION

Björntorp: This was a most interesting paper. Late onset of obesity with enlargement of fat cells seem to be the factors correlated with hypertriglyceridemia, if you will allow this small extrapolation from your data. We have also considered the possibility that a large adipose tissue mass with hypercellularity might protect against hypertriglyceridemia, for example by an increased potential for plasma triglyceride removal into adipose tissue. Compared with controls who are not obese, patients with hypertriglyceridemia have enlarged fat cells but usually no increase in fat cell number. When, on the other hand, one takes hyperplastic obese subjects, increased fat cell size does not seem to be followed by a higher triglyceride level with enlarged fat cells. Perhaps this is evidence for a protecting effect of the increased number of fat cells.

Stern: I think what is needed to answer this question is to study two groups of obese subjects – one predominantly hypertrophic and the other predominantly hyperplastic – who are well-matched with respect to other variables such as percentage adiposity, and to see if various metabolic abnormalities are found relatively more frequently among the former group than the latter.

The relationship between the adipose tissue mass and plasma lipids

G. Debry, D. Guisard*, P. Drouin and **L. Mejean****

Department of Nutrition and Metabolic Diseases and INSERM U 59 Research Group on Nutrition and Dietetics, University of Nancy, Nancy, France

Various methods of research can be used for establishing the possible types of relationship existing between the adipose tissue mass and the plasma concentration of lipid. In this work we attempted only to demonstrate the existence of such relationships and to compare them in various groups of subjects suffering from increased adipose tissue mass, generalized (as in ordinary obesity), localized (as in diffuse symmetrical lipoidosis) or associated with other conditions such as diabetes or hyperlipoproteinemia. We studied the effect of treatment on each group (dieting, dietetic treatment of hyperlipoproteinemia), but hypoglycemic treatment cannot be dealt with within the scope of this article.

METHOD

The control group consisted of patients hospitalized for benign reasons, with no cardiovascular, metabolic, endocrine, hepatic, pancreatic or digestive disorders. The experimental groups will be defined for each experiment.

The technical details concerning analytical methods, the measurement of the K_2 lipid clearance and that of post-heparin lipolytic activity in the plasma conform to those developed previously (Guisard et al., 1972). The lipid emulsion used in the perfusion for measuring the K_2 clearance consisted of 10% cotton oil emulsified in isotonic sorbitol solution (Lipiphysan Egic) and was administered by intravenous drip for 30 minutes at a dose of 0.2 g per kg body weight. The patients had been fasting for 12 hours. All tests began at 7 a.m. The statistical study was carried out using the Student *t* test.

RESULTS

Normal weight subjects, non-diabetic obese subjects and subjects with diffuse symmetrical lipoidosis (Launois-Bensaude) were compared.

The normal and obese groups were homogenous as regards age and height (Table 1). The subjects with diffuse symmetrical lipoidosis were of the same height, but older. Total lipids, triglycerides, plasma free fatty acids (Fig. 1) and K_2 clearance (Fig. 2) were significantly different in the three groups. This was not the case with regard to total cholesterol and blood sugar. It is therefore apparent that increased adipose tissue mass is accompanied by a rise in the serum concentration of most lipid constituents and by a decrease in the lipid clearance rate. However, some forms of increased adipose tissue mass with multiple preferential localization, such as diffuse symmetrical lipoidosis, are accompanied by a decrease in the serum concentration of lipids, with the exception of free fatty acids. In addition, the lipid clearance rate increases. These facts show that the nature of the relationship between the adipose tissue mass and blood lipids is variable.

* Chargé de Recherches INSERM.
** Attaché de Recherches INSERM.

TABLE 1   Characteristics of the normal weight subjects, obese subjects and subjects with diffuse symmetrical lipoidosis (Launois-Bensaude)

	No.	Age (years)	Height (cm)	Weight (kg)	Excess weight (kg)	Blood sugar (g/l)
Normal	50	32 ± 3	168 ± 2	61.5 ± 2.8		0.83 ± 0.03
Non-diabetic obese	37	37.5 ± 5	162 ± 3	97.3 ± 6.3	28.3 ± 9.0	0.88 ± 0.05
Launois-Bensaude	6	52.8 ± 2.2*	166 ± 1	71.0 ± 5.2		0.88 ± 0.07

* $0.001 < P < 0.005$.

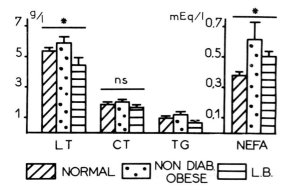

Fig. 1.   Total lipids (LT), triglycerides (TG), cholesterol (CT) and plasma free fatty acids (NEFA) in normal weight subjects, non-diabetic obese subjects and subjects with diffuse symmetrical lipoidosis (L.B.). ns = not significant. * $0.001 < P < 0.005$.

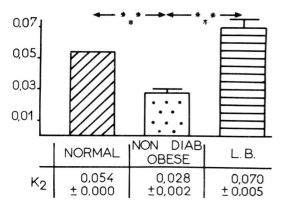

Fig. 2. K_2 clearance in normal weight subjects, non-diabetic obese subjects and subjects with diffuse symmetrical lipoidosis (L.B.). Arrows indicate significant differences ($P < 0.0005$).

In a further study normal weight subjects, non-diabetic obese subjects and diabetic obese subjects were compared. The characteristics of the three groups are shown in Table 2 and the results obtained in Figures 3 and 4.

The normal and obese subjects were homogenous as regards age and height (Table 2). Non-diabetic and diabetic obese subjects were homogenous as regards height, weight and excess weight. Obese diabetic subjects were a little older. Their diabetes was discovered during the test and invariably consisted of moderate chemical diabetes.

Except for total cholesterol, the two groups of obese subjects had significantly higher lipid concentrations than normal subjects (Fig. 3). Total lipids and triglycerides were significantly higher in diabetic obese subjects than in non-diabetic obese subjects. The K_2 lipid clearance was decreased in the latter and greatly decreased in the former (Fig. 4). The association of obesity with a decrease in cellular glucose assimilation aggravates the lipid anomalies that accompany obesity.

TABLE 2   Characteristics of the normal weight subjects, obese subjects and obese diabetic subjects

	No.	Age (years)	Height (cm)	Weight (kg)	Excess weight (kg)	Blood sugar (g/l)
Normal	50	32 ± 3	168 ± 2	61 ± 2.8		0.82 ± 0.02
Obese						
Diabetic	49	48 ± 4*	162 ± 2	99 ± 4.8	40 ± 6.2	1.23 ± 0.10
Non-diabetic	72	38 ± 4*	162 ± 2.6	96 ± 4	35 ± 6.6	0.91 ± 0.05

* $0.001 < P < 0.005$.

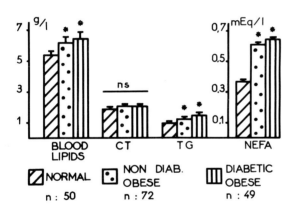

Fig. 3.   Total lipids, triglycerides (TG), cholesterol (CT) and plasma free fatty acids (NEFA) in normal weight subjects, non-diabetic obese subjects and diabetic obese subjects. * $0.001 < P < 0.005$.

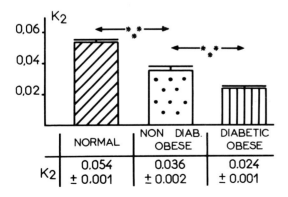

Fig. 4. K_2 clearance in normal weight subjects, non-diabetic obese subjects and diabetic obese subjects. Arrows indicate significant differences ($P < 0.0005$).

Effect of dieting on non-diabetic and diabetic obese subjects

The subjects were given a low-calorie diet varying from 500 to 1,000 cal daily, depending on the case, for a minimum period of 30 days. The lipid clearance test performed at the end of the course utilized the same quantity of exogenous lipids as the first test, despite the decrease in

TABLE 3   Characteristics of normal subjects and non-diabetic and diabetic obese subjects before and after dieting

	No.	Age (years)	Height (cm)	Weight (kg)	Excess weight (kg)	Weight loss (kg)	Blood sugar (g/l)
Normal	50	32 ± 3	168 ± 2	61 ± 2.8			0.82 ± 0.02
Obese							
Non-diabetic							
Before	36	37.5 ± 5.3	160 ± 2	97 ± 6	39.5 ± 6		0.87 ± 0.05
After	36			87 ± 5	29.5 ± 5	10 ± 4	0.90 ± 0.07
Diabetic							
Before	19	39 ± 4	163 ± 2	96 ± 4	36 ± 6		1.20 ± 0.12
After	19			84 ± 6	24 ± 8	12 ± 4	1.01 ± 0.16

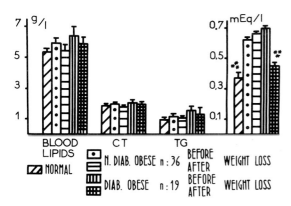

Fig. 5.   Total lipids, triglycerides (TG), cholesterol (CT) and plasma free fatty acids in normal subjects and non-diabetic and diabetic obese subjects before and after weight loss by dieting. *** P < 0.0005.

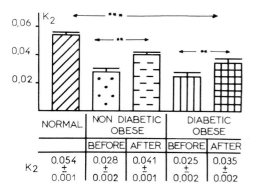

Fig. 6. K_2 clearance in the same population as in Figure 5. Arrows indicate significant differences: **0.0005 < P < 0.001; ***P < 0.0005.

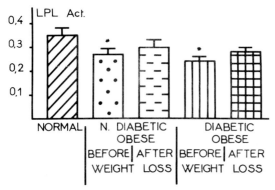

Fig. 7. Plasma post-heparin lipolytic activity in the same population before and after weight loss by dieting. * $0.001 < P < 0.005$.

body weight in obese subjects. On the other hand, in the determination of plasma post-heparin lipolytic activity, the amount of heparin injected was in proportion to the true weight. The conditions adopted were therefore the most unfavourable for demonstrating the possible effects of dieting.

The two groups of obese subjects and the control group were homogenous with regard to age and height, and both obese groups were homogenous as regards age, height, weight, excess weight and the loss of weight (about 10% of the original weight) (Table 3). Dieting, although moderate, tended to normalize lipid disturbances in both obese groups, and plasma free fatty acids and glycemia in the diabetic group (Fig. 5). In the non-diabetic obese group plasma free fatty acids were not modified by the moderate loss of weight. In both groups dieting tended to normalize plasma post-heparin lipolytic activity and significantly increased the K_2 clearance rate of exogenous lipids (Figs. 6 and 7).

Adipose tissue mass and circulating lipids in normal and obese hyperlipoproteinemic subjects

Effect of normolipemic dietetic treatment 132 hyperlipoproteinemic subjects were classified both by electrophoresis and according to their dietetic sensitivity by the sequential method we perfected seven years ago. Each patient had a glucose tolerance test to ensure that none had chemical diabetes (Table 4).

The frequency of obesity was similar to that in the adult population of economically developed countries, but appeared to be higher in the subjects with glucose-sensitive hyperlipoproteinemia.

TABLE 4   Frequency of obesity in subjects with hyperlipoproteinemia

	Non-obese		Obese		Total
	No.	%	No.	%	
Hypercholesterolemia	37	73	14	27	51
Mixed hyperlipidemia	13	62	8	38	21
Hypertriglyceridemia sensitive to alcohol	29	72	11	28	40
Hypertriglyceridemia sensitive to carbohydrates	9	53	8	47	17
Hypertriglyceridemia sensitive to carbohydrates and alcohol	2		1		3

It has been shown that dieting decreases the concentrations of circulating lipids in hyperlipo-proteinemic patients. This fact is too well known for us to communicate our own results in this respect. On the other hand, it may be asked whether this decrease is due to loss of weight or to reduction in the quantity of the particular food responsible for hyperlipoproteinemia. To answer this question, we submitted the 132 patients to 6 successive diets, each lasting 3 weeks. The isocaloric and isoprotein sequences maintained the weight at a constant level, according to the method we have already described (Debry and Drouin, 1970; Debry et al., 1970).

TABLE 5   Characteristics of obese and non-obese subjects with primary hypertriglyceridemia

Hyperlipidemias sensitive to:	No.	Age (years)	Height (cm)	Weight (kg)	Excess weight (kg)	Blood sugar (g/l)
Alcohol						
Non-obese	29	47 ± 2	168 ± 10	67 ± 4		1.07 ± 0.04
Obese	11	50 ± 2	166 ± 12	91.1 ± 4.6	29.1 ± 5	1.22 ± 0.04
Carbohydrates						
Non-obese	9	45 ± 2	173 ± 4	70.6 ± 1.8		0.93 ± 0.07
Obese	8	44 ± 3	167 ± 13	83.9 ± 4.2	21.8 ± 4	1.15 ± 0.07

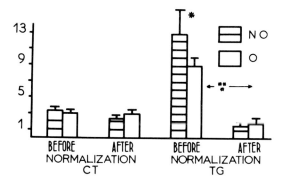

Fig. 8.   Cholesterol (CT) and triglycerides (TG) before and after normalization in non-obese (NO) and obese (O) subjects with hypertriglyceridemia sensitive to alcohol. * $0.001 < P < 0.005$. *** $P < 0.0005$.

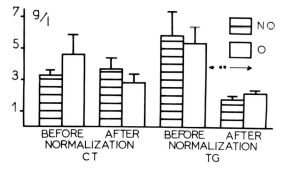

Fig. 9.   Cholesterol (CT) and triglycerides (TG) before and after normalization in non-obese (NO) and obese (O) subjects with hypertriglyceridemia sensitive to carbohydrates. ** $0.0005 < P < 0.001$.

There was no reduction in the concentration of circulating lipids in subjects with hypercholesterolemia and mixed hyperlipidemia. Subjects with primary hypertriglyceridemia (Table 5), however, normalized concentrations of circulating lipids during the diet with considerable reduction of the particular food to which they were sensitive (Figs. 8 and 9). This result was obtained without modification in weight either in subjects with normal weight or obese subjects.

Normalization of triglyceridemia without weight variation, both in obese subjects and subjects with normal weight, proves that in these disorders the concentration of circulating lipid is not connected with obesity. The reduction in hypertriglyceridemia observed during weight reduction in these patients was not due to the reduction in the adipose tissue mass, but to the reduction in the food responsible for hypertriglyceridemia.

CONCLUSION

The relationship existing between the adipose tissue mass and the circulating lipid is variable. Hyperlipidemia and decreases in the lipid clearance rate and plasma post-heparin lipolytic activity are connected with the adipose tissue mass in common obesity and diabetic obesity. They are modified by dieting, but this modification does not occur in cases of diffuse symmetrical lipoidosis and hyperlipoproteinemia with dietetic sensitivity to a particular food. These few facts open up a new pathway for the study of the relationship between the adipose tissue mass and the circulating lipid.

REFERENCES

Debry, G. and Drouin, P. (1970): Diététique des hyperlipémies et des hypercholestérolémies. Rev. Prat. (Paris), 20, 849.

Debry, G., Laurent, J., Guisard, D., Gonand, J.P., Mejean, L. and Drouin, P. (1970): Etude de la glycorégulation au cours des hypertriglycéridémies essentielles. In: Comptes Rendus, Journée Annuelle de Diabetologie de l'Hôtel Dieu, p. 229. Flammarion, Paris.

Guisard, D., Bigard, M.A. and Debry, G. (1972): Etude de l'épuration plasmatique des lipides chez des sujets atteints de cirrhose hépatique d'origine éthylique avec Ascite. Nutr. Metab., 14, 193.

Metabolic defects in adipose tissue in hypertriglyceridemia: new aspects of the pathogenesis of hypertriglyceridemia*

G. Walldius, A.G. Olsson, P. Rubba[1] and L.A. Carlson

King Gustaf V Research Institute, Stockholm and Department of Geriatrics, University of Uppsala, Uppsala, Sweden

Hypertriglyceridemia (HTG) is often found in patients with ischemic heart disease (IHD). Elucidation of the mechanisms causing HTG is thus important. In the development of HTG two mechanisms or a combination of both may operate: (1) increased inflow of triglycerides (TG) into the blood; (2) decreased removal (clearance) of plasma TG from the blood.

There is increasing evidence to suggest that impaired removal of TG is the important mechanism underlying HTG (Havel et al., 1970; Boberg et al., 1972). Two major processes regulate the removal of plasma TG from blood: (1) hydrolysis of TG-rich lipoprotein molecules (chylomicrons, very low density lipoproteins) by lipoprotein lipase (LLA) in the capillaries forming fatty acids (FA); (2) local incorporation and esterification of these FA by adipose and other tissues.

The first process has been studied in detail. Deficiency of LLA has as yet only been found in the rare Type I HTG. In some, but not all, studies low mean values for post-heparin LLA have been found in groups of patients with HTG (Fredrickson and Levy, 1972). Persson (1973) found lower mean LLA activity per g adipose tissue in a group of subjects with Type IV hyperlipidemia (HL). This difference was, however, not obvious when LLA activity was expressed per total body fat.

The second process has not been studied until recently. In a case of massive HTG (Type V) with normal post-heparin LLA we found a subnormal fatty acid incorporation into adipose tissue (FIAT), suggesting this as the cause of HTG (Carlson et al., 1973a).

To investigate whether low FIAT is common in HTG, the FIAT mechanism has been studied in subjects with different types of HTG. FIAT was determined by an isotopic method later modified to take into consideration FA release into the incubation medium (Dole, 1961). The results with both methods suggest that low FIAT may be one important cause of HTG.

MATERIALS

Apparently healthy persons with normolipidemia and hyperlipidemia from a health control centre (Group A) and patients undergoing ordinary clinical examinations both as outpatients and as inpatients (Group B) were included in this study. No patients suffered from malignant, endocrine or acute disease or were on insulin or tablets because of diabetes.

METHODS

Adipose tissue was obtained by a needle biopsy technique from lower abdominal subcutaneous fat. Biopsy specimens (20-40 mg) in triplicate were incubated in 1 ml 2% albumin buffer solution (493 nmol FA/ml) with [³H] palmitate and [¹⁴C] glucose added. After incubation FIAT and glucose incorporated (GLIAT) were determined in an extract containing adipose tissue glycerides. In Group B the release of FA during the incubation was determined by the Ni-method to

* Supported by grants from Nordisk Insulinfond and the Swedish Medical Research Council (19x-204).
[1] Visiting scientist from Naples, Italy.

correct the specific activity for dilution during the incubation. Glycerol was measured as an indicator of lipolysis. Fat cell size was determined by a microscopic technique. Total body fat was calculated from anthropometric data. All methods used are described in detail in Walldius and Rubba (1974). The type of HL was determined according to the WHO classification. TG and cholesterol were determined as described above.

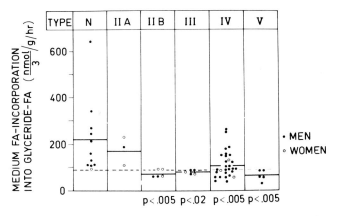

Fig. 1. Fatty acid incorporation into adipose tissue in a normolipidemic group (N) and 5 groups with different types of hyperlipidemia. The mean value for each group is indicated by the solid line. The broken line indicates the lowest value of the control group. The P-values indicate the significance versus group N.

RESULTS

In Group A, FIAT was determined assuming constant specific activity during the incubation procedure. It was found that FIAT was significantly lower in subjects with HTG when expressed per g adipose tissue (Fig. 1). Part of these preliminary data have been presented elsewhere (Carlson et al., 1973b). In these subjects there was also, however, an increased lipolysis. This suggested that a concomitant release of FA during the incubation could partly explain the difference between the FIAT values in the different types of HL, since FA release was probably higher in HTG.

This prompted further studies with Group B using the same method, including determination of FA release to the incubation medium, to obtain corrected FIAT values. In this group of patients total body fat was also calculated using anthropometric data. Age, blood lipids and total body fat in Group B are given in Table 1. Subjects with Type IV HL had significantly more body fat than normals. The mean values in Type IIB and Type III patients were larger than, although not significantly different from, those in the normal group.

When 20-30 mg biopsy specimens were incubated in 1 ml buffer solution for 2 hours there was often a lower total amount of FA after the incubation, indicating a net uptake of FA in those subjects with normolipidemia and low lipolysis. However, in subjects with HTG there was often a net release of FA, sometimes as much as 50% of the initial FA concentration in the albumin buffer, indicating a significant dilution of the isotopic medium. The mean FA release in subjects with normolipidemia was 116 ± 304 nmol/g/hour and the corresponding release in Type IV subjects was 1355 ± 214 nmol/g/hour (P < 0.01). When the correction method of Dole (1961) was applied in Group B the FIAT values in normals were elevated by 4% and the values in Type IV subjects were elevated by 9% compared to data uncorrected for FA release.

FIAT, GLIAT and glycerol expressed per g adipose tissue

The corrected FIAT values per g in the different types of HL are given in Table 2. In all types of

TABLE 1 Fasting concentrations of plasma triglycerides and cholesterol and age and body fat (mean ± S.E.M.)

	Normal	Type of hyperlipidemia				
		IIA	IIB	III	IV	V
No.	24	9	13	4	39	3
Triglycerides (mmol/l)	1.55 ± 0.09	2.17 ± 0.12	3.29 ± 0.37	6.07 ± 1.79	4.44 ± 0.44	11.6 ± 3.2
Range	0.99 – 2.3	1.44 – 2.4	2.1 ± 6.7	3.7 – 11.4	1.3 – 9.7	5.4 – 15.5
Cholesterol (mg/100 ml)	221 ± 6	345 ± 17	346 ± 9	361 ± 50	279 ± 11	260 ± 56
Age (years)	51 ± 2	52 ± 3	55 ± 2	61 ± 6	51 ± 1	52 ± 11
Range	35 – 66	43 – 65	41 – 73	50 – 74	35 – 66	41 – 73
Body fat (kg)	13.6 ± 0.9	16.3 ± 1.7	15.8 ± 0.9	17.4 ± 1.5	17.1 ± 0.8	11.1 ± 2.4
Range	5.1 – 23.8	8.6 – 24.0	9.6 – 19.4	13.8 – 20.8	8.3 – 28.7	7.3 – 15.5
P		n.s.	n.s.		<0.01	

n.s. = not significant.

HL there was a lower FIAT than in normals. In Type IIA there was a significantly lower FIAT value than in normals. This fact will be discussed later. GLIAT was significantly lower in all types of HL. FIAT and GLIAT values were significantly correlated within each HL group. Glycerol release was significantly increased in Type IV HL. Fat cell size was only significantly increased in Type IIB (91 μm versus 81 μm in the normal group, P < 0.025).

TABLE 2   Incorporation of fatty acid ($\frac{nmol}{3}$/g/hr) corrected for fatty acid release during incubation, and glucose into adipose tissue and glycerol release (nmol/g/hour) (mean ± S.E.M.)

	Normal	Type of hyperlipidemia				
		IIA	IIB	III	IV	V
No.	24	9	13	4	39	3
Fatty acid	156 ± 11	97 ± 7	74 ± 9	82 ± 4	88 ± 6	59 ± 23
Range	79 – 250	57 – 126	51 – 163	69 – 89	24 – 239	14 – 95
P		< 0.01	< 0.001	< 0.01	<< 0.001	
Glucose	199 ± 12	170 ± 17	141 ± 19	89 ± 10	135 ± 13	83 ± 60
Range	107 – 293	89 – 239	77 – 269	70 – 107	5 – 394	5 – 202
P		n.s.	< 0.05	< 0.01	< 0.01	
Glycerol	468 ± 35	513 ± 90	413 ± 42	542 ± 110	655 ± 51	578 ± 332
Range	225 – 904	221 – 970	161 – 692	303 – 818	211 – 1555	222 – 1243
P			n.s.	n.s.	< 0.01	

n.s. = not significant.

FIAT, GLIAT and glycerol expressed per total body fat

The low values for FIAT expressed per g adipose tissue in HTG might only be an expression of a greater adipose tissue mass in these subjects, who in general have more body fat (Table 1), i.e. the total FIAT expressed per body fat would be equal to normal. It is, however, evident from Table 3 that the calculated total body FIAT is lower in all types of HL and significantly lower in Types IIB, IV and V, whereas in Type IIA values are normal. Total body GLIAT was not significantly different from normal in any type of HL. However, total body fat glycerol release was about 45% higher (P < 0.001) in Type IV than in normal subjects. The total number of fat cells was in the range of 1-10 × 10^{10} and was not significantly different in any type of HL. FIAT per fat cell was lower in Type IV (P < 0.01).

DISCUSSION

The first observation of low FIAT increased by treatment with nicotinic acid (Carlson et al., 1972) at the same time as HTG was reduced prompted the development of a micro-method for the determination of FIAT from small specimens easily obtained by needle biopsy. By using such a micro-method (Group A in this report) lower FIAT values were found in subjects with HTG than in normal subjects. Lipolysis was also increased in HTG. Thus the release of FA into the incubation medium had to be determined to establish that isotopic dilution was not the cause of the difference obtained. It was found that there was a higher FA release in HTG, leading to dilution of the incubation medium. FIAT values were corrected by applying the method of Dole (1961) in Group B. By this method the same general finding of low FIAT in HTG was found as by the micro-method originally used (Carlson et al., 1973b).

   If the assumption that no major difference in FIAT is found in different body sites is valid, it might be more physiological to study total body fat FIAT, reflecting the total body fat capacity

TABLE 3   Incorporation of fatty acid corrected for fatty acid release during incubation and glucose and glycerol release (μmol/body/hour) (mean $\pm$ S.E.M.)

	Normal	Type of hyperlipidemia				
		IIA	IIB	III	IV	V
No.	24	9	13	4	39	3
Fatty acid	2157 ± 242	2158 ± 243	1168 ± 165	1416 ± 152	1423 ± 93	539 ± 161
Range	765 – 4903	798 – 2640	365 – 2331	1104 – 1810	329 – 2526	217 – 707
P		n.s.	< 0.01	n.s.	< 0.01	
Glucose	2756 ± 271	2588 ± 447	2005 ± 207	1503 ± 90	2339 ± 266	833 ± 1104
Range	950 – 6400	1560 – 4710	670 – 2580	1320 – 1750	690 – 7060	80 – 2100
P		n.s.	n.s.	n.s.	n.s.	
Glycerol	6303 ± 677	8446 ± 1873	6230 ± 713	10980 ± 2640	11116 ± 979	6107 ± 3438
Range	2300 – 16540	3090 – 17950	2300 – 10930	4180 – 17010	2840 – 23930	1960 – 12930
P		n.s.	n.s.	n.s.	< 0.001	

n.s. = not significant.

to incorporate hydrolysed FA. Since it is generally known that subjects with HTG are somewhat obese, their total body fat FIAT may be of the same magnitude as in normal subjects. In an earlier study (Persson, 1973) LLA activity expressed per total body fat was the same in all types of HL. We have found, however, that total body FIAT was significantly lower in Types IIB, IV and V HL than in normolipidemia. Thus adipose tissue cannot incorporate exogenous FA in HTG at the same rate as in normolipidemia at one and the same FA concentration. Adipose tissue in HTG may have low enzyme activities or low concentrations of coenzymes or important metabolites, as previously discussed (Carlson et al., 1973a). These findings may suggest that FIAT is one important factor in the pathogenesis of HTG. A combined defect with high lipolysis and low FIAT was found in Type IV HL.

REFERENCES

Boberg, J., Carlson, L.A., Freyschuss, U., Lassers, B.W. and Wahlqvist, M.L. (1972): Splanchnic secretion rates of plasma triglycerides and total and splanchnic turnover of plasma free fatty acids in men with normo- and hypertriglyceridaemia. Europ. J. clin. Invest., 2, 454.

Carlson, I.A., Eriksson, I. and Walldius, G. (1973a): A case of massive hypertriglyceridaemia and impaired fatty acid incorporation into adipose tissue glycerides both corrected by nicotinic acid. Acta med. scand., 194, 363.

Carlson, L.A., Walldius, G., Olsson, A.G. and Rubba, P. (1973b): Adipose tissue defects in hypertriglyceridemia (abstract). Europ. J. clin. Invest., 3, 219.

Dole, V.P. (1961): The fatty acid pool in adipose tissue. J. biol. Chem., 236, 3121.

Fredrickson, D.S. and Levy, R.I. (1972): Familial hyperlipoproteinemia. In: Metabolic Bases of Inherited Disease, 3rd ed., p. 545. Editors: J.B. Stanbury, J.B. Wyngaarden and D.S. Frederickson. McGraw Hill, New York.

Havel, R.J., Kane, J.P., Balasse, E.O., Segel, N. and Basso, L.V. (1970): Splanchnic metabolism of free fatty acids and production of triglycerides of very low density lipoproteins in normotriglyceridemia and hypertriglyceridemic humans. J. clin. Invest., 49, 2017.

Persson, B. (1973): Lipoprotein lipase activity of human adipose tissue in different types of hyperlipidemia. Acta med. scand., 193, 447.

Walldius, G. and Rubba, P. (1974): A micromethod for simultaneous determination of fatty acid incorporation into adipose tissue (FIAT) and lipolysis in relation to adipose tissue characteristics. In preparation.

Blood lipid regulation in various types of human lipodystrophy

J.L. de Gennes, G. Turpin and **J. Truffert**

Clinique Endocrinologique, Centre Hospitalier Universitaire Pitié Salpêtrière, Paris, France

Although several scattered observations of blood lipid abnormalities in human lipodystrophy have been reported in the medical literature, very few systematic studies have been devoted to this peculiar and important field. We have been able to collect a series of various types of human lipodystrophy and to investigate their relationships with blood lipid abnormalities.

MATERIAL AND METHODS

Selection of patients was on a clinical basis using the classical criteria of several types of lipodystrophy.

Total or diffuse lipoatrophy (also called lipoatrophic diabetes)

In this group there were 4 patients, 3 adults aged 30, 32, and 50 years, and 1 6-month-old girl with a congenital type of lipodystrophy. In all these cases the generalized loss of adipose tissue was obvious and was anatomically checked by biopsy of subcutaneous adipose tissue. All the other criteria of the disease, namely a rise in basal metabolism, insulin-insensitive diabetes without ketosis and marked hyperlipidemia, were also present in each case.

Segmental lipoatrophy

In this special type of lipodystrophy, two different groups of patients were defined. The first was the classical syndrome of upper segment lipoatrophy, the so-called Barraquer-Simmons syndrome, with total loss of adipose tissue in the upper part of the body and very often a tendency to obesity in the lower part of the body. Four cases (all females) of this syndrome were investigated. All were adults aged 22, 30, 39 and 46 years.

The second was a new syndrome recently described by ourselves from 5 similar cases representing the 'mirror syndrome' of the previous one, i.e. loss of adipose tissue in the lower segment of the body with excessive accumulation of fat in the upper part of the body. All 5 cases were adult women (aged 34, 47, 49, 63 and 72 years), though their history strongly suggested a constitutional type of adipose tissue distribution, with absence of adipose tissue in the lower limbs since early infancy.

Launois-Bensaude syndrome

This is a more common type of symmetrical lipodystrophy, with symmetrical lipomatosis, known as the Launois-Bensaude syndrome in the French literature and the Magdelung syndrome in the English literature. 26 cases of this syndrome were studied. There were 23 men and 3 women and their ages ranged from 31 to 68 years (mean age 48 years).

BIOLOGICAL STUDIES

These were focused primarily on blood lipid chemical studies. Chemical measurements of various lipid fractions were done according to the modified technique of Grigaut (1910) for total cholesterol, the method of Van Handel and Zilversmit (1957) for triglycerides, the method of Zilversmit and Davis (1950) for phospholipids, and the method of Chabrol and Charonnat (1937) for total lipids. Lipoprotein studies were chiefly carried out by the paper electrophoresis method of Lees and Hatch (1963) for typing according to the classification of hyperlipoprotein-emias by Fredrickson et al. (1967). Flotation tests of fasting sera were also done at 4° C for 24 to 48 hours in order to identify the collection of chylomicrons at the top, and lipomicrons remaining throughout the sera. Preparative ultracentrifugation studies were only done in 1 case of lipoatrophic diabetes, and 2 cases of Launois-Bensaude syndrome.

Post-heparin lipoprotein lipase activity (PHLA) was measured by the technique of Fredrickson et al. (1963) on Intralipid substrate, and free fatty acids (FFA) according to the colorimetric technique of Duncombe (1964). Sensitivity to dietary fat, carbohydrate and alcohol were studied in a longitudinal study in a permanent dietetic survey with controlled diets providing either specific deprivation or specific load of one of these nutrients.

Besides specific blood lipid studies, other biological studies included uric acid determination (Folin and Denis, 1920), glucose tolerance tests, radioimmunological measurement of insulin after glucose or tolbutamide stimulation, glucagon test and exogenous insulin sensitivity tests.

RESULTS

Total lipoatrophy (Lawrence, 1946)

Massive and grossly lactescent hyperlipidemia was found in 2 out of 3 adult cases and in the 1 congenital case. Hyperglyceridemia was constant, with very high, though variable, levels of 944, 4,220, 320 and 1,300 mg/100 ml. Total cholesterol was also generally high and total lipids reached very high levels of 9,600, 8,240, 1,260 and 1,800 mg/100 ml.

The lipoprotein electrophoretic pattern on a free diet was very unstable and variable, even in the same patient. Type V, i.e. simultaneous presence of chylomicrons and lipomicrons, was

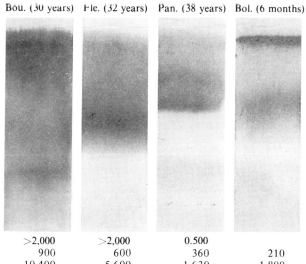

	Bou. (30 years)	Fle. (32 years)	Pan. (38 years)	Bol. (6 months)
Optical density	>2,000	>2,000	0.500	
Total cholesterol (mg/100 ml)	900	600	360	210
Total lipids (mg/100 ml)	10,400	5,600	1,630	1,800
Triglycerides (mg/100 ml)	7,100	3,800	500	1,300
PHLA (μmol/ml/minute)	0.26	0.50	0.30	0.03

Fig. 1. Lipoprotein electrophoretic pattern in 4 subjects on a free diet. PHLA = post-heparin lipoprotein activity.

found at least once in all cases. The pre-β-lipoprotein component was more constant than the chylomicron component and spontaneous variation from Type V to Type IV was frequently found (Fig. 1). On a fat-free diet disappearance of chylomicrons both on paper electrophoresis and in the flotation test was observed consistently with a partial fall of lipid concentration (triglycerides, total cholesterol and total lipid). However, complete correction of hyperlipidemia could not be achieved with this measure alone because of the persistence of the pre-β-lipoproteinemia of Type IV. The sensitivity to fat reloading either in normal diets (35% of calories as fat) or high-fat diets (70% of calories as fat) was demonstrated in the 3 adult cases, but not in the congenital case.

TABLE 1   Lipoatrophic diabetes

	Subject			
	Bou.	Fle.	Pan.	Bol.
Age (years)	30	32	38	0.5
Weight (kg)	36	44	51	5.5
Height (cm)	160	150	160	61
Free diet				
Triglycerides (mg/100 ml)	940–4600	1500–4200	300–320	260–1300
Total cholesterol (mg/100 ml)	120–940	320–640	220–260	190–230
Total lipids (mg/100 ml)	2550–9600	4000–7600	1050–1260	1000–1800
Chylomicrons	+++	++	0	+
Pre-β-lipoprotein	++	++	+	++
PHLA (μmol/ml/minute)		30–0.50	0.30	0.05
Glucose (mg/100 ml)		200	64–90	80–190
Basal insulin (μU/ml)	60	19	49–120	158
Fat-free carbohydrate-rich diet				
Triglycerides (mg/100 ml)	110–3100	1500–4200	820	190
Total cholesterol (mg/100 ml)	180–380	320–640	360	230
Total lipids (mg/100 ml)	740–4400	4000–7600	2000	930
Chylomicrons	0	0	0	0
Pre-β-lipoprotein		++	+	+
Glucose (mg/100 ml)	120–150	600–650	84–80	
Basal insulin (μU/ml)			39–45	
Fat-rich diet				
Triglycerides (mg/100 ml)	3100–6000		520–1200	240
Total cholesterol (mg/100 ml)	150–540		360–400	280
Total lipids (mg/100 ml)	4000–10,800		1660–2600	1080
Chylomicrons	+++		+	0
Pre-β-lipoprotein	++		+	±
PHLA (μmol/ml/minute)				0.07
Glucose (mg/100 ml)	150		86–98	79
Basal insulin (μU/ml)	280		70–78	21
Calorie-adjusted diet				
Triglycerides (mg/100 ml)	110–260	200		84
Total cholesterol (mg/100 ml)	210–250	210		210
Total lipids (mg/100 ml)	700–900	820		640
Chylomicrons		±		
Pre-β-lipoprotein		±		
PHLA (μmol/ml/minute)	0.26			
Glucose (mg/100 ml)	100–150	250		

PHLA = post-heparin lipoprotein lipase activity.

TABLE 3 'Mirror' lipodystrophy syndrome (De Gennes syndrome)

	Subject				
	Gai.	Hei.	Per.	Goi.	Par.
Triglycerides (mg/100 ml)	170–420	108–1150	164–194	34–950	58–250
Total cholesterol (mg/100 ml)	270–370	310–460	325–375	420–880	411–306
Total lipids (mg/100 ml)	900–1440	940–2110	940–1130	1610–3880	820
Electrophoresis	Type III	Type IV	Type II + IV	Type IV	Type II + IV
PHLA (μmol/ml/minute)	0.24	0.27		0.50	
Oral glucose tolerance test	Diabetic (4)	Diabetic (4)	Diabetic (1)	Normal	Diabetic (1)
IRI (μU/ml)					
Basal	24	27			
Max	100	56			
Weight (kg)	70	67	63	60	83
Height (cm)	151	171	165	171	173

PHLA = post-heparin lipoprotein activity.

According to our criteria of classification of hyperlipidemias (De Gennes, 1971) all those cases could be classified as mixed or combined hyperlipidemia. PHLA could be measured in only 3 cases, and was found to be low in 2 cases (0.24 and 0.27 μmol/ml/minute) and high normal in the case of Type IV massive endogenous hyperglyceridemia. Abnormalities of blood glucose regulation were found constantly associated with a diabetic glucose tolerance test and with increased basal and/or reactive insulin levels during this test. Exogenous insulin sensitivity was low. Also the inhibition of in vivo lipolysis, as tested during the glucose tolerance test by the amplitude of fall of FFA, was impaired, suggesting decreased efficiency of endogenous insulin. Marked alimentary sensitivity due chiefly to sucrose (in 1 case to ethanol) and to fat was shown in all these cases. Finally, severe risk of vascular thrombosis in small vessels, especially in the eye, was observed, with loss of vision in 2 of the 5 cases.

Launois-Bensaude syndrome

Of our 26 patients, 13 showed blood lipid abnormalities (Table 4). 6 had massive endogenous hyperglyceridemia of Type IV, 4 mixed or combined hyperlipidemia of Type III or II + IV on electrophoresis only, and 3 pseudo essential hypercholesterolemia or pseudo Type II, i.e. having the appearance of that type after a 12-hour overnight fast, but clearly showing the pattern of mixed hyperlipidemia after an 8-hour fast.

PHLA was measured in only 3 cases, and was normal in 2, and slightly decreased in 1 (0.26 μmol/ml/minute). Ethanol sensitivity of hyperlipidemia was demonstrated in 5 out of the 13 cases, although chronic ethanol intoxication was very common in the whole group (22 out of 26). In the 8 other cases restriction of carbohydrate was necessary to correct the hyperglyceridemia. Disturbances of blood glucose regulation were generally slight and inconstant. No case of overt diabetes was found in the whole group. However, abnormal glucose tolerance tests were found in 11 cases, 5 normolipidemic patients and 6 hyperglyceridemic patients. In 10 cases immunoreactive insulin was checked and no increase could be detected either in basal levels or in reactive post-glucose load levels.

DISCUSSION

An unusually high incidence of pathologic hyperlipidemias was found in these different types of human lipodystrophy: 100% (4/4) in total lipoatrophy, 40% (2/5) in Barraquer-Simmons

TABLE 4   Launois-Bensaude syndrome

Subject	Sex	Age (years)	Weight (kg)	Height (cm)	Triglycerides (mg/100 ml)	Total cholesterol (mg/100 ml)	Total lipids (mg/100 ml)	Electrophoresis classification	PHLA (μmol/ml/min)	Oral glucose tolerance test	IRI (μU/ml) Basal	Max
Mas.	M	43	84	168	3000	830	6400	Type IV	0.26	Normal	9	60
Best.	M	31	73	159	1380	370	3800	Type IV	0.40	Normal	8	29
Cal.	M	41	64	161	1500	540	3520	Type IV	—	Normal	—	—
Mes.	M	56	55	160	1900	380	3200	Type IV	—	Normal	—	52
Kal.	F	52	65	172	896	380	1980	Type IV	—	Normal	14	—
Cas.	M	47	59	164	440	250	1520	Type IV	—	Diabetic (3)	—	—
Che.	M	45	88	165	500	150	1450	Type IV	—	—	—	—
Des.	M	39	73		495	210	1340	Type IV	0.33	Diabetic (3)	18	60
Gra.	F	45	50	151	216	330	1070	Type II + IV	—	Diabetic (2)	9.2	8.1
Gue.	M	36	68	170	195	200	870	Normal	—	Normal		
Ali.	M	45	69	176	75	340	1070		—	—		
Dau.	F	64	63	144	50	300	900	Normal	—	Normal	11	110
Gar.	M	36	68	167	84	290	890		—	Diabetic (1)	8	36

PHLA = Post heparin lipoprotein activity.

syndrome, 100% (5/5) in our new 'mirror' syndrome of segmental lipodystrophy, 50% (13/26) in the Launois-Bensaude or Magdelung syndrome. These figures are far above the expected frequency of blood lipid increases in the normal population (around 4%) or even in common obesity (around 20%). A relation between lipodystrophy and disturbances of blood lipid regulation is thus suggested. The common feature of the different types of hyperlipidemia is the constant finding of endogenous hyperlipidemia in the various electrophoretic patterns of Types II, II + IV and III (electrophoretic).

As well as idiopathic hyperlipidemias of Type IV, III or II + IV, a frequent association with disturbances of blood glucose regulation was observed. All these facts suggest that the adipose tissue occupies a key position in regulation and that the normal physiological function of adipose tissue could induce by itself a definite disturbance in lipid or glucose blood levels. Adipose tissue is the storage organ for any excess of energy of exogenous alimentary origin, from fat (chylomicrons) and from glucose as well as from hepatic fat secretion (lipomicrons). Any condition in which the storage capacity of adipose tissue is impaired can be the origin of blood lipid abnormalities. In total impairment as created by total lipoatrophy, hyperlipidemia must be massive and constant, as also must diabetes. This was confirmed by our observations and those of others. Partial or total correction of the pathological disorder, by reducing and adjusting total fat and carbohydrate intake, provides another demonstration of the truth of this interpretation.

In partial or segmental impairment of adipose tissue, the occurrence of blood lipid disturbances seems closely related to the saturation of still present and functioning adipocytes. In the Barraquer-Simmons syndrome, in contrast to previous publications (Rifkind et al., 1967; Piscatelli et al., 1970) which claimed the constancy of hyperglyceridemia of Type IV with normal PHLA, our experience was that only a minority of cases had permanent and definite rises of blood lipids. Results in our small series of 5 cases were similar, however, to those of Senior and Gellis (1964) with more than 100 cases. Of these only 3 were associated with permanent blood lipid disorders. However, in these cases sensitized screening tests for blood lipid abnormalities, such as those recently introduced by us, seem to be able to detect very slight disorders which are not apparent at the conventional screening time after a 12-hour fast. The usual lack of blood lipid disorders and of blood glucose disturbances in the Barraquer-Simmons syndrome could be related to the larger storage capacity of the lower segment of the body due to the larger number of adipocytes, as recently described by Vague and Vague (1973). So, without enormous obesity of the lower part of the body, it is conceivable that the time required to saturate all these adipocytes can never be reached and this protects a large proportion of cases belonging to this syndrome from major blood lipid disorders. In contrast, in our 'mirror' syndrome of lipodystrophy, the loss of adipocytes in the lower segment of the body allows easy and quick saturation of the fat storage capacity of adipocytes in the upper segment of the body, which are much less numerous. Once saturation is achieved, it leads inevitably to accumulation of circulating lipids and glucose directly dependent on excess calorie intake. Thus, obesity is more easily produced in the upper than in the lower body. Concurrently the metabolic abnormalities are permanent, with a constant association of blood lipid disorders with endogenous hyperglyceridemia, abnormal, diabetic-type glucose tolerance tests, hyperinsulinemia both basal and reactive to glucose and decrease of exogenous insulin sensitivity. In this type of segmental lipodystrophy the entire spectrum of total lipoatrophic diabetes is reproduced. Lastly, very serious vascular complications occur in this syndrome, especially in the ocular and vertebrobasilar circulation.

The fact that in all three types of lipodystrophy a common metabolic syndrome more or less identical to lipoatrophic diabetes can be reproduced argues in favor of the primary responsibility of impairment of adipose tissue storage capacity. The whole spectrum of metabolic abnormalities, including the disturbance of insulin secretion, is completely identical to the metabolic spectrum of hypertrophic obesity with full saturation of the storage capacity of adipose tissue.

The special group of Launois-Bensaude or Magedelung syndrome could be slightly more difficult to fit exactly into such a general comprehensive theory. However, a careful analysis of adipose tissue redistribution in this syndrome indicates most often definite alternations of scattered lipoatrophic areas (namely the very frequent loss of the Bichat fat pad) and scattered areas of symmetrical lipomatosis. A restricted capacity of fat storage in the adipose tissue of such patients is also conceivable as soon as lipomatous areas reach saturation level. An interesting case of Boulet et al.(1961) in which complete disappearance of a lipoatrophic diabetes-like metabolic syndrome followed the surgical removal of the most prominent lipoma and a recurrence followed regrowth of the same lipoma argues in favor of such an interpretation. However, in this syndrome, besides partial impairment of adipose tissue storage capacity, there could exist in a proportion of cases an accumulation of several hyperlipidemic factors, in particular excessive ethanol intake, which is very common among these patients and can occasionally strikingly amplify otherwise rather limited blood lipid disorders. In each case, the respective contribution of exogenous factors and of saturation of adipose tissue to the hyperlipidemic state has to be carefully analysed because, in contrast to the case of Colwell and Cruz (1972) with marked and persistent hyperinsulinism, in 10 cases we tested we found no definite basal or reactive hyperinsulinism.

REFERENCES

Boulet, P., Mirouze, J., Barjou, P. and Mion, C. (1961): Eclosion synchrône d'un diabete sucré, d'une lipomatose diffuse et d'une hépatostéatose massive chez une acromegale. Le Diabète, March, 3, 12.

Chabrol, E. and Charonnat, R. (1937): Détermination des lipides totaux sanguins. Presse méd., 45, 1713.

Colwell, J.A. and Cruz, S.R. (1972): Effect of resection of adipose tissue on the diabetes and hyperinsulinism of benign symmetric lipomatosis. Diabetes, 21, 13.

De Gennes, J.L. (1971): Les hyperlipidémies idiopathiques. Proposition d'une classification simplifiée. Presse méd., 79, 791.

De Gennes, J.L., Croisier, J.C., Menage, J.J. and Truffert, J. (1973): Analyse de l'hyperlipidémie et du diabète et de leurs relations réciproques dans 4 cas de diabète lipo-atrophique. In: Comptes Rendus, Journées de Diabétologie de l'Hôtel Dieu, p. 69. Flammarion, Paris.

De Gennes, J.L., Turpin, G. and Truffert, J. (1972): Dépistage et identification des hyperlipidémies idiopathiques. Un nouveau test. Nouv. Presse méd., 80, 1627.

Duncombe, W.G. (1964): The colorimetric micro-determination of non-esterified fatty acids in plasma. Clin. chim. Acta, 9, 122.

Folin, G. and Denis, C. (1920): Dosage de l'acide urique dans le sérum sanguin. C.R. Soc. Biol. (Paris), 83, 1273.

Fredrickson, D.S., Levy, R.I. and Lees, R.S. (1967): Fat transport in lipoproteins. An integrated approach to mechanisms and disorders. New Engl. J. Med., 276, 32, 94, 148, 215, 273.

Fredrickson, D.S., Ono, K. and Davis, L.L.(1963): Lipolytic activity of post-heparin plasma in hyperglyceridemia. J. Lipid Res., 4, 24.

Grigaut, A. (1910): Détermination du cholestérol sanguin. C.R. Soc. Biol. (Paris), 68, 781.

Lawrence, R.D. (1946): Lipo-dystrophy and hepatomegaly with diabetes, lipaemia and other metabolic disturbances. Lancet, 1, 724.

Lees, R.S. and Hatch, F.T. (1963): Sharper separation of lipoprotein species by paper electrophoresis in albumin-containing buffer. J. Lab. clin. Med., 16, 518.

Piscatelli, R.L., Vieweg, W.V.R. and Havel, R.J. (1970): Partial lipodystrophy. Metabolic studies in three patients. Ann. intern. Med., 73, 963.

Rifkind, B.M., Boyle, J.A. and Gale, M. (1967): Blood lipid levels, thyroid status and glucose tolerance in progressive partial lipodystrophy. J. clin. Path., 20, 52.

Senior, B. and Gellis, S.S. (1964): The syndromes of total lipodystrophy and of partial lipodystrophy. Pediatrics, 33, 593.

Vague, J. and Vague, P.(1973): L'obésité diabétogène – Etat actuel de la question. Journées de Diabétologie de l'Hôtel Dieu, p. 53. Flammarion, Paris.

Van Handel, E. and Zilversmit, D.B. (1957): Micromethod for the direct determination of serum triglycerides. J. Lab. clin. Med., 50, 152.

Zilversmit, D.B. and Davis, A.K. (1950): Microdetermination of plasma phospholipids by trichloro-acetic acid precipitation. J. Lab. clin. Med., 35, 155.

DISCUSSION

Harter: It is suggested that the lipoatrophy of the lower part of the body described by Dr. De Gennes may correspond to extreme cases of android biotypes with or without obesity described by Dr. J. Vague. In this latter biotype, hypertriglyceridemia and abnormalities of carbohydrate metabolism are also frequently observed.

De Gennes: All our cases of this new syndrome were carefully studied.

The intravenous fat tolerance test in obese subjects: a study before and after jejuno-ileal shunt operation*

S. Rössner, L. Backman and **D. Hallberg**

Departments of Internal Medicine and Surgery, Karolinska Hospital, King Gustaf V Research Institute, and Nutrition Unit, Karolinska Institute, Stockholm, Sweden

In the intravenous fat tolerance test (IVFTT) (Boberg et al., 1969; Carlson and Rössner, 1972) the elimination rate of the fat emulsion Intralipid, (AB Vitrum, Stockholm, Sweden) after a single intravenous injection is followed. The elimination of the emulsion follows first order kinetics below a so-called critical concentration and thus gives a straight line in a semilogarithmic plot. The elimination rate of Intralipid expressed as the rate constant K_2 (percentage per minute) is related to endogenous plasma triglycerides (TG) in several ways. The fractional turnover of endogenous plasma TG (Boberg et al., 1972) is well correlated to the K_2 values, suggesting that the latter reflect the fractional removal rates of endogenous TG (Rössner et al., 1974). Furthermore K_2 is correlated to plasma TG concentration in a hyperbolic way, subjects with low removal rates having high plasma TG concentration and vice versa (Boberg et al., 1970). The IVFTT value is influenced by several hormonal and metabolic factors such as noradrenaline, glucagon, various lipid-lowering drugs, starvation in combination with surgical trauma and acute myocardial infarction. It has been pointed out that one reason for these observed differences in K_2 may be the fact that the removal sites are not evenly distributed in the organism.

The role of the adipose tissue in the removal of the Intralipid emulsion is not known in detail. Olivecrona and Belfrage (1965) have demonstrated that chyle TG is removed as intact particles stored in the liver or incorporated into tissue esters as chyle TG fatty acids.

The present study was undertaken to investigate whether subjects with marked obesity eliminated Intralipid differently from normal subjects. The IVFTT was therefore carried out in highly obese subjects before and during weight reduction after a jejuno-ileal shunt operation.

METHODS

Subjects

The obese patients studied had all been referred to the Department of Surgery of the Karolinska Hospital because of advanced obesity. They all had a weight/height index (kg/ (cm – 100)) above 1.41, a history of more than 5 years, unsuccessful results of conservative medical treatment and social and/or psychiatric complications of their obesity. During the preoperative evaluation program 60 patients were studied with the IVFTT. After operation the IVFTT was repeated in 23 patients (2 men and 21 women) 1 to 3 times during the period of weight reduction.

In order to compare the preoperative IVFTT of these highly obese patients with values from subjects of more normal weight, data from a control group were obtained. The control subjects were selected from earlier studies on the IVFTT. For every obese subject a control of the same sex and of approximately the same age and plasma TG concentration was chosen. The controls were not on any drugs or diets affecting lipid metabolism and had fairly normal body weights. The values of the control group were within the normal limits for IVFTT at the same age and plasma lipid concentrations as has been found previously.

* Supported by King Gustaf V 80-årsfond and Vitrum AB, Stockholm, Sweden.

Analytical methods

All blood samples were taken after an overnight fast. Plasma TG and cholesterol were determined by auto-analyzer methods. The IVFTT has recently been described in detail (Carlson and Rössner, 1972). A basal blood sample is taken and then 1 ml 10% Intralipid/kg body weight is given intravenously as a single dose injection. During the ensuing 40 minutes blood is taken and light scattering of the plasma is determined by nephelometry. The values are plotted in a semilogarithmic system; a straight line is obtained and the slope is calculated by means of the least squares. The elimination rate K_2 is expressed as percentage per minute.

RESULTS

In obese males the weight/height index was 1.70 ± 0.04 (S.E.M.) and in obese females the corresponding figure was 1.90 ± 0.04. Male controls had an index of 1.01 ± 0.01 and female controls 0.93 ± 0.02. No significant differences in age or plasma TG were found between male and female obese subjects and their corresponding control groups. Plasma cholesterol tended to be lower in obese patients than in the controls. In each group K_2 was higher in females than in males. Control females had an 83% higher mean value, which was highly significant. In the obese group females had 48% higher IVFTT values than males, but this difference did not reach statistical significance.

Plasma volume (PV) was determined preoperatively in 14 obese patients. Turnover of plasma TG determined as $K_2 \times TG \times PV$ mmol/minute did not correlate to body weight or the weight/height index. Plasma TG clearance determined as $K_2 \times PV$ l/minute did not correlate significantly to the weight/height index or body weight.

In Table 1 the values before operation have been compared with those after weight reduction. When this latter study was performed some patients had reached a steady weight, whereas others were still losing weight. Concomitant with a mean weight reduction from 129 to 104 kg

TABLE 1 Clinical data and weight reduction pattern in obese patients after jejuno-ileal shunt operation

Sub-ject	Age	Sex	Preoperation		Postoperation					
			Height (cm)	Weight (kg)	1st IVFTT		2nd IVFTT		3rd IVFTT	
					Months postop.	Weight (kg)	Months postop.	Weight (kg)	Months postop.	Weight (kg)
1	35	F	167	156	2	146	9	111		
2	46	M	173	138	7	110				
3	50	F	160	142	10	113				
4	47	F	170	146	3	117	5	100		
5	31	F	162	94	4	79				
6	43	F	172	160	5	151	18	134		
7	45	F	170	108	4	104	7	100		
8	24	F	160	150	2	132	12	105		
9	18	F	169	131	3	120				
10	27	F	155	101	8	86				
11	32	F	158	111	9	84				
12	45	F	172	115	7	92				
13	34	F	164	129	5	107	8	94	13	87
14	33	F	176	137	2	119				
15	28	F	174	115	2	110				
16	47	F	163	134	10	106				
17	18	F	178	143	14	80				
18	24	F	174	111	16	83				
19	60	F	170	130	11	111				
20	48	F	162	158	6	122				
21	37	M	180	160	3	149				
22	62	F	151	110	3	94				
23	53	F	171	100	2	96				

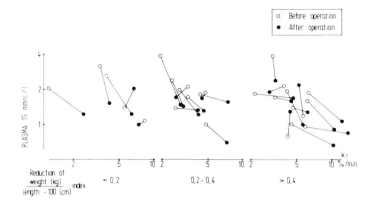

Fig. 1. Relationship between the intravenous fat tolerance and plasma TG concentration in obese subjects before and after operation. The subjects were divided into 3 groups on the basis of the weight loss between studies. Logarithmic scales.

there was a significant reduction of plasma TG from 1.92 to 1.40 mmol/1 and an increase of the IVFTT from 3.99 to 5.83%/minute, which is also significant. Plasma cholesterol concentration was significantly reduced from 216 mg/100 ml to 145 mg/100 ml. The relationship between K_2 and plasma TG concentration during weight reduction is illustrated in Figure 1, where the changes of K_2 and plasma TG are given in 3 groups with different body weight reduction patterns. In the operated subjects the weight reduction rate can be expressed as kg/month. This measurement did not correlate to the observed changes of K_2 during the corresponding time.

DISCUSSION

When the regression lines for obese males and females plotting log K_2 against log TG were compared to the control subjects there was no significant difference between the slopes. This finding suggests that the removal sites for the fat emulsion were not increased in proportion to the amount of adipose tissue. The development of obesity seems to have increased the plasma TG pool, but if the number of removal sites does not increase the final result is a plasma TG rise. The K_2 value after the jejuno-ileal shunt operation was improved concomitant with a fall of plasma TG and this finding could indicate that an improved elimination in the periphery, rather than a change in plasma TG secretion rate, may at least partly account for the reduced TG concentration in plasma. The fact that the weight loss after operation did not seem to be related to K_2 changes supports the concept that removal sites for the Intralipid emulsion are not in proportion to the amount of adipose tissue.

The pattern in Figure 1 is in principle the same as may be seen in other situations where plasma TG has been lowered by, for instance, specific drugs (Boberg et al., 1970, 1971; Olsson et al., 1974; Carlson et al., 1974). Concomitant with a plasma TG reduction an increase of the K_2 was seen.

Several cases showed a marked improvement of their K_2 and a marked plasma TG reduction although the weight loss might be small. Others reduced their body weight considerably without any major effect on plasma TG or K_2. In this connection it should be pointed out that the K_2 value is highly reproducible when repeated in the same subject at different intervals (Carlson and Rössner, 1972). This inconstant pattern raises the question of whether the effects of the jejuno-ileal shunt operation on plasma TG and K_2 may depend, at least in part, on other factors than the loss of adipose tissue. It can be speculated that the operation mechanically caused changes of the blood flow in the splanchnic region and that removal sites in this area may be perfused more or less depending on the location.

REFERENCES

Boberg, J., Carlson, L.A., Freyschuss, U., Lassers, B.W., Rössner, S. and Wahlqvist, M.L. (1972): A comparison of the removal rates of a fat emulsion and of endogenous plasma triglyceride in man. Europ. J. clin. Invest., 2, 123.

Boberg, J., Carlson, L.A. and Hallberg, D. (1969): Application of a new intravenous fat tolerance test in the study of hypertriglyceridemia in man. J. Atheroscler. Res., 9, 159.

Boberg, J., Carlson, L.A., Fröberg, S., Olsson, A.G., Orö, L. and Rössner, S. (1971): Effects of chronic treatment with nicotinic acid on intravenous fat tolerance and postheparin lipoprotein lipase activity in man. In: Metabolic Effects of Nicotinic Acid and its Derivatives, p. 465. Editors: K.F. Gey and L.A. Carlson. Hans Huber, Bern, Stuttgart, Vienna.

Boberg, J., Carlson, L.A., Fröberg, S.O. and Orö, L. (1970): Effect of a hypolipidemic drug (CH 13, 437) on plasma and tissue lipids, and on the intravenous fat tolerance in man. Atherosclerosis, 11, 353.

Carlson, L.A., Olsson, A.G., Orö, L. and Rössner, S. (1974): Clinical and metabolic effects of dextro-3-iodothyronine in hyperlipidaemia. Acta med. scand, in press.

Carlson, L.A. and Rössner, S. (1972): A methodological study of an intravenous fat tolerance test with Intralipid emulsion. Scand. J. clin. Lab. Invest., 29, 271.

Olivecrona, T. and Belfrage, P. (1965): Mechanisms for removal of chyle triglyceride from the circulatory blood as studied with [14C] glycerol- and [3H] palmitic acid-labeled chyle. Biochim. biophys. Acta (Amst.), 98, 81.

Olsson, A.G., Orö, L. and Rössner, S. (1974): Effects of oxandrolone on plasma lipoproteins and the intravenous fat tolerance in man. Atherosclerosis, in press.

Rössner, S., Boberg, J., Carlson, L.A., Freyschuss, U. and Lassers, B. (1974): Comparison between fractional turnover rate of endogenous plasma triglycerides and Intralipid (intravenous fat tolerance test) in man. Europ. J. clin. Invest., 4/2.

Concluding remarks

J. Vague

Endocrinology Clinic, Hôpital de la Timone, Marseilles, France

The number and the quality of the papers presented during this meeting, as well as the discussions they have produced, prevent any attempt to summarize them briefly. I believe they deserve a thorough reading by all physicians and biologists who want to keep in touch with the progress of our knowledge on adipose tissue. That is a reason to express our warmest thanks to all colleagues who contributed to the Congress. Yet, among much important data, a few main ideas have emerged from what we have heard during these three days.

The regulation of a determined adipose mass, which remains steady under constant internal and external conditions, is a very precise physiological phenomenon. Its neuro-hormono-enzymatic mechanism becomes more and more understood, but the signal which controls the adipose mass (either increasing or decreasing it) is still controversial.

The mechanism by which adipose mass increases, decreases or remains at a determined level is not as yet entirely known. The relation between adipocyte surface and adipocyte volume, in other words the contact surface of triglycerides with the cytoplasm and further with adipocyte membrane, seems to be an important regulating factor.

The number of adipocytes, which appears to be definitively fixed when growth is finished and which is probably decreased by testosterone during puberty, especially in the lower part of the body, is one factor determining adipose mass.

As adipose mass seems to be determined as a function of internal and external conditions, it is difficult to act on it both quickly and with a lasting effect. In most cases adipose mass tends to spontaneously recover its initial value. Early diet and physical exercise influence the adipocyte number. The avoidance of overfeeding and muscular exercise during childhood can reduce the multiplication of adipocytes. In obese people, muscular training decreases adipose mass, adipocyte volume and hyperinsulinemia. Thus from the beginning to the end of life the competition between muscular and adipose mass is the result of the past and remains more or less influenced by the environment. Real improvement of the disturbances in regulation is easier in the young than in adults. We believe that the metabolic complications of adipose mass excess are rather specifically the consequence of excess in the upper part of the body, with enlarged deltoid adipocytes and a relatively reduced number of trochanteric adipocytes.

Numerous points regarding the mechanism of the regulation of adipose mass still remain unexplained. They will certainly stimulate further research to extend our knowledge in a field of biology which probably still holds the secret of widespread metabolic diseases.

Author index

Subject index